Essential Forensic Biology
Second Edition

Essential Forensic Biology

Second Edition

Alan Gunn
Liverpool John Moores University, Liverpool, UK

A John Wiley & Sons, Ltd., Publication

Library of Congress Cataloguing-in-Publication Data

Gunn, Alan.
  Essential forensic biology / Alan Gunn. – 2nd ed.
    p. ; cm.
  Includes bibliographical references and index.
  ISBN 978-0-470-75804-5 (HB) – ISBN 978-0-470-75803-8 (PB)
  1. Forensic biology. I. Title.
  [DNLM: 1. Forensic Medicine. W 700 G976e 2009]
  QH313.5.F67G86 2009
  363.25–dc22
                                                    2008040263

ISBN: 978-0-470-75804-5 (HB)
         978-0-470-75803-8 (PB)

A catalogue record for this book is available from the British Library.

Set in 10/12 pt Sabon by SNP Best-set Typesetter Ltd., Hong Kong.
Printed in Singapore by Ho Printing Singapore Pte Ltd.

6   2013

To Sarah, who believes that no evidence is required in order to find a husband guilty.

Contents

Acknowledgements

Thanks to Sarah and to all of the academic and technical staff at the School of Biological & Earth Sciences, Liverpool John Moores University who helped me along the way.

Introduction

The word 'forensic' derives from the Latin *forum* meaning 'a market place': in Roman times this was the where business transactions and some legal proceedings were conducted. For many years the term 'forensic' had a restricted definition and denoted a legal investigation but it is now commonly used for any detailed analysis of past events i.e. when one looks for evidence. For example, tracing the source of a pollution incident is now sometimes referred to as a 'forensic environmental analysis', determining past planetary configurations is referred to as 'forensic astronomy', whilst historians are said to examine documents in 'forensic detail'. For the purposes of this book, 'forensic biology' is defined broadly as 'the application of the science of biology to legal investigations' and therefore covers human anatomy and physiology, organisms ranging from viruses to vertebrates and topics from murder to the trade in protected plant species.

Although forensic medicine and forensic science only became specialised areas of study within the last 200 or so years, their origins can be traced back to the earliest civilisations. The first person in recorded history to have medico-legal responsibilities was Imhotep, Grand Vizier, Chief Justice, architect and personal physician to the Egyptian pharaoh Zozer (or Djoser). Zozer reigned from 2668–2649 BC and charged Imhotep with investigating deaths that occurred under suspicious circumstances. The codification of laws was begun by the Sumerian king Ur-Nammu (ca 2060 BC) with the eponymous 'Ur-Nammu Code' in which the penalties of various crimes were stipulated whilst the first record of a murder trial appears on clay tablets inscribed in 1850 BC at the Babylonian city of Nippur.

In England, the office of coroner dates back to the era of Alfred the Great (871–899) although his precise functions at this time are not known. It was during the reign of Richard I (1189–1199) that the coroner became an established figure in the legal system. The early coroners had widespread powers and responsibilities that included the investigation of crimes ranging from burglary to cases of murder and suspicious death. The body of anyone dying unexpectedly had to be preserved for inspection by the coroner, even if the circumstances were not suspicious. Failure to do so meant that those responsible for the body would be fined, even though it might have putrefied and created a noisome stench by the time he arrived. It was therefore not unusual for unwanted bodies to be dragged away at night to become another village's problem. The coroner's responsibilities have changed considerably over the centuries but up until 1980 he was still expected to view the body of anyone dying in suspicious circumstances.

Essential Forensic Biology, Second Edition   Alan Gunn
© 2009 John Wiley & Sons, Ltd

Although the coroner was required to observe the corpse he did not undertake an autopsy. In England and other European countries, the dissection of the human body was considered sinful and was banned or permitted only in exceptional circumstances until the nineteenth century. Most Christians believed that the body had to be buried whole otherwise the chances of material resurrection on Judgement Day were slight. The first authorized human dissections took place in 1240 when the Holy Roman Emperor Frederick II decreed that a corpse could be dissected at the University of Naples every five years to provide teaching material for medical students. Subsequently, other countries followed suit, albeit slowly. In 1540, King Henry VIII became the first English monarch to legislate for the provision of human dissections by allowing the Company of Barber Surgeons the corpses of four dead criminals per annum and in 1663, King James II increased this figure to six per annum. Subsequently, after passing the death sentence, judges were given the option of permitting the body of the convict to be buried (albeit without ceremony) or to be exposed on a gibbet or dissected. Nevertheless, the lack of bodies and an eager market among medical colleges created the trade of body snatching. Body snatchers were usually careful to leave behind the coffin and the burial shroud because taking these would count as a serious criminal offence – which was potentially punishable by hanging. Removing a body from its grave was classed as merely a misdemeanour. The modern day equivalent is the Internet market in human bones of uncertain provenance (Huxley & Finnegan, 2004; Kubiczek & Mellen, 2004). A recent notorious case arose when it was discovered that the body of the eminent journalist Alistair Cooke had been plundered whilst 'resting' in a funeral parlour in New York. Alistair Cooke died on March 30th 2004 and despite the fact that he was 95 years-old at the time of his death and had been suffering from cancer, his arms, legs and pelvis were surreptitiously removed and sold to a tissue processing company. There is a perfectly legal market for bones and other body tissues for use in surgery or as dental filler but it is also highly lucrative and some people have been tempted into criminal behaviour.

Although the ancient Greeks are known to have performed human dissections, Julius Caesar (102/100 – 44 BC) has the dubious distinction of being the first recorded murder victim in history to have undergone an autopsy. After being assassinated, his body was examined by the physician Antistius who concluded that although Julius Caesar had been stabbed 23 times, only the second of these blows, struck between the 1st and 2nd ribs, was fatal. The first recorded post mortem to determine the cause of a suspicious death took place in Bologna in 1302. A local man called Azzolino collapsed and died suddenly after a meal and his body very quickly became bloated whilst his skin turned olive and then black. Azzolino had many enemies and his family believed that he had been poisoned. A famous surgeon, Bartolomeo de Varignana was called upon to determine the cause and he was permitted to undertake an autopsy. He concluded that Azzolino had died as a consequence of an accumulation of blood in veins of the liver and that the death was therefore not suspicious. Although this case set a precedent, there are few records from the following centuries of autopsies being undertaken to determine the cause of death in suspicious circumstances.

The first book on forensic medicine may have been that written by the Chinese physician Hsu Chich-Ts'si in the 6th century AD but this has since been lost. Subsequently, in 1247, the Chinese magistrate Sung Tz'u wrote a treatise entitled '*Xi Yuan Ji Lu*' that is usually translated as 'The Washing Away of Wrongs', and this is generally accepted as being the first forensic textbook (Peng & Pounder, 1998). Sung Tz'u would also appear to be the first person to apply an understanding of biology to a criminal investigation as he relates how he identified the person guilty of a murder by observing the swarms of flies attracted to the bloodstains on the man's sickle. In Europe, medical knowledge advanced slowly over the centuries and forensic medicine really only started to be identified as a separate branch of medicine in the 1700s (Chapenoire & Benezech, 2003). The French physician Francois-Emanuel Foderé (1764–1835) produced a landmark 3 volume publication in 1799 entitled *Les lois éclairées par les sciences physiques: ou Traité de médecine-légale et d'hygiène publique* that is recognised as a major advancement in forensic medicine. In 1802, the first chair in Forensic Medicine in the UK was established at Edinburgh University and in 1821 John Gordon Smith wrote the first book on forensic medicine in the English language entitled '*The Principles of Forensic Medicine*'.

Today, forensic medicine is a well-established branch of the medical profession. Clinical forensic medicine deals with cases in which the subject is living (e.g. non-accidental injuries, child abuse, rape) whilst forensic pathology deals with investigations into causes of death that might result in criminal proceedings (e.g. suspected homicide, fatal air accident). Pathology is the study of changes to tissues and organs caused by disease, trauma and toxins etc. Theoretically, any qualified medical doctor can perform an autopsy but in practise, at least in the UK, they are conducted by those who have received appropriate advanced training.

The majority of deaths are not suspicious so an autopsy is unlikely to take place. Indeed, even if a doctor requests an autopsy, the relatives of the dead person must give their permission. Some religious groups are opposed to autopsies and/or require a person to be buried within a very short period of death so this may be refused. For example, many Muslims, orthodox Jews and some Christian denominations remain opposed to autopsies. Some doctors are concerned about how few autopsies take place since it is estimated that 20–30% of death certificates incorrectly state the cause of death (Davies *et al.* 2004). The errors are seldom owing to incompetence or a 'cover-up' but a consequence of the difficulty of diagnosing the cause of death without a detailed examination of the dead body. Unfortunately, there are rogue elements in all professions and Dr Harold Shipman is believed to have murdered over 200 mostly elderly patients over the course of many years through the administration of morphine overdoses and then falsified their death certificates (Pounder, 2003). Dr Shipman's victims suffered from a range of chronic ailments and because of their age and infirmities nobody questioned the certificates he signed. In addition, he also falsified his computer patient records so that it would appear that the patient had suffered from the condition that he claimed had led to their death. He would sometimes do this within hours of administering a fatal dose of morphine. Ultimately, suspicions were aroused and several of his victims who had been buried were disinterred and subjected to an autopsy. The findings indicated that although they may have been infirm they had not died as a consequence of disease. They did, however, contain significant amounts morphine: morphine residues can be detected in buried bodies for several years after death. Dr Shipman had therefore, surprisingly for a doctor, chosen one of the worst poisons in terms of leaving evidence behind. Dr Shipman was found guilty of murdering 15 of his victims in January 2000 and subsequently committed suicide whilst in prison.

In England and Wales, when a body is discovered in suspicious circumstances the doctor issuing the death certificate or the police will inform the coroner and they can then request that an autopsy is performed regardless of the wishes of the relatives. In this case, the autopsy will usually be undertaken by one of the doctors on the Home Office List of pathologists. As of 1 April 2006, there were 38 of these each of whom covered one of 8 regions of England and Wales. The name is a bit of a misnomer because although they are accredited by the Home Office, they are not employed by the Home Office. Scotland has its own laws and the Procurator Fiscal is the person who decides whether a death should be considered suspicious and also whether one or two pathologists should conduct the autopsy. [In England and Wales the pathologist usually works on their own.]. The situation in Northern Ireland is slightly different again with pathology services provided by The State Pathologist's department. Other countries have their own arrangements and there are calls for a thorough overhaul for the provision of forensic services in England and Wales and of the coroner system in particular (e.g. Whitwell, 2003).

Animals and plants have always played a role in human affairs, quite literally in the case of pubic lice, and have been involved in legal wrangles ever since the first courts were convened. Disputes over ownership, the destruction of crops and the stealing or killing of domestic animals can be found in many of the earliest records. For example, Hammurabi, who reigned over Babylonia during 1792–1750 BC, codified many laws relating to property and injury that subsequently became the basis of Mosaic Law. Amongst these laws it was stated that anyone stealing an animal belonging to a freedman must pay back ten fold whilst if the animal belonged to the court or a god, then he had to pay back thirty fold. Animals have also found themselves in the dock accused of various crimes. In the Middle Ages there were several cases in which pigs, donkeys and other animals were executed by the public hangman following their trial for murder or sodomy. The judicial process was considered important and the animals were appointed a lawyer to defend them and they were tried and punished like any human. In 1576, the hangman brought shame on the German town of Schweinfurt by publicly hanging a pig in the custody of the court before due process had taken place. He never worked in the town again and his behaviour is said to have given rise to the term 'Schweinfurter Sauhenker' (Schweinfurt sow hangman) to describe a disreputable scoundrel (Evans, 1906). However, the phrase has now fallen out of fashion. Today, it is the owner of a dangerous animal who is prosecuted when it wounds or kills someone, although it may still find itself facing the death penalty.

During the nineteenth century, a number of French workers made detailed observations on the sequence of invertebrate colonisation of human corpses in cemeteries and attempts were made to use this knowledge to determine the time since death in murder investigations (Benecke, 2001). Thereafter, invertebrates were used to provide evidence in a sporadic number of murder investigations but it was not until the 1980s that their potential was widely recognised. Part of the reason for the slow development is the problem of carrying out research that can be applied to real case situations. The body of the traditional experimental animal, the laboratory rat, bears so little resemblance to that of a human being that it is difficult to draw meaningful comparisons from its decay and colonisation by invertebrates. Pigs, and in particular foetal pigs are therefore the forensic scientists' usual choice of corpse although America (where else?) has a 'Body Farm' in which dead humans can be observed decaying under a variety of 'real life (death?) situations' (Bass & Jefferson, 2003). Leaving any animal to decay inevitably results in a bad smell and attracts flies – so it requires access to land far from human habitation. It also often requires the body to be protected from birds, dogs, and rats that would drag it away. Consequently, it is difficult both to obtain meaningful replicates and to leave the bodies in a 'normal' environment. Even more importantly, these types of experiments conflict with European Union Animal By Products Regulations that require the bodies of dead farm and domestic animals to be disposed of appropriately to avoid the spread of disease – and leaving a dead pig to moulder on the ground clearly contravenes these.

The use of animals other than insects in forensic investigations has proceeded more slowly and that of plant-based evidence has been slower still. The first use of pollen analysis in a criminal trial appears to have taken place in 1959 (Erdtman, 1969) and although not widely used in criminal trials since then its potential is now being increasingly recognised (Coyle, 2004). By contrast, the use of plants and other organisms in archaeological investigations has been routine for many years. Microbial evidence has seldom featured in criminal trials although this is likely to change with the development of new methods of detection and identification and the concerns over bioterrorism.

By contrast to the slow progress in the use of animal and plant-based evidence, the use of molecular biology in forensic science is now well established and it is an accepted procedure for the identification of individuals. This is usually on the basis of DNA recovered from blood and other body fluids or tissues such as bone marrow and Jobling & Gill (2004) provide a thorough review of current procedures and how things may develop in the future. The use of molecular biology for forensic examination of non-human DNA is less advanced, although this situation will probably improve in the near future as DNA databases become established (Coyle, 2007). When this happens, animals and plants can be expected to play a larger part in legal proceedings.

One of the major stumbling blocks to the use of biological evidence in English trials is the nature of the legal system (Pamplin, 2004). In a criminal prosecution case, the court has to be sure 'beyond all reasonable doubt' before it can return a guilty verdict. The court therefore requires a level of certainty that science can rarely provide. Indeed, science is based upon hypotheses and a scientific hypothesis is one that can be proved wrong – provided that one can find the evidence. Organisms are affected by numerous internal and external factors and therefore the evidence based upon them usually has to have qualifications attached to it. For example, suppose the pollen profile found on mud attached to the suspect's shoes was similar to that found at the site of the crime: this suggests a possible association but it would be impossible to state beyond reasonable doubt that there are not other sites that might have similar profiles – unlikely perhaps, but not beyond doubt. Lawyers are, quite correctly, experts at exploiting the potential weaknesses of biological evidence because it is seldom possible for one to state there is no alternative explanation for

Table A   Questions arising when a body or stains are found in suspicious circumstances

Are the remains or stains of human origin?
Who is the victim?
What was the cause of death?
How long ago did the victim die?
Did the victim die immediately or after a period of time – and if so, how long?
Did the person die at the spot where their body was found?
Did the person die of natural causes, an accident or a criminal act?
If the person was killed as a result of a criminal act who was responsible?

the findings or an event would never happen. Within civil courts, biological evidence has greater potential since here the 'burden of proof' is based upon 'the balance of probabilities'.

Although all biological evidence has its limitations, it can prove extremely useful in answering many of the questions that arise whenever a body is found under suspicious circumstances. The first question is, of course, are the remains human? This might be obvious if the body is whole and fresh or even if there is just a skull but sometimes there may be no more than a single bone or some old bloodstains. Assuming that the remains are human, biological evidence can also help to answer the subsequent questions (Table A).

Similar sorts of questions arise in the cases of wildlife crime (e.g. killing of / trade in protected species), neglect of humans and domestic animals, miss-selling of animal products, and food contamination. This book is intended to demonstrate how an understanding of biology can answer all these questions and is designed for undergraduates who may have a limited background in biology and not the practicing forensic scientist. I have therefore attempted to keep the terminology simple whilst still explaining how an understanding of biological characteristics can be used to provide evidence. Descriptions of potential sources of biological evidence and tests that could be performed upon it continue to grow at a bewildering rate. However, to be truly useful any test / source of evidence should be accurate, simple, affordable, and deliver results within an acceptable time period (Table B). With such a large subject base, it is impossible to cover all topics in depth and readers wishing to identify a maggot or undertake DNA analysis should consult one of the more advanced specialist texts in the appropriate area. Similarly, those wishing more detailed coverage of individual cases would be advised to consult the excellent books by Erzinclioglu (2000), Goff (2000), Greenberg & Kunich (2002) and Smith (1986). Where information would not otherwise be easily accessible to undergraduate students, I have made use of web-based material although the usual caveats apply to such sources.

At the start of each chapter, I have produced a series of 'objectives' to illustrate the material covered. They are written in the style of examination essay questions, so that the reader might use them as part of a self-assessment revision exercise. Similarly, at the end of each chapter I have produced a number of questions to test knowledge and recall of factual information. Also at the end of each chapter, I have made some suggestions for undergraduate projects. Because the usefulness of bio-

Table B Characteristics of an ideal forensic test

Accurate: The results must stand up to intense scrutiny in court

Sensitive: Many forensic samples are extremely small and are finite (i.e. one cannot collect more material once it used up)

Specific: If the test also cross-reacts with other materials then its accuracy will be compromised.

Quick: Investigations cannot be allowed to drag on. If there is a chance that a criminal might offend again they must be apprehended and charged as soon as possible. It is also unfair to deprive a suspect of their liberty for long periods whilst time-consuming tests are conducted.

Simple: The more complex a test becomes the more opportunity there is for mistakes to be made. It also becomes expensive to train people to conduct the tests.

Reliable and repeatable: It is essential that a test can be replicated by other workers at other laboratories.

Affordable: Financial considerations are important. Exceedingly expensive tests cannot be used on a routine basis.

Equipment and reagents are readily available: The effectiveness of the test will be compromised if equipment becomes unusable through lack of spare parts or reagents are difficult to obtain.

---

logical material as forensic evidence depends on a thorough understanding of basic biological processes and the factors that affect them, there is plenty of scope for simple projects based upon identifying species composition or that measure growth rates. Obviously, for the majority of student projects cost, time and facilities will be serious constraints; DNA analysis can be extremely expensive and requires specialist equipment. Similarly, the opportunities to work with human tissues or suitably sized pigs may not exist. However, worthwhile work can still be done using the bodies of laboratory rats and mice or meat and bones bought from a butcher as substitute corpses with plants and invertebrates as sources of evidence.

PART A

Human Remains: Decay, DNA, Tissues and Fluids

1 The decay, discovery and recovery of human bodies

Chapter outline

The Dead Body
The Stages of Decomposition
Factors Affecting the Speed of Decay
Discovery and Recovery of Human Remains
Determining the Age and Provenance of Skeletonized Remains
Future Developments
Quick Quiz
Project Work

Objectives

Compare the chemical and physical characteristics of the different stages of decomposition.

Explain how a body's rate of decomposition is affected by the way in which death occurred and the environment in which it is placed.

Compare the conditions that promote the formation of adipocere and of mummification and how these processes preserve body tissues.

Compare and contrast the various techniques by which a dead body may be located and retrieved.

Evaluate the potential and limitations of radiocarbon dating and stable isotope analysis as means of determining the age and geographical origin of human remains.

The dead body

The time before a person dies is known as the ante mortem period whilst that after death is called the post mortem period. The moment of death is called the 'agonal

---

Essential Forensic Biology, Second Edition   Alan Gunn
© 2009 John Wiley & Sons, Ltd

period' – the word being derived from 'agony' because it used to be believed that death was always a painful experience. Either side of the moment of death is the peri mortem period although there is no consensus about how many hours this should encompass. It is important to know in which of these time periods events took place in order to determine their sequence, the cause of death and whether or not a crime might have been committed. Similarly, it is important to know the length of the post mortem period, referred to as the post mortem interval (PMI) because by knowing exactly when death occurred it is possible, amongst other things, to either include or exclude the involvement of a suspect. The study of what happens to remains after death is known as 'taphonomy' and the factors that affect the remains are called 'taphonomic processes'. Thus, burning, maggot feeding, and cannibalism are all examples of taphonomic processes.

When investigating any death it is essential to keep an open mind as to the possible causes. For example, if the partially clothed body of a woman is found on an isolated moor, there are many possible explanations other than she was murdered following a sexual assault. First of all, she may have lost some of her clothes after death through them decaying and blowing away or from them being ripped off by scavengers (Chapter 8). Secondly, she may have been a keen rambler who liked the open countryside. Most people die of natural causes and she may have suffered from a medical condition that predisposed her to a heart attack, stroke, or similar potentially fatal condition whilst out on one of her walks. Another possibility is that she may have committed suicide: persons with suicidal intent will sometimes choose an isolated spot in which to die. Another explanation for the woman's death would be that she had suffered an accident, such as tripping over a stone, landing badly and receiving a fatal blow to her head. And, finally, it is possible that she was murdered. All of these scenarios must be considered in the light of the evidence provided by the scene and the body.

The stages of decomposition

After we die our body undergoes dramatic changes in its chemical and physical composition and these changes can provide an indication of the PMI. The changes also influence the body's attractiveness to detritivores – organisms that consume dead organic matter – and their species composition and abundance can also be used as indicators of the PMI. Furthermore, the post mortem events may preserve or destroy forensic evidence as well as bring about the formation of artefacts that need to be recognized for what they are. An understanding of the decay process and the factors that influence it is therefore essential for the interpretation of dead human and animal remains.

The stages of decomposition in terrestrial environments can be loosely divided into four stages: fresh, bloat, putrefaction, and putrid dry remains. However, these stages merge into one another and it is impossible to separate them into discrete entities. In addition, a body seldom decays in a uniform manner. Consequently, part of the body may become reduced to a skeleton whilst another part continues to retain fleshy tissue.

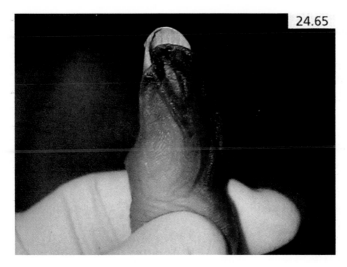

24.65

Figure 1.1 Mummified fingertip. The drying and retraction of the surrounding skin makes the fingernail appear longer and hence the common perception that after death nails continue to grow. The drying of the skin can make taking fingerprints impossible. (Reproduced from Dolinak, D. *et al.*, (2005) *Forensic Pathology Theory and Practice*. Copyright © 2005, Elsevier Academic Press.)

Fresh

Owing to the blood circulation ceasing and the settling of blood to dependent regions (see later), the skin and mucous membranes appear pale immediately after death. Because the circulation has ceased, the tissues and cells are deprived of oxygen and begin to die. Different cells die at different rates, so, for example, brain cells die within 3–7 minutes while skin cells can be taken from a dead body for up to 24 hours after death and still grow in a laboratory culture. Contrary to folklore, human hair and fingernails do not continue to grow after death, although shrinkage of the surrounding skin can make it seem as though they do (Fig. 1.1).

Temperature changes

Because normal metabolism ceases after death our body starts to cool – this cooling is referred to as *algor mortis*: literally, the coldness of death. For many years measurements of body temperature were used as the principal means of determining the PMI but it is now recognized that the technique suffers from a variety of shortcomings. To begin with, the skin surface usually cools rapidly after death and the mouth often remains open so measurements recorded from the mouth or under the armpits would not accurately reflect the core body temperature. The core body temperature must therefore be measured using a long rectal thermometer. However, inserting a rectal thermometer often involves moving the body and removing the clothing and it could also interfere with evidence collection in cases where anal intercourse before or after death might have occurred. It has therefore been suggested that it might be

Table 1.1 Factors affecting the rate at which a body cools after death

Factors that enhance the rate of cooling
Small body size
Low fat content
Body stretched out
Body dismembered
Serious blood loss
Lack of clothes
Wet clothes
Strong air currents
Low ambient temperature
Rain, hail
Cold, damp substrate that conducts heat readily (e.g. damp clay soil)
Body in cold water
Dry atmosphere

Factors that delay the rate of cooling
Large body size
High fat content
Foetal position (reduces the exposed surface area)
Clothing – the nature of clothing is important because a thin, highly insulative layer can
   provide more protection than a thick poorly insulative material.
Insulative covering (e.g. blanket, dustbin bags, paper etc)
Protection from draughts
Warm ambient temperature
Warm microclimate (e.g. body next to a hot radiator)
Exposed to the sun
Insulative substrate (e.g. mattress)
High humidity

feasible to measure temperature changes in the external auditory canal (Rutty, 2005). A second major problem with using temperature as a measure of the PMI is that the rate of cooling depends upon a host of complicating factors starting with the assumption that the body temperature at the time of death was 37 °C. In reality the body temperature may be higher (e.g. owing to infection, exercise or heat stroke) or lower (e.g. hypothermia or severe blood loss). In addition, the rate of temperature loss depends upon numerous factors (Table 1.1). For example, the body of a fat man who dies inside a car on a hot sunny day may not lose heat to any appreciable extent; indeed, his body temperature may even increase.

Various formulae have been developed to relate body temperature to time since death but these are mostly too simplistic to be reliable. Clauss Henßge has designed a sophisticated nomogram (Fig. 1.2) that accounts for body weight and environmental temperature and allows for corrective factors to be applied according to the individual circumstances of the case (Henßge & Madea, 2004; Henßge et al., 2002). A nomogram is a graphical calculator that usually has three scales. Two of these scales record known values (rectal and environmental temperature) and the third scale is the one from which the result is read off (time since death). Unfortunately, even this approach has limitations – for example, it is not reliable if the body was

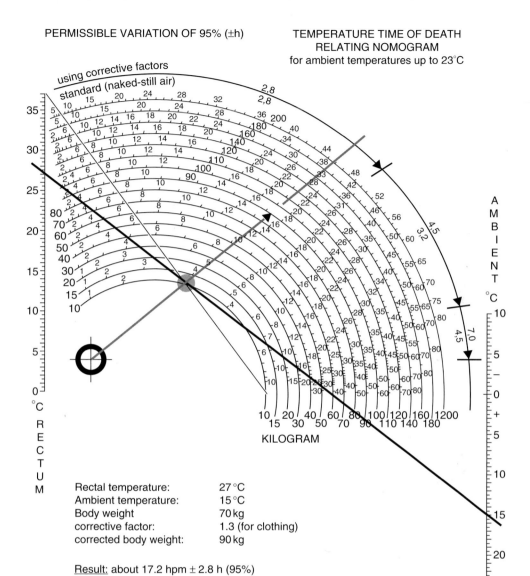

Figure 1.2 Clauss Henßge's nomogram for the determination of time since death from body temperature. (Reproduced from Henßge, C. and Madea, B. (2004) Estimation of time since death. *Forensic Science International*, **144**, 167–175. With permission from Elsevier.) The nomogram works as follows (a) A straight line is drawn between the rectal temperature and the ambient temperature. In the case illustrated here the line is therefore drawn from 27 °C to 15 °C. (b) The 'standard' is a naked body lying in an extended position in still air and therefore 'corrective factors' need to be applied for any situations other than this. These factors are listed by Henßge et al. (2002). In this example, the body was found wearing three thin layers of dry clothes in still air and therefore the corrective factor is 1.3. The weight of the body is now multiplied by the corrective factor. The body weighed 70 kg and therefore 70 × 1.3 = 91 kg. The nomogram goes up in units of ten and therefore 91 kg is rounded down to 90 kg. (c) A second straight line is drawn from the centre of the circle that is found at the left-hand side of the nomogram so that it hits the intersection of the nomogram's diagonal line and that drawn between the rectal temperature and the ambient temperature in step (a). The line is then continued until it hits the outermost circle. (d) Where the line drawn in step (c) hits the 90 kg semicircle is the time since death (17.2 hours). Where the line hits the outermost circle one can read off the 95% confidence limits (2.8 hours). Therefore, the person is judged to have been dead for 17.2 + 2.8 hours (95% CI).

left exposed to the sun or if there is reason to believe that it was moved after death. In the latter situation, the body would have been exposed to at least two different environments and could therefore have spent time cooling at two very different rates. This is not to say that temperature measurements are pointless but one should be aware of possible complicating factors.

Body temperature, like most biological measurements of the PMI can be classed as a 'rate method'. Rate methods are those in which events are initiated or stopped at the time of death and the subsequent rate of change provides an estimate of elapsed time. Other examples include the increase in the potassium ion concentration in the vitreous humor of the eye, the development of rigor mortis and the growth of maggots on the dead body. Rate methods become increasingly inaccurate the longer the PMI because they suffer from being influenced by a wide variety of biotic and abiotic factors but as long as their limitations are recognized they can be extremely useful and if there is concordance between several different methods then the time of death can be predicted with a fair degree of confidence. Furthermore, in the absence of any other evidence an indication is more useful to a police investigation than nothing at all. The other methods of determining the time since death are known as 'concurrence methods' and they work by evaluating the occurrence of events that happened at known times at or around the time of death. Typical concurrence events would be finding that the victim's watch had stopped at a particular time as a consequence of being smashed (e.g. following a fall or during a struggle) or that mobile phone records indicated that the victim must have been alive until at least a certain date and time.

Chemical changes

Owing to the lack of oxygen, after death cellular processes switch from aerobic to anaerobic and there are dramatic increases and decreases in specific metabolites. Furthermore, as membrane integrity is lost metabolites redistribute within and between tissues. These changes do not take place uniformly throughout the body at the same time. For example, energy metabolism ceases more rapidly in the blood than it does in the vitreous humor of the eye. A number of workers have attempted to estimate the PMI by measuring chemical changes after death (e.g. Vass et al., 2002). Unfortunately, few comparative studies have been made between different chemical measurements or between chemical measurements and other existing techniques. In addition, most studies to date lack field data and their reliability could potentially be adversely affected by environmental factors such as temperature and ante mortem factors such as age, drug use and disease (Henßge & Madea, 2004; Madea & Musshoff, 2007). The most commonly used chemical measurement of PMI is the determination of potassium ion concentration in the vitreous humor of the eye although there are marked discrepancies between authors concerning its reliability (Chapter 2).

Hypostasis

Between 20 and 120 minutes after death hypostasis (also called *livor mortis* and post mortem lividity) is usually seen – it can be found in all bodies but may be difficult to observe. Hypostasis is a purple or reddish purple discoloration of the skin caused by the blood settling in the veins and capillaries of the dependent parts of the body. Blood plasma also settles to the dependent regions and this causes oedema (fluid accumulation) and the formation of blisters on the surface of the skin. If the person is lying on their back, hypostasis will develop in the back and those body surfaces adjacent to the ground whilst if the person is hanging by their neck, pronounced hypostasis will develop in their hands, forearms and lower legs. It starts as a series of blotches that then spread and deepen in colour with time. Initially, the blood remains in the blood vessels but eventually the blood cells haemolyse (break down and rupture) and the pigment diffuses out into the surrounding tissues, where it may be metabolized to sulphaemoglobin that gives rise to a greenish discoloration. Sulphaemoglobin is not present in normal blood although it may be formed after exposure to drugs such as sulphonamides. This emphasizes the need to be aware that normal decomposition processes may mimic those that are induced before death or by the action that induced death.

The rate of development of hypostasis varies from body to body and is also influenced by underlying medical conditions, such as circulatory disease. Consequently, there is some variation in the literature about when events begin and when they reach their maximal effect. Indeed, hypostasis may not develop at all in infants, the elderly or those suffering from anaemia. Some of the literature suggests that after about 10–12 hours of a body remaining in a set position, the discoloration caused by hypostasis becomes 'fixed' (Fig. 1.3). Furthermore, if the body is then moved and left in a different position a second area of discoloration forms. Two or more distinct patterns of discoloration therefore indicate movement of the body. However, according to Saukko & Knight (2004) there is so much variation in the time it takes for 'fixation' to develop, if it develops at all, that it is not a particularly reliable forensic indicator of the PMI or evidence of movement after death.

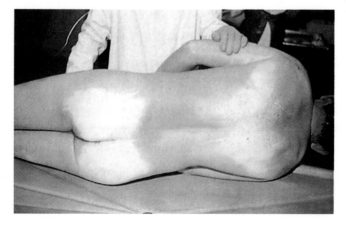

Figure 1.3 Characteristic pattern of hypostasis and pressure pallor resulting from a dead body lying on its back. The reddening results from the settling of blood in the veins whilst the pale regions are where the pressure of the body against the underlying substrate has constricted the vessels. (Reproduced from Shepherd, R. (2003) *Simpson's Forensic Medicine*, 12th edn. Copyright 2003, Hodder Arnold, London.)

Pressure, whether from tight fitting clothes such as belts and bra straps, a ligature around the neck, ropes used to bind hands together, or corrugations in the surface on which the body is resting, will prevent the underlying blood vessels from filling with blood and therefore these regions will appear paler than their surroundings – this is known as 'pressure pallor' or 'contact pallor'. Whilst the body is fresh, it is possible to distinguish between ante mortem bruising and hypostasis because bruising results from the leakage of blood out of damaged blood vessels into the surrounding tissues and the consequent formation of clots. By contrast, in hypostasis the blood is restricted to dilated blood vessels although as time passes and tissues decay, blood begins to leak out of the vessels and it becomes more difficult to distinguish between the two.

Initially, blood remains liquid within the circulatory system after death, rather than coagulating, because of the release of fibrinolysins from the capillary walls. These chemicals destroy fibrinogen and therefore prevent clots from forming. However, wounds inflicted after death do not bleed profusely because the heart is no longer beating and therefore blood pressure is not maintained. Blood from even a severed artery therefore trickles out as a consequence of gravity rather than being spurted out as it might if inflicted during life. A common question that arises when a person's body is found at the bottom of a building after suffering a great fall is whether or not they were still alive when they hit the floor. This is important because it is possible for a murderer to attempt to mask the wounds caused by a violent assault within the much greater trauma that would result from a fall – especially if the fall could be construed as an accident or suicide. If the victim was already dead then their body might bleed a lot less than if they were still alive at the time of impact. Furthermore, if the person was bleeding before being thrown it would be expected that bloodstains would be found near the point from which the body fell and/or cast from it during the fall (Chapter 2). Unlike the situation on land, in the case of drowning or a dead body disposed of in a lake or river, there may be a considerable loss of blood from wounds. After initially sinking, a dead body tends to rise to the surface owing to the accumulation of gas from the decay process and then floats face downwards. Consequently, the blood pools in the facial and dependent regions and wounds affecting these areas after death may bleed profusely.

Changes in muscle tone

Immediately after death, the muscles usually become flaccid and the joints relax such that a person's height may increase by as much as 3 cm. Furthermore, the body may be found in a posture that would be highly uncomfortable in life. Once consciousness is lost, a standing individual collapses without making any attempt to break their fall whilst a seated individual slumps forwards (usually) and may fall to the floor unless supported. Consequently, the body may receive injuries which might themselves have been life-threatening had the person not already been dead. The relaxation of muscles can lead to the sphincters loosening, and the release of urine and faeces or the regurgitation of gut contents at or shortly after the moment of death. Suffocation can lead to the victim urinating involuntarily but this may also happen naturally at the time of death. Therefore, it would be unwise to make

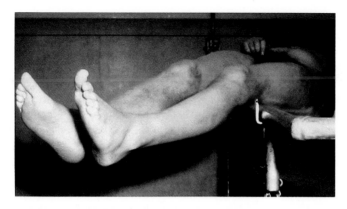

Figure 1.4 *Rigor mortis* in the lower limbs. Note how the legs remain in a fixed, rigid position despite the lack of support. (Reproduced from Saukko, P. and Knight, B. (2004) *Knight's Forensic Pathology*, 3rd edn. Copyright 2004, Hodder Arnold.)

too much of such findings unless there was other evidence to indicate that criminal activity may have been involved. By contrast, when a person is in a coma the volume of urine in the bladder can increase markedly because they are not responsive to stimuli that would normally wake them up. Consequently, an unusually distended bladder is an indication that a person was comatose for several hours before they died.

Approximately 3–4 hours after death, *rigor mortis*, the stiffening of muscles and limbs becomes noticeable and the whole body becomes rigid by about 12 hours (Fig. 1.4). The condition can, however, be broken by pulling forcefully on the affected limbs. Rigor is usually first noticeable in the small muscles of the face and those being used most actively prior to death. Rigor affects both the skeletal and the smooth muscles. When it affects the *arrector pili* muscles it can result in the scalp and body hairs standing on end – this can make it look as though the person died in a state of shock. The *arrector pili* are smooth muscles that run from the superficial dermis of the skin to the side of the hair follicles. Normally our hair emerges at an angle to the skin surface but when the *arrector pili* are stimulated to contract – for example as a consequence of the body's response to cold or stress – the hair is pulled into a more upright position. This also gives rise to the phenomenon of 'goose bumps'. The rigor that follows death can give rise to a similar appearance.

Rigor mortis is brought about by the rise in the intracellular concentration of calcium ions in muscle cells that follows death, as the membranes around the sarcoplasmic reticulum and the cell surface become leaky and calcium ions are therefore able move down their concentration gradient into the cytoplasm of the muscle cells. This rise causes the regulatory proteins troponin and tropomyosin to move aside, thereby permitting the muscle filaments actin and myosin to bind together to form cross bridges. This is possible because the head of a myosin molecule would already be charged with ATP before death. However, actin and myosin, once bound, are unable to detach from one another because this process requires the presence of ATP – and this is no longer being formed. Thus, the actin and myosin filaments remain linked together by the immobilized cross bridges, resulting in the stiffened condition of dead muscles. Subsequently, *rigor mortis* gradually subsides as the proteins begin to degrade and it disappears after about 36 hours. The speed of development of *rigor mortis* and its duration are both heavily influenced by environmental temperature with onset commencing earlier and duration shorter at high environmental temperatures. By contrast, onset is delayed at low temperatures and at a constant 4 °C may last for at least 16 days with partial stiffening still detectable up until 28 days after death (Varetto & Curto, 2004). Children tend to develop *rigor mortis* sooner than adults whilst onset is said to be delayed if death was owing to asphyxiation or poisoning with carbon monoxide. The extent and degree of *rigor mortis* is therefore not an especially accurate measure of the PMI.

Heat stiffening is distinct from rigor mortis and results from the body being exposed to extreme heat. It causes the body to exhibit what is known as a 'pugilistic posture' (Chapter 5) and evidence of severe burning will inevitably be apparent. Exposure to very low temperatures will also cause the body to stiffen but can prevent the onset of *rigor mortis* entirely. In this case, the body will become flaccid when it is warmed up and may then subsequently exhibit *rigor mortis*. In this way, a murderer may confuse a police investigation by storing his victim in a freezer immediately after death before disposing of the body some time later. There is a considerable literature in the food science sector on means of distinguishing between fresh meat and that which has been frozen but there are far fewer studies on human tissues. Miras *et al.*, (2001) have suggested that it would be possible to distinguish muscle tissue that had previously been frozen by its higher levels of the enzyme short-chain 3-hydroxyacyl-CoA dehydrogenase but it is uncertain how effective this would be in practise and would presumably rely on the body being discovered within a few hours of defrosting.

Unlike *rigor mortis*, 'cadaveric rigidity' (also called 'cadaveric spasm') sets in immediately after death and according to Shepherd (2003) is a 'forensic rarity'. It may affect part or all of the body and is said to be associated with individuals who were extremely stressed, emotionally and physically, immediately before they died. However, one would have thought that this would include most murder victims and also many who die of painful medical conditions, so there must be some other reason why it is not found more frequently. Nevertheless, its occurrence can provide useful indications of a person's last actions such as their hands may be found firmly grasping hair from their attacker or an object in a vain attempt to prevent themselves from drowning (Fig. 1.5). Persons who commit suicide by shooting themselves may be found with the gun so tightly held that it would have been impossible for a second person to have arranged the corpse in this manner after death. However, there is no evidence to suggest that majority of people who kill themselves in this way exhibit this trait.

Indications of poisoning

Sometimes the cause of death may result in striking changes to normal skin coloration. For example, deaths from carbon monoxide poisoning often result in a cherry red / pink coloration to the skin, lips and internal body organs (Fig. 1.6) although if the body is not discovered until several hours after death the coloration may not be immediately apparent owing to the settling of the blood to the dependent regions.

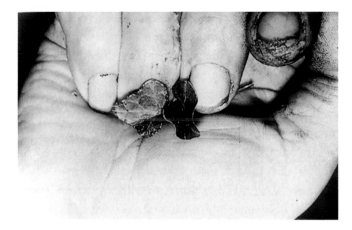

Figure 1.5 Cadaveric rigidity. This person grasped at vegetation before falling into water. (Reproduced from Shepherd, R. (2003) *Simpson's Forensic Medicine*, 12th edn. Copyright 2003, Hodder Arnold, London.)

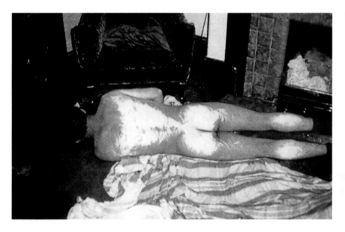

Figure 1.6 Cherry-red coloured hypostasis as a consequence of carbon monoxide poisoning causing the formation of carboxyhaemoglobin. (Reproduced from Shepherd, R. (2003) *Simpson's Forensic Medicine*, 12th edn. Copyright 2003, Hodder Arnold, London.)

Carbon monoxide gas forms during the combustion of many substances and poisoning is a common feature of accidental deaths in which people are exposed to fumes from a faulty gas boiler or during fires and suicides in which the victim breaths in vehicle exhaust fumes. Carbon monoxide poisoning may also be the cause of death in homicides resulting from arson or where the flue to a fire or gas boiler is deliberately blocked. Carbon monoxide has much greater affinity than oxygen for the haeme molecule of haemoglobin and therefore, even at very low atmospheric concentrations it will rapidly replace it and thereby reduce the oxygen carrying capacity of the blood. When carbon monoxide binds with haemoglobin in the blood or myoglobin in the muscles it forms carboxyhaemoglobin and carboxymyoglobin respectively and they are responsible for the pink coloration. There are cases in which carbon monoxide poisoning does not result in the formation of a cherry pink coloration (Carson & Esslinger, 2001) and it can be difficult to spot when the victim is dark skinned – though it may be apparent in the lighter regions such as the palms of the hands or inside the lips or the tongue. There are big differences in susceptibility to carbon monoxide poisoning and this is at least partly a consequence of age, size and general health. For example, children tend to be more susceptible owing to their higher respiration rate.

Cyanide poisoning also results in the skin developing cherry red coloration although it is said to be somewhat darker than that caused by carbon monoxide. Cyanide ingestion is sometimes used as means of suicide and homicide but cyanide is also a potentially lethal component of the smoke formed during the combustion of many substances (e.g. wool, plastics) and its effect in conjunction with carbon monoxide is additive since they work by different mechanisms. Indeed, a person inhaling smoke may die of cyanide poisoning before there is marked rise in the levels of carboxyhaemoglobin. Cyanide affects a variety of enzymes and cell processes but has its principal effect through the inhibition of cytochrome oxidase and thereby prevents the production of ATP via oxidative phosphorylation. The cherry red coloration results from the increased oxygenation of the blood in the veins as a consequence of the inability of cells to utilize oxygen for aerobic metabolism.

Cyanide poisoning can also cause cyanosis – a bluish tinge to the skin, fingernails and mucous membranes – although the term is derived from the blue–green colour cyan rather than the chemical cyanide. Cyanosis may be localized or more widespread and be found on its own or in conjunction with the cherry red skin coloration. It is caused by a reduction in the level of oxygen in the blood and therefore darker deoxygenated blood imparts colour to the tissues, blood vessels, and capillaries rather than the normal bright red oxygenated blood. Cyanosis is therefore a common symptom of a whole range of conditions that interfere with the supply of oxygenated blood to the tissues including carbon monoxide poisoning, a heart attack and asphyxia from hanging. Cyanide has the reputation for causing rapid, near instantaneous death, but although this can occur a lot depends on the nature of the cyanide and its means of delivery (e.g. breathing in gaseous hydrogen cyanide, ingestion of a salt in solid or liquid form or absorption through the skin) and the dose. Death may occur within minutes or hours of acquiring a lethal dose and involve a long period of struggling to breathe so cyanosis is to be expected.

Bloat

The intestines are packed with bacteria and these do not die with the person. These micro-organisms break down the dead cells of the intestines, while some, especially the *Clostridia* and the enterobacteria, start to invade the other body parts. At the same time, the body undergoes its own intrinsic breakdown, known as autolysis, that results from the release of enzymes from the lysosomes (subcellular organelles that contain digestive enzymes) thereby causing cells to digest themselves and chemicals, such as the stomach acids, from the dead cells and tissues. The pancreas, for example, is packed with digestive enzymes, and so rapidly digests itself. Autolysis may also occur on a more restricted scale in a living person as a consequence of certain diseases.

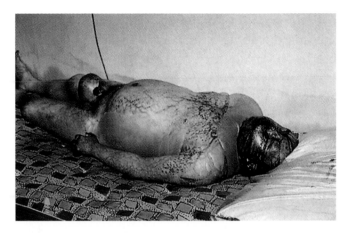

Figure 1.7 Late bloat stage of decomposition. The body is about 7 days old and exhibits pronounced swelling owing to accumulation of gas. Note discoloration of the skin and exudates from the mouth and nose. (Reproduced from Shepherd, R. (2003) *Simpson's Forensic Medicine*, 12th edn. Copyright 2003, Hodder Arnold, London.)

The decomposing tissues release green substances and gas which make the skin discoloured and blistered, starting on the abdomen in the area above the caecum (Fig. 1.7). The front of the body swells, the tongue may protrude and fluid from the lungs oozes out of the mouth and nostrils (Fig. 1.8). This is accompanied by a terrible smell as gasses such as hydrogen sulphide and mercaptans, sulphur-containing organic molecules, are produced as end products of bacterial metabolism. Methane (which does not smell) is also produced in large quantities and contributes to the swelling of the body. In the UK, this stage is reached after about 4–6 days during spring and summer but would take longer during colder winter weather. The accumulation of gas can become so severe that the abdominal wall ruptures and this may lead to concerns over whether the wound was caused maliciously. In 1547, the corpse of King Henry VIII underwent such extreme bloat that his coffin, which was being transported back to Windsor castle for burial, exploded overnight and dogs were found feeding on the exposed remains in the morning. This was deemed to be divine judgement on the king for his dissolution of the monasteries.

Detritivores

Blowflies and other detritivores are attracted by the odour of blood and decomposition (Fig. 1.9), and as the smell changes during the decay process so does the species of invertebrates that are attracted. Therefore, 'fresh corpses' attract different detritivores to corpses in an advanced state of decay (Table 1.2). Blowflies will lay their eggs on bodies within seconds of death occurring but do not lay their eggs on corpses once these have passed a certain state of decomposition or they have become dry or mummified. By contrast, dermestid beetles do not colonize corpses until these have started to dry out (for more details see Chapters 6 and 7).

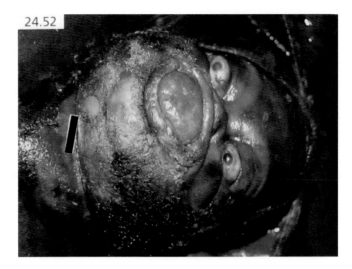

Figure 1.8   Late bloat stage of decomposition. Note how the swelling has made recognition of facial features impossible. The tongue is forced out and the eyeballs bulge as a consequence of internal pressures. These are normal decomposition features and should not be taken as an indication of asphyxiation. (Reproduced from Dolinak, D. *et al.*, (2005) *Forensic Pathology Theory and Practice*. Copyright © 2005, Elsevier Academic Press.)

Figure 1.9   Blowfly maggots developing upon a corpse. Note how mature maggots can be seen crawling over the surface and the discoloration of the skin. (Reproduced from Klotzbach, H. *et al.*, (2004) Information is everything – a case report demonstrating the necessity of entomological knowledge at the crime scene. *Agrawal's Internet Journal of Forensic Medicine and Toxicology*, **5**, 19–21. Copyright © 2004, with permission from Elsevier.)

Putrefaction

Some authors distinguish several stages of putrefaction (decay) but the usefulness of this is uncertain. As the body enters the bloat stage, it is said to be 'actively decaying' and during this time the soft body parts rapidly disappear as a result of

Table 1.2 The sequence in which insects arrive and colonize a corpse during the decomposition process. The stages of decay merge into one another and the insects may arrive or leave sooner or later than is indicated in the table depending upon the individual circumstances. For more details see Chapters 6 and 7

Stage of decomposition	Insect
Fresh	Blowfly eggs and 1st instar larvae
	Fleshfly 1st instar larvae
	Burying beetle adults
Bloat	Blowfly eggs + 1st, 2nd, 3rd instar larvae
	Fleshfly 1st, 2nd, 3rd instar larvae
	Burying beetle adults and larvae
	Histerid beetle adults and larvae
Putrefaction	No blowfly eggs once advanced putrefaction
	Blowfly 2nd, 3rd instar larvae
	Fleshfly 2nd, 3rd instar larvae
	Blowfly & fleshfly larvae leaving corpse for pupation site
	Histerid beetle adults and larvae
	Eristalid fly larvae (liquefied regions)
	Phorid fly larvae (later stages of putrefaction)
	Piophilid fly larvae (later stages of putrefaction)
Putrid dry remains	No blowfly larvae
	Stratiomyid fly larvae
	Dermestid beetle adults and larvae
	Tineid moth larvae
	Pyralid moth larvae

autolysis and microbial, insect and other animal activity. The body then collapses in on itself as gasses are no longer retained by the skin. At this point, the body enters a stage of 'advanced decay' and, unless the body is mummified, much of the skin is lost. Obese people tend to decay faster than those of average weight and this is said to be due to the 'greater amount of liquid in the tissues whose succulence favours the development and dissemination of bacteria' (Campobasso *et al.*, 2001). At first sight, this appears surprising since fat has a lower water content than other body tissues and obese individuals therefore have a lower than average water content. However, fat can act as a 'waterproofing' preventing the evaporation of water and therefore the drying out of the corpse whilst its metabolism yields large amounts of water.

Adipocere

Adipocere (grave wax or corpse wax) is formed during the decay process if the conditions are suitable and it is capable of influencing the future course of decay (Forbes *et al.*, 2004; Fiedler & Graw, 2003). It is a fatty substance that is variously described as being whitish, greyish or yellowish and with a consistency ranging from

Figure 1.10 The formation of adipocere has preserved the body of this child despite it being buried for about 3 years. (Reproduced from Shepherd, R. (2003) *Simpson's Forensic Medicine*, 12th edn. Copyright 2003, Hodder Arnold, London.)

paste-like to crumbly. Extensive adipocere formation inhibits further decomposition and ensures that the body is preserved for many years (Fig. 1.10). Adipocere formation is therefore a nuisance in municipal graveyards because it prevents the authorities from recycling grave plots but very useful to forensic scientists and archaeologists who wish to autopsy long-dead bodies.

The term 'adipocere' refers to a complex of chemicals rather than a single chemical compound and it results from the breakdown of body lipids. After death, autolysis and bacterial decomposition of triglycerides, which make up the majority of the body's lipid stores, results in the production of glycerol and free fatty acids. The free fatty acids comprise a mixture of both saturated and unsaturated forms, but as adipocere formation progresses, the saturated forms become predominant. The fatty acids lower the surrounding pH and thereby reduce microbial activity and further decomposition. Adipocere has a characteristic odour the nature of which changes with time and this is used to train cadaver dogs to detect dead bodies. Extensive adipocere formation results in the body swelling and consequently the pattern of clothing, binding ropes or ligatures can become imprinted on the body surface whilst incised or puncture wounds may be closed and become difficult to detect. Adipocere formation is not exclusive to human decomposition (Forbes, *et al.*, 2005d) and this should be borne in mind if there is a possibility that human and animal remains are mixed together. For example, the bodies of animals are often found at the bottom of disused mine shafts having stumbled in or been thrown in by a farmer looking for a quick means of disposing of dead livestock. Murderers will also make use of such facilities.

Adipocere formation has been described from bodies recovered from a wide variety of conditions including fresh water, seawater and peatbogs, shallow and deep graves, tightly sealed containers, and in bodies buried but not enclosed at all (e.g. Evershed, 1992; Mellen *et al.*, 1993). Some authors mention that warm conditions may speed its formation but adipocere has been recorded from bodies recovered from seawater at a temperature of 10–12 °C and from icy glaciers (Kahana *et al.*, 1999; Ambach *et al.*, 1992) – the preservation of the 5300-year-old 'Iceman' found in the Tyrol region appears to be at least partly a consequence of the formation of adipocere (Sharp, 1997; Bereuter *et al.*, 1997). A wide variety of durations are cited in the literature for the time taken for adipocere formation to become extensive, ranging from weeks to months to over a year. Obviously, the time will be heavily dependent upon the local conditions and it is not yet possible to use the formation of adipocere as an estimate of the PMI. However, because adipocere leaks out of the body, its presence in the soil can indicate whether a corpse was left in a particular location but then removed or if the extent of adipocere formation in the body matches that which might be expected in the surrounding soil if the body had lain there since death.

Forbes *et al.* (2005 a, b, c) conducted an extensive series of experiments on the physical and chemical factors promoting the formation of adipocere. They found that adipocere would form in soil types ranging from sandy to clayey, provided that the soils were kept moist, and also in sterile soil that was heated at 200 °C for 12 hours to remove the normal soil microbial flora. 'Bodies' buried directly in the ground tended to form adipocere more rapidly than those contained in a coffin. Interestingly, placing the 'body' in a plastic bag retarded the formation of adipocere but if the 'body' was clothed and then placed in the plastic bag adipocere formation was promoted. They suggested that this was owing to the clothing absorbing glycerol and other decay products that would otherwise inhibit the pathways through with adipocere is formed. Polyester clothing was deemed to be the most effective, probably as a consequence of its ability to retain water and, compared to cotton clothing, resistance to decay.

Mummification

Mummification occurs when a body is exposed to dry conditions coupled with extreme heat or cold, especially if there is also a strong air current to encourage the evaporation of water. It is typically seen in persons who die in deserts, such as the hot Sahara and the cold Tibetan plateau. It is also found in murder victims who are bricked up in chimneys or persons who die in well sealed centrally heated rooms. Size is important, and dead babies, owing to their large surface area to volume ratio lose water more rapidly than an adult. Newly born babies lack an active gut microbial flora and therefore not only do they lose water quickly, they may dehydrate before microbial decomposition can cause major destruction of tissues. Once a body has mummified it can remain intact for hundreds of years provided that it is in a dry environment and those insects that are capable of consuming dry organic matter (e.g. dermestid beetles and the larvae of tineid moths) do not gain access to it.

Putrid dry remains

After the skin and soft tissues are removed, the body is reduced to the hard skeleton and those structures that are more difficult to break down, such as the tendons,

Table 1.3   Summary of the stages of decomposition and their characteristic features

Stage of decomposition	Characteristics
Fresh	Body starts to cool and autolysis begins. Hypostasis and *rigor mortis* may be seen.
Bloat	Discoloration of skin surface, body swells from accumulation of gasses. Tongue protrudes, fluid expelled from orifices. Soft tissues visibly decaying. Rapid decay owing to intense microbial and invertebrate activity.
Putrefaction	Progressive loss of skin and soft tissues. Body deflates as decomposition gasses escape. Decay owing to invertebrate and microbial activity starts to slow down once soft tissues removed and body starts to dry out.
Putrid dry remains	Skin and soft tissues lost. Decay proceeds more slowly. Progressive loss of uterus/prostate gland, tendons, cartilage, fingernails, hair. Skeleton may become disarticulated through environmental and biological processes.

ligaments, fingernails and hair. Organs such as the uterus and prostate gland are also fairly resistant to decay and may last for several months if the body is kept in a well-sealed container. Because there are still traces of dead organic matter being broken down by microbes, a skeletonized body still smells of decay.

Bones also undergo a decay process referred to as diagenesis, and their chemical composition and microscopic structure changes as a consequence of microbial attack and environmental exposure. Bone decay begins soon after death and bacteria invade the bone via its natural pores and create tunnels within it (Jans *et al.*, 2004). Significant changes are therefore already apparent by the time of skeletonization and invasion by soil bacteria is not thought to be of major importance. Not surprisingly, therefore, those bones that are closest to the abdominal cavity tend to exhibit the most marked microbial attack. However, it is not yet possible to use these changes to estimate the PMI. Lasczkowski *et al.* (2002) have suggested that the rate of disappearance of the chondrocyte cells (cartilage cells that produce collagen) could be used as a measure of the PMI during the first few weeks after death but more experimental work is required to confirm how reliable this approach would be. Indeed, estimating the age of skeletal remains from morphological and biochemical characteristics can only be done very crudely. For example, the association of ligaments and other soft tissues with the bones is said to indicate that the remains are less than 5 years old whilst the presence of blood pigments within the bones indicates they are less than 10 years old.

Factors affecting the speed of decay

The rapidity and extent to which a corpse is colonized by the larvae of blowflies, along with the activities of other invertebrates, microbes and vertebrates such as dogs and birds, heavily influences the speed with which a body decays. Consequently, those factors that restrict their access or reduce their activity, such as physical exclusion, lack of oxygen or the temperature being too low or too high reduces the rate of decomposition enormously.

Geographical location

The abundance and species composition of the microbial, invertebrate and vertebrate detritivore community varies between regions and this affects the speed with which a body is located, colonized and decomposed. Decay proceeds much faster in the tropics, where conditions are both hot and humid, and slower in cold or dry conditions. In the tropics, a corpse can become a moving mass of maggots within 24 hours but in the UK it would take several days to reach this stage, even during the summer. Temperatures that are too high or climates that are too dry also restrict the activity of invertebrates and microbes and thereby reduce the rate of decay.

Time of year

Bodies decay fastest in warm damp environments. Consequently, in the UK, decay is fastest during the summer months and slowest during winter. This is partly owing to the effect of the environment on microbial decay and partly through the invertebrate detritivores showing distinct patterns of seasonal activity.

Cities and large towns offer warm microclimates and therefore some blowfly species may remain active there throughout the winter period but they would be inactive in the surrounding region. In both cities and the countryside, the adults of some blowfly species enter buildings during the autumn period and attempt to overwinter indoors (e.g. in loft spaces, garages and sheds). Should a body be placed within a building where they occur and the temperature is high enough for them to be active, then colonization of the corpse will commence even in the depths of winter.

Exposure to sunlight

The effect of exposure to sunlight is case dependent. If a body is warmed by the sun's rays bacterial decay will be promoted but if there is low humidity and a strong wind the body would desiccate and mummify thereby retaining much of the its integrity. Galloway et al. (1989) have written a detailed account of how the human body decays under arid conditions.

A body exposed to the sun is usually visible and smellable to both vertebrate and invertebrate detritivores and this leads to its rapid dismemberment and consumption and/ or colonization and consumption. Invertebrates avoid laying their eggs on and colonizing regions of a corpse that are exposed to the full sun because the combination of desiccation and UV light would kill their delicate eggs and larvae. However, eggs laid on the under surfaces or beneath clothing or other coverings (provided that they are not too tight to restrict access) will be protected from the sun and have a more humid microenvironment. This facilitates microbial and maggot growth and consequently these covered regions may decay more rapidly than exposed body parts although a lot depends on the nature of the covering material.

Wrapping and confinement

Persons disposing of a dead body often wrap or otherwise cover it up so as to make its transport easier and/or reduce the risk of its subsequent discovery. Metal foil is commonly used as is plastic sheeting and bags and cling film. Corpses are sometimes wrapped within carpets but this must make movement extremely difficult because any carpet large enough to conceal a body is already a heavy, bulky object. Corpses are also found in the boots of vehicles, placed in suitcases or similar large containers, left within locked rooms or bricked up. It is surprising how many bodies are placed under floorboards and this is usually one of the first places that police will look for concealed evidence. All of these scenarios have their own individual effects on the rate of decomposition but the common factor is that the rate of decay is slowest where the covering is most effective at excluding oxygen, vertebrates and invertebrates. Once a body starts to decay – which happens regardless of the scenario unless the temperature is below freezing – the smell attracts detritivores of various shapes and sizes; flies usually find their way into a locked car boot whilst foxes will bite and claw their way through plastic wrapping. If the detritivores are unable to gain access then the body will decay slowly but this helps to preserve evidence should the body be eventually discovered. Furthermore, it is not unusual for murder weapons and other evidence to be enclosed or entombed with the body. Encasing a body within concrete is no guarantee of successful concealment; the smell of decay may still be detectable and when someone goes missing and a suspect starts mixing concrete it isn't long before suspicions are aroused (Preuß et al., 2006). Interestingly, analysis of the composition of the concrete (a process known as petrography) can enable determination not only where and when the concrete was sourced but also how old it is and thereby, by inference, the PMI. For example, an unusual case arose in USA in which the body of a young woman was found in Nevada entombed within a home-made concrete sarcophagus (Morel, 2004). Petrographic analysis identified not only the type of concrete but that it was made within a narrow range of about 1.5–2 years previously. By contrast, anthropological analysis of the woman's remains could only place the time of death as some point between 2 and 7 years previously. It was also possible to link the sarcophagus with a concrete spattered retractable utility knife found nearby that was the probable murder weapon.

Burial underground

Buried corpses decay approximately four times slower than those left on the surface, and the deeper they are buried, the slower they decay (Dent et al., 2004). The nature of the soil affects the rate of decay directly through chemical actions and indirectly through its effect on the abundance and activity of soil organisms. Microbial density varies enormously with the soil type but figures are generally in the region of 2×10^9 per gram in the top metre and 10^8 per gram at a depth of 1–8 metres (Coleman *et al.*, 2004). Heavy clay soils are poorly aerated and therefore have low oxygen levels and this reduces microbial activity and hence the rate of decay. Very acid soil reduces microbial activity but the low pH dissolves soft tissues and bone. High soil calcium content reduces chemical dissolution of the bones but will not prevent its microbial decomposition. Even a shallow covering of soil usually prevents blowflies from colonizing a body and hence reduces the rate of decomposition. However, some fly species, such as *Muscina stabulans* lay eggs on the soil surface and their larvae then burrow down to the body whilst adult coffin flies (*Conicera tibialis*), which are very small, will crawl through cracks in the soil to locate bodies a metre or more below ground. Shallow buried bodies may, however, be detected and dug up by dogs, foxes or badgers and as soon as this occurs the rate of decay will increase and there is a high probability that the body will be dismembered. Corpses are often wrapped or enclosed in something before they are buried and this will further reduce the rate of decay dependent upon the degree to which they are airtight and/or can exclude detritivores.

Hanging above ground

Bodies are usually found suspended above ground as a result of suicidal or homicidal hanging (lynching) although they may also end up within trees as a result of being hurled there by an explosion, thrown from a vehicle after a crash or knocked there after being hit by a train or vehicle. A body that is left hanging above ground may decay more slowly than one that is lying on the surface of the ground (Wyss & Cherix, 2004). This is probably because when a body is suspended in mid air, there is not a moist, dark, under-surface where flies can lay their eggs, the circulation of air will promote drying out, and many maggots would fall off whilst crawling around or be washed off by the rain.

Burial underwater

Bodies that are disposed of in water are often said to decay twice as slowly as when the body is exposed to air. This is probably largely due to the lower temperature and the rate declines with depth because of the progressively lower temperatures and oxygen levels. Bodies are said to decay more slowly in the sea than they do in freshwater because it contains fewer marine micro-organisms. In truth there is little experimental data on decay rates in marine environments and surface seawater contains similar bacterial densities to those of lakes (10^6 per ml compared with 10^5–10^7 per ml) whilst microbial populations in sediments, even at great depths, are much higher than previously thought (Azam & Malfatti, 2007; Tranvik, 1997). The rate of decay will also depend upon the abundance of aquatic invertebrate and vertebrate detritivores.

Unless they are firmly weighted down, bodies buried underwater usually float up to the surface when gas formation occurs. Bodies floating on the surface of ponds, lakes and rivers can be colonized by blowfly larvae and this, along with the exposure to higher temperature and oxygen levels (than underwater), increases the rate of decomposition. Blowflies are not found at sea though they may colonize bodies washed up on beaches.

Wounds

Wounds, whether inflicted at the time of death or immediately afterwards, allow entry of air and invertebrates into the body and can therefore speed up the rate of decay. The smell of blood is also attractive to blowflies and therefore leads to the body being discovered and exploited more quickly. However, if there is severe blood loss this may slow the bacterial colonization of the body / body parts because the microbes are no longer able to grow rapidly through the liquid medium of the blood vessels.

Infections

Pre-existing infections such as septicaemia or infected wounds can speed up the rate of decay because there are already bacteria colonizing and breaking down the body at the time of death.

Burning

Murderers often attempt to dispose of their victim's body by burning it. However, they are seldom completely successful owing to the extremely high temperatures required – identifiable human remains may still be found among the ashes produced by a crematorium, which typically operates at over 1000 °C for 2–3 hours. The temperatures reached in typical house fires are much lower than this and although they may exceed 700 °C this tends to be for relatively short periods and to occur close to the ceiling. On the floor, the temperature may only reach 166 °C – although a lot depends on the presence of combustible material. Victims of house fires, explosions and traffic and aircraft accidents are also often badly burnt. Usually, the extremities, the limbs and the head are most badly affected and the torso is the last part of the body to be fully consumed. Bodies with a high fat content burn the best and clothing, provided it is flammable, will contribute to the extent to which the body is destroyed. Sometimes the body is placed in a tyre that is then set alight as a means of providing extra fuel.

   Some factors associated with burning reduce the rate of decomposition whilst others promote it, so it is difficult to generalize. For example, burning sterilizes the skin surface and dries the underlying tissues making them unsuitable for the growth of microbes and blowfly maggots but it also causes cracks through which they may invade the deeper tissues that are less affected. Similarly, although the skin surface may be charred, the temperature may not have been high enough to affect the gut microbial flora, so decomposition may commence here as normal. Some workers have found that burnt corpses retain their attractiveness to blowflies whilst others have found that it reduces their likelihood to lay eggs on the body (Avila & Goff, 1998; Catts & Goff, 1992). Obviously, a great deal depends on the degree of burning and the individual circumstances.

Burning induces chemical changes in proteins, carbohydrates, lipids and other organic molecules that may affect their suitability to support microbial and maggot growth but there is little published information on this in a forensic context. However, there is a lot of literature on how pre-slaughter conditions, storage and cooking affects the chemical composition and palatability of meat destined for human consumption (e.g. Varnam & Sutherland, 1995). Basically, when meat is cooked it causes the fats to melt and therefore they become susceptible to degradation (e.g. triglycerides break down to glycerol and fatty acids) and oxidation (e.g. the carboxylic acid group of a fatty acid chain is oxidized). The oxidation of fatty acids produces a range of chemicals that contribute to the smell of cooked meat and also undergo reactions with other chemicals to produce compounds that contribute to the taste of meat. Heating above 40–50 °C causes some proteins to become denatured (their three-dimensional shape is changed) and hydrolysed (i.e. broken down) and as the temperature continues to rise more and more proteins are affected. Denaturation causes muscle proteins to contract (hence heat stiffening) and in the process the cytoplasm of the muscle cells is forced out. Other proteins are also denatured causing further shrinkage, loss of fluid contents and therefore the tissues dry out as the water evaporates. As the temperature rises, pyrolytic reactions occur. (Pyrolytic reactions are those in which heat causes compounds to become converted into one or more products.) For example, collagen is converted into gelatine. The breakdown of proteins results in a rise in the concentration of amino acids and some of these will react with sugars in a complex, nonenzymic, reaction known as the Maillard reaction. This is important in the development of flavour although it must be remembered that what humans consider flavoursome and what invertebrate detritivores respond to may not be the same. If heating is prolonged the body will become desiccated and if the temperature is high enough it will be carbonized and turn black as all the organic matter is lost.

Chemical treatment

Murderers occasionally cover their victim's body with a chemical in order to enhance its speed of decay, reduce its likelihood of detection or simply to destroy potential incriminating evidence. It is a common belief that covering a body with lime will lead to its rapid and total destruction. In fact, it can actually contribute to the body's preservation. Lime (calcium oxide [CaO]) reacts with water to produce highly corrosive slaked lime (calcium hydroxide [$Ca(OH)_2$]) in a highly exothermic reaction (i.e. a lot of heat is produced). Indeed, so much heat is produced that although some surface corrosion takes place when a body is covered by lime, the temperature of the body becomes so high it is desiccated. A dry body can remain resistant to decay for many years and a corrosive covering deters detritivores from attacking it.

Placing a body in an acid bath destroys all the soft tissues as well as the bones but the speed and effectiveness are affected by the water content of the tissues; fat is hydrophobic and not readily solubilized and neither are gallstones – these are predominantly composed of cholesterol and variable amounts of calcium. Some plastics are also resistant to attack.

Case Study: The acid bath murders

John George Haigh earned himself a name in history as the 'acid bath murderer' – crimes for which he was hanged in Wandsworth prison in 1949 – by placing his victims in a bath of sulphuric acid after killing them. There were several victims but it was the death of his final one, Mrs Durand-Deacon that provided most of the evidence that led to his conviction. When the police investigated his property following the disappearance of Mrs Durand-Deacon they recovered a large quantity of body fat, three gallstones, part of a left foot, several bone fragments, an intact set of dentures, a bag handle and a lipstick container. There were also bloodstains on a wall but the technology of the time was only sufficient to identify them as being of human origin. The dentist of Mrs Durand-Deacon had kept plaster casts of her patient's upper and lower jaw and was therefore able to confirm the provenance of the dentures. A cast of the left foot was found to perfectly fit into one of Mrs Durand-Deacon's shoes whilst the pathologist who examined the foot confirmed that the victim had suffered from osteoarthritis in life – something that Mrs Durand-Deacon was known to have done. This indicates that even an acid bath has its limitations as a means of disposing of a dead body although it should also be noted that the remains of Haigh's other victims were never found.

Embalming is the process by which a dead body is treated with chemicals to delay or prevent its decomposition. The ancient Egyptians, amongst others, embalmed their dead for religious reasons and sought to prevent decay entirely. Similarly, Lenin's body was embalmed before being put on public display in his Moscow tomb and is constantly monitored to prevent decay. Nowadays, at least in Europe and USA, embalming is not done for religious or political purposes but to temporarily preserve the body so that it remains presentable and not a danger to health until it is buried or cremated. In the UK, this usually involves treating the body with a 2% solution of formaldehyde although the embalming fluid may also contain a variety of other chemicals. The forensic relevance of this is that embalming can compromise the subsequent chemical analysis of the body's fluids and tissues. Obviously, it is preferable that appropriate samples are taken before embalming occurs but this is not always possible or the need is not recognized in time. Embalming can reduce the level of benzodiazepines whilst the embalming of Princess Diana's body after her death in a car crash on 31st August 1997 prevented hormone analysis that would have proved whether or not she was pregnant with her lover Dodi Fayed's child. Needless to say, this spawned a 'cover-up' conspiracy. Embalming can also introduce artefacts, such as methanol, so it is helpful to have a sample of the embalming fluid to confirm whether or not this could have been the source. Not all chemical

Table 1.4 Summary of factors promoting or delaying the rate at which a body decays

Factors promoting decay	Factors delaying decay
Oxygen supply not restricted	Oxygen supply restricted
Warm temperature (15–37 °C)	Cold temperatures (<10 °C; decay will cease below 0 °C)
Humid atmosphere	Dry atmosphere
Presence of invertebrate detritivores (e.g. blowfly larvae)	Absence of invertebrate detritivores
Wasp, ant and other invertebrate predators feeding on corpse	Wasp, ant and other invertebrate predators feeding on detritivores
Wounds permitting invertebrates easier access to internal body tissues	Inability of detritivores to gain access to all or part of the corpse
Surface burning causing skin to crack and thereby allowing easier access of invertebrates and oxygen to internal tissues	Intense burning resulting in tissues becoming carbonized and drying out.
Obesity	
Suffering from septicaemia or myiasis before death	
Body exposed to the environment above ground	Burial on land or underwater (rate of decay declines with increasing depth)
Body resting on soil	Body suspended above ground (e.g. hanging)
	Formation of adipocere
	Mummification
	Embalming

analyses are affected by embalming, for example, the levels of morphine and strychnine are not altered. Blood ethanol levels may be reduced or impossible to assay following embalming but at least for the first 4 days afterwards a reliable estimation of the ante mortem levels can be made by measuring the levels in the vitreous humor – provided the eyes have not been injected with a mixture that compromises the analysis.

Discovery and recovery of human remains

Discovery

Most homicides are brought to the attention of the police soon after they occur owing to the disturbance caused, the confession of the culprit or the body being discovered in a public place. Sometimes an attempt is made at concealment but this is seldom very effective because human bodies, especially those of adults, are heavy and difficult to carry. Consequently, dead bodies are usually left close to where they fell or dragged a short distance from a path into nearby undergrowth – perhaps following a ride in the boot of a car – and they are often soon discovered by accident by a man walking his dog or children playing. Where the remains are buried the grave is usually relatively shallow and in the vicinity of where the homicide occurred.

Where an intentional search is employed it is essential that this is done in a careful and methodical manner – often the area is divided into grids and one section of the grid is searched at a time. Physically searching large areas of countryside is difficult, requires a lot of manpower and is time consuming – and therefore expensive. To speed up the search a variety of methods are employed. Dogs have an extremely well-developed sense of smell and the police employ specially trained cadaver dogs to help them locate dead bodies – American workers report that cadaver dogs work best at a temperature of 4–16 °C when the earth and atmosphere are moist and there is a wind speed of at least 8 km/hour. Dogs can detect bodies that have been dead for many years and those that are buried or sunk underwater.

Aerial photography will reveal disturbed soil and localized changes in plant growth patterns whilst infrared cameras will detect the heat given off by a dead body as it starts to decay. Grave sites can be detected by probing with a long metal rod fitted with a 'T-shaped' handle: graves are characterized by a covering of disturbed soil and it is possible to insert the probe to a much greater depth than in the surrounding soil for the same amount of applied pressure. By contrast, as a rule it becomes progressively harder to insert the probe into undisturbed soil beyond a depth of about 30 cm. The dimensions of a 'soft spot' can indicate the likelihood of it being a grave and additional soil chemistry tests might be done on the area before attempting an excavation. This approach works best where the soil is naturally undisturbed and requires experience, it is also time consuming and requires care if damage is not to be done to the buried object. A ground magnetometer can be used to scan a large area in a relatively short period of time. The device works by detecting minor differences in the magnetic properties of the underlying soil. The human body cannot be detected directly by magnetometry but the process of burial disturbs the soil profile and hence its magnetic properties. Magnetometry will also detect objects such as metal buckles and keys. However, the technique doesn't work unless there is sufficient iron oxide content in the soil and its effectiveness is compromised if the body is on a tip or waste dumping ground where the soil contains numerous metal objects. Ground penetrating radar works by projecting electronic pulses into the ground and then monitoring the way in which they return. This allows the operative to build up a three-dimensional map of the underlying soil including the position and shape of objects and the presence of voids (these would show up if the body was in a coffin or box). However, the equipment is expensive and works best where the terrain is level. Like all the other techniques it is not fail-safe and may indicate the presence of a clandestine grave where there is none whilst failing to identify remains where they are known to occur. A brief summary of some of the methods used to detect bodies is given in Table 1.5 and more details on the discovery and recovery of human remains are covered by Dupras *et al.*, (2006).

Recovery

Once human remains are discovered the surrounding area should be cordoned off and a tent erected over the body to protect evidence from being lost and contaminants from being introduced. The whole surrounding area should be photographed and searched methodically for evidence. Evidence from a crime scene can consist of anything from a beer can to a bone and all surrounding objects should be photographed *in situ*, their position noted and then collected, given an identifying number and stored appropriately for future analysis. The method of retrieving a buried body depends upon the local circumstances. Strictly speaking, the term 'exhumation' should be restricted to the retrieval of a body legally buried in a coffin within a graveyard but it is also commonly used where burial was unlawful. If the body was buried legally mechanical diggers can be used to expose the coffin before it is retrieved but where the body was buried unlawfully it should be retrieved in an extremely careful manner so as to gather as much evidence as possible and to avoid damaging the remains. In all cases, the soil should be removed sequentially in layers using plastic and wooden tools and the soil hand sorted and sifted for evidence before being disposed of. In the 'hole method' the soil is removed from above and around the body until the body is fully exposed and can be removed. In the 'trench method' a 60 cm wide trench is dug around 3 sides of the grave to provide room within which to work and collect evidence. The trench is extended until it is about 60 cm below the level of the body so that the soil underneath the body can be examined. In the 'table method' a trench is dug as described above except that it is extended to include all four sides of the grave. This facilitates access to all sides of the body and is therefore the preferred method where circumstances allow. Once exposed an examination of the body in its context should take place. When the body is ready to be moved the hands and feet should be enclosed in clean plastic bags to preserve any biological/chemical evidence attached to the palms/soles or underneath the nails. The bags also protect the skin from physical damage: during the later stages of decay, the surface of the skin sloughs off and if the epidermal surface is damaged it may be impossible to take prints. Afterwards, the whole body (still in its clothes, if present) should be placed in a clean body bag and removed to a mortuary where an autopsy can take place. This ensures that evidence is kept together and reduces the possibility of contamination. If the body was buried and skeletonized it may be easier to wrap it in a white sheet and place it on a wooden board for easier transportation. It is easy to damage a body that is in the late stages of decay and cause post mortem artefacts that could be mistaken for ante mortem wounds so such bodies should be handled as little as possible and treated gently.

Health and safety should be a priority both at the crime scene and within the laboratory when examining the evidence. Extreme care should be taken when handling samples collected from corpses. Decaying bodies contain many bacteria and fungi, some of which are pathogenic and there is always the possibility that the victim was suffering from a transmissible disease. The risks extend beyond the tissues of the body as, for example, both anthrax and tuberculosis spores pass unharmed through the guts of blowfly maggots and adult flies so they too can be potential sources of infection. Special care needs to be taken when handling and storing exhibits such as needles owing to the risk of HIV and Hepatitis B and Hepatitis C. Illegal cannabis grow rooms are seldom set up in accordance to building regulations and can therefore contain dangerous electrical wiring or present a fire risk.

Table 1.5 Summary of the advantages and disadvantages of the main methods of detecting dead bodies

Method	Advantages	Disadvantages
Physical search	Requires only manpower.	Manpower is expensive. Slow. Large numbers of people moving through an area may destroy evidence unless they are suitably trained.
Cadaver Dogs	Rapid, simple, cheap, nondestructive, detects old bodies and buried bodies.	Can be affected by climate. Dogs are best trained for single task (e.g. drugs or cadaver but not both).
Aerial photography	Nondestructive, provides overall picture of immediate and wider area. Databanks of the area may be available which highlight local changes.	Flight time expensive and normal cameras seldom provide sufficient resolution. Trained personnel needed for interpretation. Images may be required at different times of year.
Satellite images + spectral analysis	Nondestructive, provides overall picture of immediate and wider area. Databanks of the area may be available which highlight local changes.	Although mass graves in Iraq detected by spy satellite, the potential for locating small graves is uncertain. Suitable images not available for all regions of world and not all images are archived.
Thermal images	Nondestructive. Rapid. Can survey a wide area quickly. Will detect buried body if it is not too deep.	Most successful whilst corpse actively decaying but less effective once skeletonized. Equipment expensive and affected by wind.
Magnetometry	Nondestructive, rapid, can survey a wide area quickly. Equipment readily available. Can work through snow.	Compromised by artefacts (e.g. soil contains many metal objects) and not effective in some soil types (e.g. chalk, limestone, gravel deposits). Difficulties on rough terrain.
Ground penetrating radar (GPR)	Nondestructive, relatively rapid, can survey a wide area. Can work through snow or over water.	Expensive and requires trained operator. Usually requires smooth terrain.
Probing	Simple. Can work through snow and on rough terrain.	Potentially destructive (e.g. may damage skeletal remains), slow, requires experience.
Metal detector	Simple, cheap, quick, nondestructive. Potential if missing person contains metal medical implant.	Will only detect metal objects and to a limited depth.
Botanical survey	Simple, cheap, nondestructive. Typically used in conjunction with aerial photography.	A human corpse or burial causes the same changes to the plant community as any other dead animal or localized soil disturbance. Requires experienced botanist.

Determining the age and provenance of skeletonized remains

The longer that a person has been dead the harder it becomes to determine their time of death. Unidentified skeletons that are unearthed during construction work present particular difficulties because the police need to know whether they need to devote time and effort on an investigation and there are commercial pressures to resume building as quickly as possible. Even if the skeleton exhibits obvious signs of suspicious death (e.g. a knife still stuck in the skull) the police do not investigate crimes that were committed more than 70–75 years ago since the perpetrator would almost certainly have died and the victim's family would be several generations removed.

Radiocarbon dating

The vast majority of elements present in our bodies and the environment exist as a variety of isotopes – that is, their nuclei contain the same number of protons but a different number of neutrons. Some of these isotopes are unstable and decay through the loss of subatomic particles into other isotopic forms. For example, carbon-14 (^{14}C) decays to nitrogen-14 through the loss of beta (β) particles. The rate of decay of an element, measured as the 'half-life', depends on the instability of the isotopic form and may vary from fractions of a second to millions of years. The levels of unstable isotopes, especially ^{14}C, are useful in determining the age of both human and animal remains. Carbon-14 has a half-life of 5730 years and would have disappeared from the earth millions of years ago except for the fact that it is constantly being formed by the interaction of nitrogen in the air with cosmic rays. Once formed, the ^{14}C becomes incorporated into carbon dioxide which plants then metabolize into organic molecules when performing photosynthesis. The ^{14}C subsequently becomes incorporated into the bodies of herbivores when they eat the plants and then into the bodies of any carnivores or parasites that feed on the herbivores – consequently there is a constant cycling of ^{14}C between all living organisms. However, once an organism dies it no longer acquires new ^{14}C and the level within its body slowly declines. Because the rate of radioactive decay is constant, the level of ^{14}C present in the body provides an indication of how long the organism has been dead. Obviously, allowances have to be made through calibrations for the effects of complicating factors that influence the level of ^{14}C in the atmosphere such as solar storms and human activities (e.g. the burning of coal and other fossil fuels). Carbon-14 dates are usually expressed as 'years before present' (BP), present being the year 1950 – the year in which extensive above ground nuclear weapons testing began and thereby dramatically increased the levels of ^{14}C in the atmosphere. Carbon-14 dates are always cited as plus and minus a standard deviation (e.g. 526 ± 40 BP) to allow for errors that arise through the nature of the sample, how it was collected and the methodology used. It is therefore impossible to state the exact year in which a person died. Furthermore, traditional ^{14}C dating is insufficiently precise to accurately age the skeleton of a person who died within the last 100–200 years.

The carbon isotopes are usually extracted from collagen which is in turn extracted from the bones. This is because the mineral component of bone contains only a small amount of carbon – mostly in the form of calcium carbonate – and in buried bodies this can be affected by ion exchange with the surrounding soil.

Bomb curve radiocarbon dating

Bomb curve dating (also called bomb pulse dating) is based on the change in the ^{14}C levels in the atmosphere subsequent to the testing of nuclear weapons during the 1950s and 1960s. The nuclear explosions led to an increase in the atmospheric levels of ^{14}C followed by a decline once testing ceased – this is known as the 'bomb curve' or 'bomb pulse'. This pattern is repeated in all living organisms. The levels of ^{14}C can be measured in all tissues and the amounts reflect the carbon turnover within them. Metabolically active tissues such as the brain have high levels that mirror the atmospheric ^{14}C levels whilst those that are less active, such as the collagen content of bone, have much lower levels that reflect the year in which they were formed. It is therefore possible to determine the age of a tissue by relating its ^{14}C level to those on the atmospheric bomb curve. Tissues with a low carbon turnover that were formed before the start of nuclear weapons testing have a low level of ^{14}C whilst those formed afterwards exhibit elevated levels.

Bomb curve radiocarbon dating therefore has potential for aiding forensic investigations (Ubelaker & Buchholz, 2006). The presence of elevated ^{14}C levels in any of the tissues recovered from a dead body would suggest that the person was alive at some point after 1950. The circumstances surrounding the remains would therefore require further investigation. The precise level of ^{14}C could provide evidence of when the tissue was formed and, potentially, when the person died but there could be many years discrepancy between these two figures. Furthermore, the levels of ^{14}C in tissues are strongly affected not only by the type of tissue but also by age, growth pattern, diet, health and disease. The levels would therefore have to be considered in relation to other information. For example, whether there was evidence of osteoporosis. The use of bomb curve radiocarbon dating to determine a person's age from their teeth is described in Chapter 4.

Stable isotopes

Carbon-14 is not the only isotope that can be used to determine the age of human remains. For example, our bones contain small quantities of lead-210 (^{210}Pb) that we acquire through our food and breathing in radon-222 (^{222}Ra; this decays into ^{210}Pb) from the atmosphere. Because ^{210}Pb has a half-life of only 22.3 years its levels decline more rapidly after death than those of ^{14}C thereby making it potentially valuable for forensic studies (Swift *et al.*, 2001). Isotopes that do not undergo radioactive decay are said to be 'stable'. An element may posses both unstable and stable isotopes – for example, carbon may occur as both the unstable isotope carbon-14 and the stable isotopes carbon-12 and carbon-13. Stable isotopes have many uses in forensic studies as their ratios can provide a unique identifying feature of a specimen's provenance or as an indication of previous diet or geographical origin (Anon, 2004a). For example, strontium has four stable isotopes (strontium-84, -86, -87, and -88) and because their ratios vary between geographical locations these can act as a 'signature' indicating where a person or animal was living. Strontium enters our bodies via our diet and becomes sequestered in the bones and teeth. Once formed, the tissues that make up teeth have a low metabolic turnover and the strontium is immobilized within them. The strontium isotope ratio of the teeth therefore tends to reflect that of the environment in which a person grew up. By contrast, bones are metabolically more active with new tissues being constantly formed throughout life, although the rate of renewal varies between bones. The strontium isotope ratios found in bones therefore reflects those in the environment where a person lived in the previous ten years. Beard & Johnson (2000) analysed strontium isotope ratios to differentiate between the disarticulated skeletal remains of three American servicemen that were found mixed together in a shallow grave in Viet Nam 20 years after the conflict ended. The identity of the victims was known but the remains needed to be separated from one another so that they could be returned to the appropriate family. The three servicemen grew up in different areas of America so it was possible to identify at least some of the remains by comparing the strontium isotope ratios of the teeth and bones with those found in the three regions. Because the servicemen were not stationed in Viet Nam for long before they were killed, the strontium levels in the bones were not thought to have been affected by the food and water they consumed whilst there. Although this type of analysis offers promise for future development, its effectiveness is ultimately dependent upon the availability of databases of isotope ratio analyses for different geographical regions and is potentially subject to many complicating factors. For example, human populations are increasingly mobile and our food and water is acquired from different geographical regions to where we live.

Future developments

More effective means are required to detect dead bodies, especially buried bodies. Although satellite technology has been proposed there are relatively few of these with sufficient resolution to detect the presence of a small individual grave and their flight path cannot be altered to meet the requirements of a local police force. GPR has great potential but the technology will have to be refined and made cheaper and more robust if it is gain widespread use as a field instrument for forensic investigations.

The possibility of using the radiocarbon bomb curve dating technique for forensic studies was suggested many years ago but its potential has not been widely exploited. This is probably because we still have limited data on the factors affecting the ^{14}C levels in human tissues and how biological and environmental factors affect these after death. Similarly, the use of stable isotopes in forensic science offers considerable scope but will require agreement on standardized techniques and databases of isotope distributions within and between countries to be established.

Quick quiz

(1)   What is hypostasis?

(2)   What is meant by the term 'taphonomic process'. State two examples of taphonomic processes.

(3)   What causes the 'bloat' stage of decomposition?

(4)   What is adipocere and what is its forensic relevance?

(5)   Why do newly born babies mummify more readily than older children and adults?

(6)   Why does a buried body decay more slowly than one lying on the surface of the ground?

(7)   Why does a hanging body decay more slowly than one lying on the surface of the ground?

(8)   During winter, why might an exposed body decay faster if left in the centre of a city than in the outlying countryside?

(9)   State one means by which burning can reduce the rate of decay of a corpse and one means by which its rate of decay might be increased.

(10) Distinguish between the 'trench' and the 'hole' methods of retrieving a dead body.

Project work

Title

The effect of freezing on the rate of decay.

Rationale

Murderers sometimes store their victim's body in a freezer before disposing of it. Freezing will cause tissue damage so once the body has thawed, does it decay at the same as rate as an unfrozen body?

Method

Bodies or tissues can be frozen for varying lengths of time and then placed above or below ground and the rate of decay, speed of colonization by invertebrates etc. compared with those of a control unfrozen body. If a thermocouple is placed on the frozen body when it is exposed, it would be possible to determine whether blowflies are deterred from laying until the surface temperature has risen to near ambient levels. Histological changes could also be assessed along with biochemical assays to determine the speed with which autolysis begins.

Title

The effect of burying in concrete on the rate of decay.

Rationale

Bodies are sometimes disposed of in the concrete foundations of buildings or bridges.

Method

Bodies or tissues, which may or may not be wrapped in clothes, would be encased in concrete and then left at varying temperatures. After varying times the body would be retrieved from the concrete and its state of decay compared with those of control bodies that were not placed in concrete. The ability to extract DNA from the bodies would be assessed and structural changes to the surrounding concrete determined.

2 Body fluids and waste products

Chapter outline

Objectives

Distinguish between the different types of blood cells and explain how the presence of blood can be detected.

Describe how the deposition and composition of blood at a crime scene can be used to determine how the crime was committed and identify both the victim and the assailant.

Explain how saliva, semen, urine and faeces found at a crime scene can be used as forensic indicators.

Discuss how the composition of the vitreous humor can provide information on the time since death and the circumstances surrounding it.

Blood cells and blood typing

Blood consists of a variety of different cell types, collectively known as blood cells, which are suspended in a watery fluid called serum. Traditionally, the study of blood

Essential Forensic Biology, Second Edition   Alan Gunn
© 2009 John Wiley & Sons, Ltd

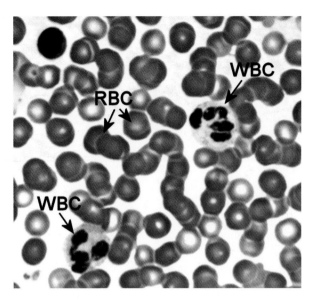

Figure 2.1 Thin blood film stained with haematoxylin and eosin showing mature red blood cells and white blood cells (arrows). Note the absence of nuclei in the red blood cells.

is referred to as haematology whilst the study of serum (and in particular the immune factors within it) is called serology. Typically, the intrusion of forensic science has complicated things and the term 'forensic serology' is sometimes taken to encompass not only the study of serum but also blood cells, saliva and semen for forensic purposes.

Mature human blood cells can be divided into those that possess a nucleus and those that do not: both types can provide forensic information (Fig. 2.1). Those lacking a nucleus are the red blood cells (RBC) (also known as erythrocytes) and the platelets. The red blood cells possess the pigment molecule haemoglobin that gives blood its red colour and is responsible for the transport of oxygen and carbon dioxide. The platelets are smaller than the red blood cells and unlike them they are capable of amoeboid-like movement. That is, they move by sending out cell processes in a similar manner to the single celled organism 'amoeba' and consequently often look star-shaped when viewed using a microscope. Platelets are responsible for the clotting mechanism and the production of chemical growth factors that maintain the integrity of the blood vessels. The cells that contain nuclei are collectively known as white blood cells (WBC) or leucocytes. There are many different sorts of white blood cell but all of them are capable of amoeboid-like movement and they are responsible for the body's immune defence capabilities. In addition to a nucleus, all white blood cells contain mitochondria – subcellular organelles that produce energy for the cell. Therefore, unlike the red blood cells, white blood cells contain both nuclear and mitochondrial DNA.

All cells within the body carry upon their outer surface an array of molecules that are called 'antigens'. These antigens serve a bit like a passport and identify the cell as a legitimate 'citizen of the body'. The antigens are recognized by a group of pro-

Table 2.1 Summary of ABO blood group interactions. Abbreviations: Ag = antigens, Ab = antibodies, RBC = red blood cells

Characteristic	Blood type			
	A	B	AB	O
Ag on RBC	A	B	Both A and B	Neither A nor B
Ab in plasma	Anti-B	Anti-A	Neither anti-A nor anti-B	Both anti-A and anti-B

teins, called 'antibodies' that are secreted by certain white blood cells and serve as 'immigration control'. If a cell does not possess the correct antigens on its surface it is deemed to be a foreigner and an immune response is mounted to destroy it. The nature of the antigens found on the surface of red blood cells is of medical importance because unless the correct blood type is used during a blood transfusion it would be rejected by the recipient's body and thereby result in fatal consequences. The most common antigens found on red blood cells are those that comprise the ABO system. The ABO system works as follows: an individual's red blood cells may possess either only class A antigens (type A), only class B antigens (type B), both classes A and B antigens (type AB), or neither class A or class B antigens (type O). Persons who are type A tolerate their own class A antigens but produce antibodies against B antigens that bind to the surface of the red blood cells and form bridges between them. This causes the red blood cells to agglutinate (clump together) – a reaction that occurs quickly and can be observed using a microscope. In a similar manner, type B persons produce antibodies against class A antigens, type AB persons produce antibodies against neither class A or class B antigens, and type O persons produce antibodies against both class A and class B antigens. These interactions are summarized in Table 2.1.

There are many other groups of antigens that are found on red blood cells, the most well known being those responsible for the Rhesus (Rh) Factor (also known as antigen-D and agglutinogen-D). The Rhesus factor is so called because the antigens responsible for it were first described in rhesus monkeys. Persons who possess these antigens are said to be Rhesus positive [Rh(+)] whilst those who do not are Rhesus negative [Rh(−)]. 83% of the UK population are Rh(+) and most belong to either blood types O (44%) or A (42%). Although we talk of a person being of blood types A, B, AB or O and being Rh(+) or Rh(−), in actual fact all of these characteristics can themselves be divided up into many more sub-combinations (e.g. O1 and O2). Certain races tend to have higher proportions of particular blood groups than others – for example type AB is more common among the Japanese (10%) than among Europeans (UK, 4%). More precise details of racial origin may be obtained from finding evidence of rare inherited disease traits or particular antigens. For example, sickle cell anaemia occurs almost exclusively among black people of African descent. Similarly, there is a variety of the Duffy antigen (called phenotype a-b-) that is extremely common among West Africans and their descendents but almost absent from white and Asian populations. Both these variations are considered to be protective against malaria because the *Plasmodium* parasite finds it harder to recognize and invade red blood cells expressing these traits. By contrast, the Kell antigen is predominantly found among white persons (Reid & Lomas-Francis, 1996). In addition to blood type and the Rhesus factor, many of the enzymes and other proteins found in red blood cells are also polymorphic – that is, they exist in more than one form. Consequently, even though an individual may have a common blood type such as type O, once all the other variables are taken into account, their combination may be shared by only a small number of the population. If a total of eight serological variables are used, it is generally estimated that the chances of two unrelated persons sharing the same profile is between 0.01 and 0.001 (i.e. between 1 in 100 and 1 in a 1000). This estimate is known as the match probability (*Pm*). This means that blood typing can provide a means of identification – although it is not as accurate as DNA profiling in which match probabilities of up to 10^{-10} are claimed. Matching a person's blood profile to traces found at a crime scene might be incriminating but on its own would be insufficient to prove guilt. It is, however, effective at excluding suspects and therefore allowing the police to concentrate their resources on more profitable lines of enquiry. Blood typing suffers from several further drawbacks as a forensic tool. Firstly, by comparison with DNA profiling, it requires relatively large amounts of sample and is therefore of limited use where only small specks of blood are available. Serological markers degrade quickly and consequently the amount of information that can be obtained from either a body or a bloodstain that is several days old is reduced. The procedure is further compromised by the interference caused by enzymes released by contaminative bacteria. Interference can also be a problem where the blood stain contains both the assailant's and victim's blood (for example, after a fight), if one of the parties has recently received a blood transfusion (in which case their blood profile may be temporarily altered), or following a rape in which semen is left inside the victim's body and therefore will be diluted by his or her serological profile. In addition, the investigation may require a suspect to provide a blood sample. This is an invasive medical procedure with which they may legitimately refuse to cooperate.

Methods for detecting blood

Bloodstains are not always obvious because of the manner in which they are formed or because the assailant cleans up the crime scene after committing an assault. However, once blood is spilt, it is extremely difficult to remove all the traces. Any attempt at cleaning up inevitably means using either the kitchen or bathroom as a 'base' – so it is good place to start looking for blood. In the cleaning process, blood can flow beneath tiles or linoleum or between the boards of wooden flooring. Similarly, if a water tap is grasped with a bloody hand, the blood may flow into the screw mechanism. Consequently, blood can be detected in even an apparently spotlessly clean room if one knows where to look. To identify the presence of this hidden or 'latent' blood an investigator uses an indicator substance. There are many of these but most work on the basis of haemoglobin catalysing the oxidation of an oxidant (e.g. hydrogen peroxide), and the products of oxidation then interacting with other chemicals to bring about a colour change. These tests can be highly sensitive but a positive reaction only suggests the presence of blood and is not proof and they are therefore called 'presumptive tests'. In addition, all of them suffer from interference from common environmental chemicals that also break down the oxidant or interact directly with the test reagent thereby inducing false positive reactions. There is some discrepancy in the literature regarding the extent to which false positives occur and the chemicals responsible although various vegetables (e.g. potatoes, horseradish, red onions) and inorganic substances (e.g. ferric sulphate) are among the most commonly cited sources of interference. In addition, in order to be sure the test is working, one should always perform a negative control (e.g. a spot of distilled water) and a positive control (e.g. a spot of dried animal blood) immediately before assessing an unknown stain. If this is not done then the results could be called into question.

Luminol

Luminol was first used to detect bloodstains in 1937 and it remains one of the most effective presumptive tests. It has the added advantage of not interfering with subsequent DNA analysis so it can be sprayed directly onto the suspect stains (Tobe et al., 2007a). Several of the other presumptive tests (e.g. Kastle–Meyer test) can destroy DNA and therefore when they are employed a sample of suspect stain should be scratched onto filter paper and the test performed on that. To perform the luminol test, luminol (5-amino-2,3,-dihydro-1,4-phthalazinedione) is mixed with sodium carbonate to form an alkaline solution whilst a separate solution of either sodium perborate or hydrogen peroxide is prepared to act as the oxidant. Immediately before use the two solutions are mixed together and sprayed onto the suspect stains. If haemoglobin is present, the oxidant is broken down and highly reactive free oxygen radicals formed; these bring about the conversion of luminol to 3-aminophthalate in a chemiluminescent reaction which is manifested as a faint bluish glow that is best seen in dim light (Fig. 2.2). Luminol is so sensitive that it can detect the presence of blood on clothing that was machine washed although after washing the stains may become diffuse and bear little resemblance to the original patterns. Consequently, caution is needed when interpreting such stains. Luminol can also demonstrate the presence of blood in soil for up to a year after the blood was initially spilt although it may be necessary first to scrape off the top layer of soil before spraying. When blood is spilt onto soil a person or animal subsequently walking over the stained area, even months afterwards, may leave tracks that stain positive with luminol. In the absence of other evidence it would not be possible to know whether the tracks were associated with the spillage of blood or a subsequent artefact (Adair et al., 2006). The luminescence does not last long and the requirement for dim light conditions can present practical difficulties in some circumstances. Further problems can arise through autoluminescence owing to degradation or contamination and spraying luminol solution can cause the formation of luminescent spots that are of a similar size to the mist-like blood stains caused by shooting injuries or severe beatings. Consequently, the use of luminol spray is not recommended where such stains are thought to be present.

Figure 2.2 The presence of (presumptive) blood is indicated following treatment with luminol. The smeared appearance is due to an attempt having been made to clean up the bloodstains. Confirmatory tests would be needed to prove that the stains were due to blood and that the blood was of human origin. (Reproduced from James, S.H. *et al.*, (2005) *Principles of Bloodstain Pattern Analysis*, Copyright © 2005, Taylor & Francis.)

Kastle–Meyer test

In this test a sample of the stain is scraped onto filter paper and treated with a drop of ethanol (to improve the sensitivity of the test). A drop of phenolphthalein (a colour indicator) is then added followed by a drop of hydrogen peroxide solution. If the phenolphthalein changes colour before the hydrogen peroxide is added the test is considered negative. The hydrogen peroxide interacts with the haem molecule of haemoglobin and is broken down into water plus free oxygen radicals – these radicals then interact with the phenolphthalein resulting in the solution changing from colourless to pink. The Kastle–Meyer reaction is slightly less sensitive than the luminol test and must not be performed directly on the stains since it interferes with subsequent DNA extraction. However, it is quick, simple and cheap and the colour change is immediately apparent.

Titanium dioxide

A 10% w/v solution of titanium dioxide in methanol sprayed onto a suspect area acts as a white dye and is therefore particularly good for demonstrating the presence of latent blood on dark surfaces. Unfortunately, the spray is flammable and poisonous and therefore not particularly suitable for work at crime scenes and once used the residues will compromise further analyses. A safer water-based solution has reportedly been developed but published details were not available at the time of writing.

Infrared imaging

Infrared imaging shows potential for demonstrating the presence of latent blood-stains without the use of chemicals. The area or object can be viewed using a camera or video recorder capable of detecting wavelengths in the 760–1500 nm region. The technique is still in the development stages and is compromised if the background material has the same infrared absorbance and reflectance characteristics as the bloodstains but otherwise exhibits high sensitivity (Lin *et al.*, 2007).

Confirming the presence of blood

Even if blood is detected, it cannot be assumed that it is human blood. Household pets are just as likely to fight, scratch and otherwise injure themselves as any human and therefore leave their bloodstains on upholstery and clothing. Similarly, when handling raw meat in the kitchen it is not unusual to leave bloodstains upon surfaces and clothing. Sometimes it is important to specifically look for animal blood, for example, when investigating crimes such as badger baiting or when a pet was injured or killed during the course of a break-in or homicide. To determine whether blood comes from a human or an animal one usually undertakes a precipitin test which involves reacting antigens in the blood sample with anti-human antibodies – these are available commercially and are raised in rabbits. The blood samples and the antisera (anti-human antibodies) are placed in wells punched into an agar gel that is spread over a glass dish or slide. The samples move towards one another through the agar by diffusion or the process can be speeded up using an electric current. If a white line – called the precipitin line – is formed at the point at which the two samples meet, this is indicative of an interaction between the antigens in the blood and antibodies in the rabbit antisera and therefore the blood is human. If no pre-cipitin line forms and there is a suspicion that the blood belongs to an animal, then the procedure can be repeated using antibodies raised against the appropriate animal sera.

DNA-based techniques are now available to differentiate human and non-human blood and tissue samples and will probably supersede the precipitin test in the near future. For example, Matsuda *et al.*, (2004) describe a highly specific PCR-based protocol that utilizes primers for the human mitochondrial cytochrome b gene. They state that following the agarose electrophoresis step of the PCR process, human DNA produces a single band whilst blood from other vertebrates fails to produce any bands at all.

Bloodstain pattern analysis

Bloodstain pattern analysis is a specialized branch of forensic science in which the investigator deduces evidence from the shape and distribution of bloodstains. From careful analysis it is possible to determine whether or not a crime was actually com-mitted and if so, how the conflict developed and how the wounds were inflicted. The earliest record of bloodstains being used in court proceedings relate to the trial of the unfortunate Richard Hunne in 1514. He was being held in Lollard's Tower in London on five charges of heresy but before he could be examined he was found hanged in his prison cell. This obviously upset the authorities because he was promptly charged with 13 more counts of heresy and suicide. (Suicide was a serious offence because it meant that the victim's body could not be buried in consecrated ground. Suicide remained a crime in England and Wales until 1961.) However, bloodstains were found in his cell and after reviewing these and medical evidence it was concluded that 'whereby it appearth plainly to us all, that the neck of Hunne was broken, and the great Plenty of Blood was shed before he was hang'd. Wherefore all we find by God and all our Consciences, that Richard Hunne was murder'd. Also we acquit the said Richard Hunne of his own Death' (Forbes, 1985).

Types of bloodstain

An adult human contains about 5 litres of blood; loss of approximately 30% blood volume (~1.5 litres) usually results in loss of consciousness or incapacitation whilst the loss of 40% (~2 litres) blood volume can be fatal. It should be remembered that bleeding takes place both internally and externally so the amount of blood surrounding a body may not reflect the amount actually lost. Arterial blood (with the exception of that going from the heart to the lungs) is bright red in colour owing to its high oxygen content whilst venous blood is darker in coloration as it is deoxygenated. However, once shed, blood darkens and begins to clot within about 3 minutes and it is impossible to tell whether it originated from an artery or vein from its colour. There are no reliable tests to determine the age of dry bloodstains although studies on the extent of DNA decomposition have shown some potential.

Many texts classify bloodstains as low-, medium- and high-velocity spatter based upon their size distribution and hence the force necessary for their production. Briefly and very crudely, these causes could be given as dripping, beating and shooting i.e. the higher the velocity the smaller the stains. However, there is a move away from this approach because it is difficult to be so prescriptive. For example, a severe beating and certain gunshot injuries can generate similar bloodstain size distributions although these two causes would normally be categorized as medium- and high-velocity respectively whilst other mechanisms such as sneezing or coughing can produce stains within the same size category as medium- to high-velocity. Perhaps the best classification system is that set out in detail by James *et al.* (2005) in which they recognize three main categories: passive, spatter and altered and these are then subdivided on the basis of the mechanism most likely to have produced them.

Passive stains: These stains are subdivided into:

(a) *Drops*: These stains occur when blood drips passively under the influence of gravity (Fig. 2.3). They may occur singly, in clusters or provide a trail and thereby indicate movement and its direction. However, direction may be difficult to determine if the victim moves slowly and therefore the drops fall at 90° resulting in the stains being more or less oval. A classic situation is when

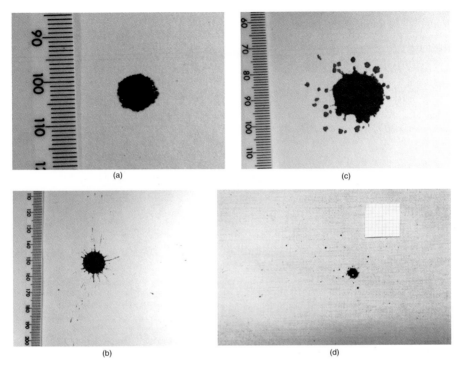

Figure 2.3 Factors affecting the spatter pattern of passively falling droplets of blood. (a) A droplet falling a short distance (5 cm in this case) horizontally onto a smooth hard surface tends to form a circular-shaped stain. Scale = mm. (b) Droplets falling from a greater height (100 cm in this case) form a sunburst-shaped bloodstain (Scale = mm). (c) When a second droplet falls upon a bloodstain that is still wet it causes the formation of satellite spatter (Scale = mm). (d) The texture of the surface a droplet impacts on can markedly affect the shape of the resultant bloodstain. This droplet fell 10 cm onto a piece of clean dry calico cloth and promptly disintegrated to form satellite spatter. Square scale = 20 mm.

you cut yourself and leave a drip trail from the place where the accident occurred to the nearest washbasin.

(b) *Flows*: These occur when blood flows passively to produce small rivers or streams of blood. These can be useful for determining orientation at the time of attack or movement after death. For example, if a person is stabbed in the chest whilst standing a trickle of blood will head vertically down the body (Fig. 2.4). If, however, the person was lying on the floor the flow will veer to whichever side of the body is angled to the floor. Similarly, if the body is moved after death, multiple flow paths may be seen although this is most obvious when the skin is uncovered since clothing can diffuse and complicate flow patterns.

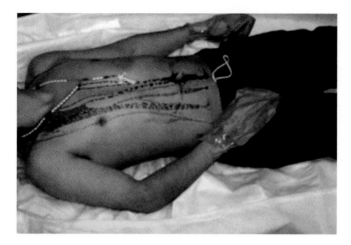

Figure 2.4   The flow of blood vertically down the chest and abdomen from a neck wound indicates that this person was standing upright when they were injured. (Reproduced from James, S.H. *et al.*, (2005) *Principles of Bloodstain Pattern Analysis*, Copyright © 2005, Taylor & Francis.)

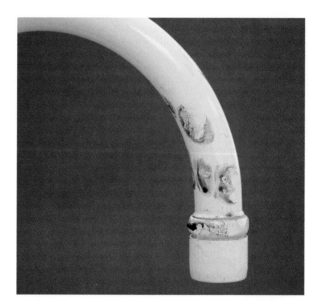

Figure 2.5   Transfer bloodstains on a tap. Transfer stains tend to smear if the fingers/hands/feet/ object are 'overloaded' with liquid, even when placed on an absorbent surface.

(c)   *Transfers*: These occur when wet blood is transferred from one object to another (Fig. 2.5). For example, when a bloody knife is wiped onto clothing, a bloody hand leaves prints on doorknobs or weapons and when a person stepping in blood leaves behind impressions of their footwear.

(d)   *Large volumes*: A dead person may be found lying in a large volume of blood that has formed a pool around the body and started to seep underneath it

Figure 2.6 This man was subject to a severe beating with a blunt object. Note the pool of blood on the floor, impact spatter and cast-off stains from the weapon on the wall above the man's left shoulder. (Reproduced from James, S.H. *et al.*, (2005) *Principles of Bloodstain Pattern Analysis*, Copyright © 2005, Taylor & Francis.)

(Fig. 2.6). If the volume is considered in relation to the injuries received it can provide and indication of the time the body was resting in that position. Persons dying of pulmonary TB and a variety of other diseases may cough up or vomit large amounts of blood shortly before or at the time of their death (Fig. 2.7) and this can lead to them being found in a pool of blood that at first sight may appear suspicious. This is a not unusual situation in the deaths of vagrants and those who have become cut off from society.

Spatter stains: James *et al.* (2005) subdivide spatter stains into three subcategories based upon their cause: secondary mechanisms, impact mechanisms, and projection mechanisms.

(a)   *Secondary mechanisms* are those that are secondary to the cause of the initial wound. For example, if blood drips onto an existing wet stain it will distort the original stain's shape and form a patchwork of surrounding spatter (Fig. 2.3c).

(b)   *Impact mechanisms*, as the name suggests, indicate the means by which the wounds were caused. Although separate categories, such as gunshot, beating (sharp/blunt implements) and industrial tools are recognized it is not always easy to 'work backwards' and establish the cause solely from the bloodstain evidence. For example, bloodspatter resulting from gunshot wounds is affected by the type of firearm, the type of ammunition, the distance of the victim from the firearm, the presence of clothing and the location of the wound. Gunshot wounds often produce back spatter (i.e. towards the firearm) (Fig. 2.8) and forward spatter (i.e. in the direction of the projectile leaving the body). Both may include small fragments of flesh and bone. Suicidal gunshot injuries can

Figure 2.7 Bloodstains found on the toilet of a person who subsequently died of haemorrhaging as a consequence of lung cancer. Air bubbles can be seen within the blood indicating it came from the lungs. In addition, one can see elongated projected spatter stains on the toilet seat and splashes of blood on the floor where large volumes of blood have dropped almost vertically. (Reproduced from James, S.H. *et al.*, (2005) *Principles of Bloodstain Pattern Analysis*, Copyright © 2005, Taylor & Francis.)

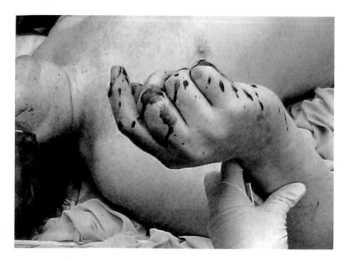

Figure 2.8 Backspatter stains on the right hand of a man who committed suicide using a shotgun. (Reproduced from James, S.H. *et al.*, (2005) *Principles of Bloodstain Pattern Analysis*, Copyright © 2005, Taylor & Francis.)

therefore potentially be distinguished from homicidal ones by the presence of both blood spatter and gunshot residues on one of the hands of the victim and/or within the cuffs of long-sleeved clothing. The absence of both would lead to the suspicion that the gun was placed in the victim's hand after they were shot. However, it is worth noting that in experimental work by Bernd *et al.* (2002), in which calves were shot in the head with a 9 mm Luger, and a survey of human suicide incidents by Betz *et al.* (1995) there were always cases in which back spatter was not observed. As the speed of impact increases, the dimensions of the blood droplets decreases and bullets from a powerful gun cause the formation of a mist-like array of tiny stains 1 mm in diameter or less. Even when formed by high velocity impact, such small droplets of blood seldom travel far.

Beatings with a blunt instrument, whether by fists, feet or an iron bar, can produce a wide variety of bloodstain patterns. Where the beating is prolonged and brutal the blows are inflicted upon already open bleeding wounds and this generates large volumes of blood spatter over the surrounding area (Fig. 2.6). In extreme cases, mists of small droplets may be formed similar to those caused by gunshots. With so much blood being spilt, anything getting in the way of the projected blood, such as the assailant or an item of furniture, will result in a clean region or 'void' on the surrounding vertical surfaces (e.g. wall or door). Such voids are useful for determining the position of people during the course of an assault. When large amounts of blood are projected it also means that it is highly unlikely that the assailant would be able to leave the scene without substantial staining to their body, clothes, and footwear. It can also indicate whether doors were open of closed at the time of the assault. For example, if a door was open, projected blood might be found on the hinges or parts of the door frame that would be covered if the door was shut.

(c) *Projection mechanisms* are divided into spurt patterns, sneezed, coughed or breathed blood and cast off stains.

Arterial blood flows under high pressure and will spurt out over a considerable distance should the vessel be cut. Furthermore, it will continue to spurt out in bursts owing to the beating of the heart thereby producing a characteristic undulating 'arterial spurt stain' (Fig. 2.9). By contrast, when a vein is cut, the blood usually seeps out rather than spurts because of the comparatively lower blood pressure. However, restricting the terminology to arteries could give rise to interpretation errors since in certain circumstances pulse-type stains result when veins are severed (Brodbeck, 2007). This is particularly the case in venous insufficiency syndrome, in which blood pools in the lower legs and often involves the formation of varicose veins. Varicose veins result from the valves within the vein becoming leaky – this disrupts the normal flow of the blood allowing it to pool and therefore cause localized swelling. Where the varicose veins are close to the skin surface, they are vulnerable to being knocked and ruptured. The resultant pulse stains are therefore most commonly found on the floor and lower vertical surfaces. Although not a common cause of death, elderly sufferers of varicose veins may stagger around after breaking

Figure 2.9 This stain was caused when the victim slit their wrist. It exhibits the typical undulating pattern caused by arterial bleeding. Note the accompanying spray of smaller spatter stains. (Reproduced from James, S.H. *et al.*, (2005) *Principles of Bloodstain Pattern Analysis*, Copyright © 2005, Taylor & Francis.)

open a vein spreading large amounts of blood and giving the impression that they were physically attacked.

Coughing and sneezing generates a stream of fast moving air that carries a mixture of spatter consisting of small droplets just visible to the naked eye and an aerosol of microscopic droplets <100 μm in diameter. Wounds to the chest, mouth or nose will result in the coughing and sneezing of blood droplets of a similar range of sizes. Stains formed from these droplets will often be contaminated with strings of mucus thereby indicating their probable origin. Droplets >100 μm lack the mass and kinetic energy to travel far but aerosolized droplets (<100 μm) can remain airborne for long periods and the distance travelled will depend, in part, on the local air currents. Small droplets may also be projected when a person is breathing face down through a pool of blood (e.g. after a beating or being shot in the head). The absence of such stains would indicate that the person was possibly dead by the time their head hit the floor. Bloodstains containing tiny air bubbles are an indicator that the blood was expired (Fig. 2.7).

Cast off stains result from blood being thrown from a weapon or limb as it is being wielded (Fig. 2.6). For example, when a knife is pulled outwards and upwards from the body there will be projected a series of bloodstains from the knife blade. The second and any further blows tend to be the ones causing the most blood splatter. Axes and similar hefty weapons are most frequently used with a downward chopping action and the returning upswing leaves a cast off pattern of blood on the ceiling. Weapons that pick up a lot of blood may produce cast off stains on both their outward and return stroke. Typically, in an arc of bloodstains, the stains become longer and more elliptical at the end of the stroke and more circular as the blow centres above the assailant's head. If there was only one victim and cast off bloodstains were found in more than one room it indicates that the attack was prolonged and there was a chase. However, if there were cast off bloodstains in

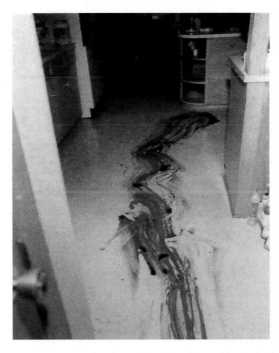

Figure 2.10 Smeared bloodstain pattern formed by dragging a bloody body across the floor. The presence of large blood clots among the smear pattern indicates the victim was not moved for some time after they were wounded. (Reproduced from James, S.H. *et al.*, (2005) *Principles of Bloodstain Pattern Analysis*, Copyright © 2005, Taylor & Francis.)

only one room but passive or transfer bloodstains in other rooms it is possible that the attack took place in one room after which the victim staggered or was dragged elsewhere.

Altered Stains: Any blood stain can become altered at some point after its initial formation. During the course of an assault it is not unusual for bloodstains to fall upon one another and for a bleeding person to fall across a spatter pattern. Changes can also occur as the blood dries out and clots, diffuses through clothing or as result of attempted cleaning up operations or the movement of insects through the bloodstains. Although blood tends to remain fluid within the body after death, clots will form at wound sites and shed blood also clots. Consequently, if a severely wounded body is dragged away some hours after the assault took place it will leave large clots of blood among the drag pattern (Fig. 2.10). The absence of such clots would suggest the body was moved shortly after death. Similarly, attempts at cleaning up bloodstains will result in smears and smudges, as would the movement of a bleeding body across the floor. The pattern of a blood smear can indicate the direction in which a person or bloodstained object was dragged: the smear usually begins as a series of drops and these then become ragged along one edge indicating the direction of travel. The initial spots may be disrupted by the passage of the head (especially if it has long hair) through them – for example, if the victim is wounded in the chest and dragged by the feet. However, the direction of travel will remain obvious. A 'thinking murderer' wishing to move their victim's body would be expected to drag it by its feet because this reduces the risk of blood being transferred onto their clothes. If a dead body is held under the armpits and then dragged, a smear might be expected from one or both heels passing through the initial bloodstains.

Interpreting bloodstains

In the initial stages of an investigation, the characteristics of all suspicious stains at a crime scene are noted. This will include the exact position of every stain along with its size and shape and the nature of the material on which it was formed (clothing, plastic, wood etc): these records are made using photography and a written report so that every spot that is to form part of the evidence is given an identifying number for future reference. From the distribution and shape of the stains one can determine how the blood was shed and hence how a wound was inflicted. For example, large (4 mm or more in diameter) circular drops of blood on the floor would indicate that the blood was travelling slowly and that the victim was stationary or hardly moving. This is indicative of blood dripping passively from a wound or a weapon. The shape of these 'passive stains' would indicates how far it had fallen – blood falling vertically onto the floor from 1–50 cm tends to form circular drops with slightly frayed edges whilst blood falling from a greater height would form a sunburst pattern (Fig. 2.3). However, above a certain height it is impossible to determine how far a blood droplet has fallen because once it reaches terminal velocity it cannot impact with any greater force. Larger droplets develop a greater terminal velocity than smaller ones but the distance a droplet would have to fall to reach terminal velocity is best determined by experimentation since it is affected by local factors such as air currents. Indeed, the shape of a bloodstain is much more strongly affected by the nature of the surface it impacts upon (Fig. 2.3). For example, blood falling onto a textured surface such as concrete is far more distorted than if it falls onto a smooth surface such as glass. It is therefore important to verify any conclusions drawn from a bloodstain pattern by experimentation.

If more than one person is present when a violent crime is committed it may be possible to distinguish their roles from the types of bloodstains upon them. For example, the person pulling the trigger or wielding the murder weapon may have only blood spatter stains on their clothes whilst the one shifting the heavily bleeding corpse may acquire only transfer bloodstains. The absence of bloodstains on a suspect is not necessarily an indication of innocence since it is possible that they may have had time to wash and change their clothes. Alternatively, they may have removed their clothes before committing the crime or they were sufficiently far away from any flying blood not to be hit or shielded from it in some way.

Determining the area of haemorrhage

Most texts use the terms 'area of origin' or 'point of origin' to refer to site from which bloodstains originated but I find 'area of haemorrhage' more apposite because it clearly indicates 'blood loss began here'. Its determination is notoriously difficult and relies upon both knowledge of physics and a great deal of practical experience. It is therefore not unusual for analysts to contradict one another's interpretations in complicated cases.

There are five principle methods of determining the area of haemorrhage and hence the position of a person at the time they were struck: Observation, Trigonometry, Stringing, Graphics, and Computer Programmes.

Observation

Simple observation of the crime scene cannot provide the numerical data that most forensic investigations require but it can sometimes provide an initial indication of the approximate area of haemorrhage/origin without the need for time consuming analysis. For example, if a pattern of aspirated bloodstains is found 1.8 metres up a wall this would indicate that the victim must have been standing upright and must have been very close to the wall since such small droplets travel only very short distances.

Trigonometry

When a drop of blood falls, it is held together by surface tension and behaves as though it had an elastic skin. Surface tension is caused by cohesive forces between molecules at the surface of a liquid and results in the drop pulling into its smallest possible area. Consequently, a drop of blood will not fall unless it is subjected to a force greater than its surface tension and once in flight the drop will tend to stay together in a sphere (rather than break up or form a teardrop shape) until it comes into contact with another object. This assumption of a spherical shape is crucial to the behaviour of the blood droplet whilst in flight and also when it comes into contact with a solid object.

If a blood droplet hits a smooth clean flat surface at 90° (i.e. 'head on') it will form an oval shape with about the same diameter as the droplet and with a number of spines dependent upon the velocity with which it impacted and the nature of the surface it hit. If, however, the droplet strikes the surface at an angle then it will form an ellipse the length of which will be dependent on both the diameter of the original droplet and the angle with which it hit the surface. Basically, as the angle of impact decreases the length of ellipse increases, i.e. a droplet impacting at an angle of 15° produces a longer ellipse than one impacting at 45° (Fig. 2.11). The ellipses will seldom be perfectly symmetrical and the narrower end and/or where tails are formed indicate the direction in which the blood droplet was travelling and hence the direction of the force that was propelling it. Because the shape of the ellipse is dependent upon the angle of impact it is possible to calculate this angle using basic trigonometry (Figs 2.12 and 2.13).

Angle θ = angle of impact (θ = Greek symbol 'theta')

$$\text{Sine of angle } \theta = \frac{\text{Width of stain}}{\text{Length of stain}}$$

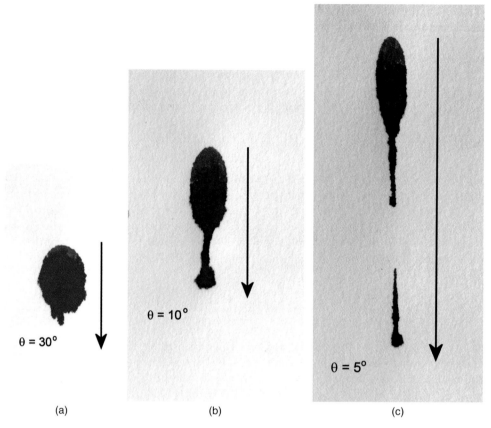

Figure 2.11 The influence of angle of impact (θ) on the shape of bloodstain forming on white card. For a blood droplet falling onto a smooth hard surface, as the angle of impact decreases the length of the ellipse increases. At low impact angles the blood droplet may break up on impact to form two or more separate stains. If the impact surface has a rough texture the chances of the blood droplet breaking up on impact are increased.

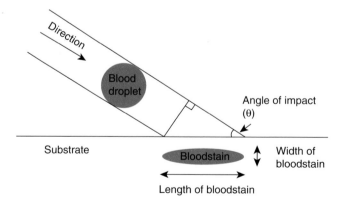

Figure 2.12 Diagrammatic representation of how the length of an elliptical bloodstain is related to its impact angle.

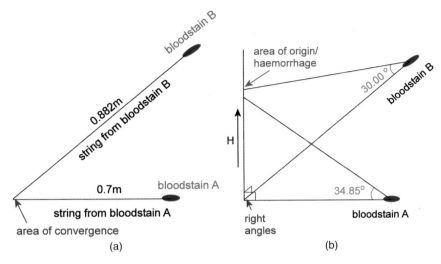

Figure 2.13 Diagrammatic representation of how the point of convergence and area of origin/haemorrhage can be determined using trigonometry. For details, see the text.

$$\text{Therefore } \theta = \text{Sine}^{-1}\frac{(\text{Width of stain})}{\text{Length of stain}}$$

If bloodstain A is 4 mm wide and 7 mm long then the calculation would be

$$\theta = \text{Sine}^{-1}(4/7)$$

$$\theta = \text{Sine}^{-1}(0.5714) = 34.85°$$

However, it is impossible to determine the point of haemorrhage from this one calculation so one needs measurements from at least one other stain. Assume therefore that a second bloodstain B lies to one side of bloodstain A and has the dimensions 5 mm wide and 10 mm long. The calculations would be:

$$\theta = \text{Sine}^{-1}(5/10)$$

$$\theta = \text{Sine}^{-1}(0.5000) = 30.00°$$

Provided that both stains result from the same event one can calculate their area of convergence and thereby estimate the distance of the stains from the area of haemorrhage. This is done by drawing lines (sometimes called 'strings') through the centres of the long axis of the stains and where the two lines meet is the area of convergence (Fig. 2.13a). The distances between the centres of the bloodstains and the area of convergence can now be measured. Let us assume that in our case the distance between bloodstain A and the area of convergence is 0.70 metres and for bloodstain B it is 0.882 metres (bigger droplets tend to travel further than smaller ones).

A line, we can call it 'H', is now drawn at 90° from the area of convergence so creating two right angled triangles: the one formed from bloodstain A and the one formed by bloodstain B and both of them will share H as one of their sides (Fig. 2.13b). This is done by drawing a line with an angle of 34.85° from bloodspot A and another with an angle of 30.00° from bloodspot B where they meet on line H will be the area of haemorrhage. This can also be calculated as follows:

For bloodstain A

$$H = \text{Tangent } 34.85° \times 0.70 = 0.487 \text{ metres}$$

For bloodstain B

$$H = \text{Tangent } 30.00° \times 0.882 = 0.509 \text{ metres}$$

Note that the two H values are not identical. This is because in real life there is a lot of ambiguity, not least because blood droplets travel in parabolic flight paths rather than straight lines and the further a droplet travels the more it assumes a parabolic arc. In addition, when we are hit by something our body is sent into motion and when we bleed from a serious wound we do not (usually) lose blood one drop at a time from a single point source (Fig. 2.9). Hence, the source of any two spots of blood in three-dimensional space may not be exactly the same even if they were caused by the same event. Consequently, some workers use the terms 'area of convergence' and 'area of origin/haemorrhage' rather than 'point of con-vergence' or 'point of origin/haemorrhage' because it is seldom possible to establish these sites with pinpoint precision. Accuracy can be increased by carefully choosing which stains to use – those with the lowest impact angles are thought to provide the greatest accuracy (Willis *et al.*, 2001) – and whilst there are no recommended minimum numbers it is obviously prudent to use more than two or three if many more suitable stains are available. Several workers have developed more compli-cated equations for the determination of the area of haemorrhage that provide a more realistic account of droplet's flight path (e.g. Knock & Davison, 2007) but these remain at the experimental stage. Absolute accuracy is not always needed anyway since the investigator often only needs to know the general position of a person when they shed their blood. For example, if a suspect claims that he was defending himself but the bloodstains indicate that the blood was shed from a fatal wound to the head that was at a height of about 14 cm from the ground when struck – then this is evidence that the victim was already on the floor at the time and therefore hardly a threat.

Stringing

After labelling and measuring all the stains that are to be used in the analysis, the scene is recorded using a camera and a scale is included with each photograph along with directional indicators. On a horizontal surface these take the form of the points of the compass and an indication whether they are on the upper/lower surface or ceiling/floor as appropriate. On a vertical surface an arrow indicating up/down is required and the arrow should be drawn at 90° vertically downwards (i.e. as a 'plumb line'). Alternatively, a spirit level can be used to draw a 'level line' underneath the stain. This enables a record to be made of the stains' orientations in relation to the surface they are found on. Next, an elastic cord (or 'string') is placed through the mid-line of each stain and run backwards in the direction the blood came from. The individual cords are secured at either end and where they overlap one another is the point of convergence. The distance from the leading edge of each stain to the area of convergence is then recorded. Let us assume that our stains are on a vertical surface, in this case one should extend the level line so that it cuts across the elastic cords that are running to the area of convergence. The angle formed between the plumb line and individual cords can be recorded using a protractor and is called the 'direction of flight angle'. In order to work out the distance from the vertical surface and hence the area of haemorrhage/origin a stick or similar object needs to be secured so that it projects at 90° from the area of convergence (Fig. 2.14). The angle of impact can be determined for each stain from its dimensions, as described above, and a cord with this angle to the surface (as determined using a protractor) run back to the stick. Where the cords cross on the stick is the area of haemorrhage. Needless to say, the whole exercise takes a long time,

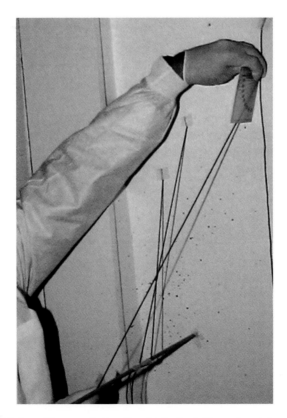

Figure 2.14 Bloodstain pattern analysis using 'stringing'. (Reproduced from James, S.H. *et al.*, (2005) *Principles of Bloodstain Pattern Analysis*, Copyright © 2005, Taylor & Francis.)

especially where there are numerous separate 'events' to record and the tendency of the cords to droop adds to the difficulties.

Graphics

The angle of impact and distance to the area of convergence is calculated for each stain as described in the trigonometry method. This information is then plotted on an X–Y graph in which the distance to the area of convergence is the horizontal X axis and the Y axis represents the distance to the area of haemorrhage/origin. For example, let us assume that there are four blood stains whose characteristics are as follows:

	Distance to area of convergence (cm)	Angle of impact
Stain A	72.4	50.9°
Stain B	99.0	41.6°
Stain C	128.0	35.8°
Stain D	160.0	30.0°

As can be seen from Fig. 2.15, by drawing a line with the appropriate angle from each stain to the Y axis, the point at which they meet represents the area of haem-

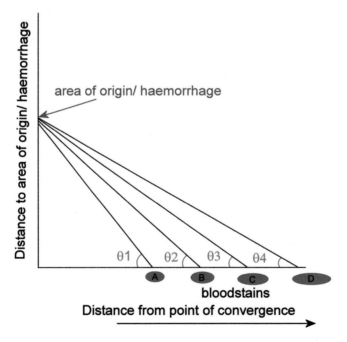

Figure 2.15 Determination of the area of origin/haemorrhage by the graphics method. For details, see the text.

orrhage, i.e. the distance above/below/to the side (as appropriate) of the point of convergence.

Computer programmes

The use of computer programmes (e.g. Hemospat®, BackTrack®) to analyse the distribution of bloodstains is rapidly replacing manual procedures such as stringing. Not only do these programmes permit rapid analysis but they also provide numerical estimates of the error associated with individual measurements. This makes the recording of data more scientifically rigorous, the reconstruction of the events easier to understand and also facilitates the subsequent presentation of evidence in court.

The crime scene is recorded using a pre-calibrated digital camera and a scale included with each photograph along with a directional indicator. Ideally, this should indicate the points of the compass and up/down if the stains are on a vertical surface. In some programmes, the inclusion of a calibrated pattern such as a chequerboard in the photographs enables the angled shots to be rectified by the computer to appear as though they were taken from overhead. The computer superimposes an ellipse over any chosen bloodstain and through the use of algorithms identifies its angle of impact. Then, following the analysis of several bloodstains, the computer can generate lines to identify their area of origin and illustrate this in either two or three dimensions.

The current computer programmes require numerous photographs of the crime scene because it is seldom possible to take just one or two photographs that both show the bloodstains with sufficient resolution to facilitate measurement and display them in the wider context of the room. There is therefore an interest in developing the use of three-dimensional laser scanners which can rapidly map a whole room and its contents in three dimensions (www.DelftTech.com). The computer can then use digital pattern analysis to identify the oval and elliptical shapes of typical bloodstains within the room and analyse them as described above. This would permit the crime scene to be recorded for bloodstain evidence far more quickly than is currently the case. Computer analysis software can also be particularly useful in complex situations in which two or more blood patterns overlap. This is not unusual – for example, if blows were delivered in quick succession or if a conflict progressed back and forth over the same area. Through the use of statistical techniques such as fuzzy cluster analysis the probability that an individual stain resulted from one blow or another can be calculated.

Case Study: Shooting by bow and arrow, the Peter Rattya Case

On 15th January, 2005 Peter Rattya and a group of other people were drinking and smoking cannabis in the backyard of a boarding house in Melbourne, Australia. Subsequently, everyone apart from Rattya and another man, Amer Alihromic, drifted away and the two had an argument about religion. The evening ended with Alihromic being found fatally shot through the chest with a bow and arrow and with a knife clasped in his hand whilst Rattya suffered from a knife wound to his leg. Rattya initially claimed that Alihromic's death was an accident (presumably the bow 'just went off') and then claimed self-defence. As is not unusual in such circumstances there were no witnesses willing to either confirm or deny his account of events at the start of the investigation.

Bloodstain pattern analysis of the backyard and the boarding house coupled with DNA analysis to confirm which stains belonged to which person indicated a very different sequence of events. To begin with the pattern of staining indicated that Alihromic was in a prone position when he was shot. Indeed, a witness (some persons did eventually agree to testify) subsequently stated that Alihromic was lying on the ground begging for his life when he was shot from a distance of about 2 metres. The autopsy indicated that the arrow had entered at the solar plexus, pierced the liver and punctured a lung causing fatal haemorrhage. It would be difficult for the arrow to take such a trajectory unless the victim was lying down and angled away from the shooter. A trail of drip stains indicated that after being shot Alihromic staggered towards the house before finally collapsing at the point where he was found. Rattya, meanwhile had gone into the house and forced one of the witnesses to stab him in the leg with the knife so that it would look like he acted in self defence. He then forced the same man to go outside and put the knife in Alihromic's hand and return the bow to Rattya's room. This, however, meant that Rattya's blood was found in the house, well away from the scene of conflict. Although Rattya went outside and stood over Alihromic's body so that his blood was found on the corpse the pattern would be one of passively dripped blood onto a non-moving object – hardly symptomatic of a violent conflict.

At the trial it transpired that Rattya (who had psychiatric problems) became upset about Alihromic making 'devil/hyena-like noises'. He therefore went to his room, took down his bow and arrows from the wall where they were stored, went outside and shot the unarmed Alihromic from close range. Rattya then went indoors and forced those there to do as he said and keep quiet. He also told them to phone the emergency services and inform them that he had been stabbed – there was to be no mention of Alihromic. Rattya had a long history of violent offences and people were afraid of him but once it became clear that the evidence against him was so strong they started to co-operate with the investigation. Ultimately, Rattya was jailed for murder.

Case Study: Were bloodstains acquired during an assault or after it had taken place? The Sion Jenkins Case

On February 15th, 1997 13-year-old Billie-Jo Jenkins was bludgeoned to death with an 18-inch iron tent peg outside her family home in Hastings, East Sussex. At the first trial, her foster-father, Sion Jenkins, was convicted of her murder and sentenced to life in prison. However, Mr Jenkins has repeatedly stated his innocence and the case contains many anomalies that were given another airing at the retrial (April–July, 2005). The prosecution claimed that the distribution of 158 microscopic bloodstains found on Mr Jenkins' clothes were acquired when, for reasons that have never been satisfactorily explained, he attacked his foster-daughter with the tent peg, before driving off on an unnecessary shopping trip with his other daughters to provide an alibi. By contrast, the defence maintained that, whilst Mr Jenkins was away on his shopping trip Billie-Jo was attacked by an unknown intruder who then fled before Mr Jenkins returned. In this scenario, the stains represent blood droplets exhaled from Billie-Jo's airways onto Mr Jenkins' clothing when he went to her aid.

One of the first points of dispute was why, if Mr Jenkins really was the killer, his clothing was stained with only microscopic spots of blood when Billie-Jo had been struck at least eight times with a heavy metal object in a frenzied attack and the surrounding area was coated with blood. Attempts to replicate the attack using a pig's head covered in blood always resulted in the striking arm becoming covered in bloodstains yet Mr Jenkins, who is right-handed, had only three microscopic spots of blood on his right arm. The prosecution countered that this was not a serious issue and at the most recent trial they claimed to have identified a white substance that they took to be human flesh on three of the bloodspots on Mr Jenkins' trousers. This was suggested to be evidence that Mr Jenkins had not only been spattered with blood but also with flesh – indicating that he was the person delivering the blows. However, from the ferocity of the attack, tissue fragments might be also be expected on Billie-Jo's body and clothing so it is conceivable that these might be passively transferred onto someone who subsequently held her.

In the first trial, the forensic scientists called by the prosecution claimed that, in the scenario presented by Mr Jenkins, Billie-Jo would have been dead by the time he arrived back from his shopping trip and therefore incapable of spluttering blood over him. If a person is decapitated, there is little doubt about when they cease to breathe but on the basis of head injuries, even horrendous ones such as those received by Billie-Jo, it will remain a matter of conjecture. Not surprisingly, therefore, medical experts have failed to agree on this issue. The defence argued that Billie-Jo was still alive – albeit only just – when Mr Jenkins arrived on the scene. Histological examination of Billie-Jo's lungs has supported Mr Jenkins' version of events by demonstrating evidence of interstitial emphysema and also that Billie-Jo could have been alive for at least 20 minutes after the attack. Interstitial emphysema is a condition in which air escapes through a tear or other injury into the connective tissue of the lungs. It is a common occurrence when the airways are obstructed (as was argued by the defence in the case of Billie-Jo) or there is a penetrating wound to the chest. It was highly unfortunate that at the initial post mortem a histological examination of the lungs was not performed even though this is a recommendation in the guidelines provided by the Home Office and the Royal College of Pathologists. The defence went on to argue that, when Mr Jenkins saw Billie-Jo, he ran to her and placed his hand on her body – this slight movement was then sufficient to dislodge a blockage in her larynx caused by blood seeping from her head wounds. As a consequence, of the blockage, the air in Billie-Jo's lungs would have been under pressure. By removing the blockage, the pressure would force air out of the lungs along with a fine mist of blood droplets. Hence anyone who was near to her mouth would have been coated with tiny blood particles without even noticing it. Further support for this scenario was provided in the retrial when forensic scientists called by the prosecution admitted for the first time under cross-examination, that it was a reasonable possibility that Billie-Jo had projected out the blood droplets over Mr Jenkins. It is always possible that the lack of extensive blood-stains on his clothes and body might have been because he cleaned himself up after killing Bille-Jo. However, according to the defence counsel, the timings and witness statements would suggest that in this scenario Mr Jenkins would have had only about 3 minutes in which to get angry with Billie-Jo, kill her and clean himself up. He would then need to calm down and take his other daughters out shopping without them thinking he was unduly upset.

To add to the uncertainties associated with much of this case, at the 2005 re-trial the QC acting for Mr Jenkins claimed that important evidence had gone missing. For example, only approximately 40 of the blood spots remained for future analysis and the hospital storing Billie-Jo's lung tissue samples appeared to have lost them. There are further strange features associated with this case that do not 'add up', such as the testimony of the other members of Mr Jenkins' family. There is also the mysterious finding of a scrap of black plastic bin liner that had been forced deep inside Billie-Jo's left nostril and the role (if any) of 'Mr X' – a paranoid schizophrenic who was known to have been in the neigh-bourhood and whom police had observed pushing plastic bags into his own mouth and nose. Ultimately, at the end of the second trial, the jury failed to reach a verdict and was discharged. More background information on the Billie-Jo Jenkins case, albeit biased towards the defence, can be found at www.innocent.org.uk/cases/sionjenkins/.The case of Sion Jenkins is provided in some detail to illustrate how 'messy' and confusing the evidence can be in a forensic investigation and therefore how experts not infrequently arrive at completely different conclusions when analysing the same material.

Artificial blood

Where the conclusions drawn from bloodstain analysis are ambiguous or likely to be controversial then it is always a good idea to verify them through experimenta-tion. The ideal test material is, obviously, human blood, or failing that equine or pig blood. However, working with real blood presents Health and Safety problems and is not necessary for demonstration or training purposes. Millington (2004) has provided an excellent discussion of a variety of artificial bloods the formula of one of which is given in Table 2.2.

All the ingredients are mixed together thoroughly to a smooth consistency and the 'blood' may then be stored at room temperature. Millington's data are based on 'blood' used at room temperature – and the same is true of many other labora-

Table 2.2 Formula for artificial blood (After Millington 2004)

Superfine self-raising flour	13 g
Sodium chloride	1 g
Glycerol	1 ml
Strawberry sugar syrup (Lyles)	1 ml
Scarlet food colouring	1 ml
Distilled water	183 ml

tory studies, regardless of the nature of the blood used. Temperature will affect viscosity and it would therefore be interesting to know whether this might affect the results obtained.

Post mortem toxicological analysis of blood

Most toxicological studies involve the analysis of blood samples but the collection and analysis of post mortem blood presents far more difficulties than it does when the donor is still living (Skopp, 2004). Once a person dies the blood stops circulating and settles to the dependent regions whilst chemical and microbial decay begins. Consequently, many drugs become redistributed and their concentration at a particular site may increase, decrease, or remain more or less the same as it was at the time of death. All drugs behave differently and there are also differences between their behaviour in different individuals whilst the post mortem environment and time since death are also important factors. For example, drug addicts often develop a tolerance to their drug of choice but there is no way of estimating this once they are dead. Consequently an otherwise healthy drug addict may have much higher drug levels in their body after death than a person not addicted to that drug but in the drug addict the levels would not have been toxic. Similarly, post mortem changes mean that the blood drug concentration can differ markedly in different regions and there is little consensus about how many regions should be sampled or whether the highest, lowest, or mean value of those samples should be used in any calculations. For example, if a drug is prone to redistribution (e.g. morphine), its concentration tends to be much higher in blood taken from within one of the heart's chambers (cardiac blood) or the surrounding pericardial sack than it does in the peripheral blood – which is usually taken from the left and right femoral veins. Microbial fermentation causes the formation of ethanol and could therefore, potentially, lead to a false assumption of alcohol intake prior to death. Although microbial ethanol production is not usually marked until 3–10 days after death it might, presumably, occur sooner under conditions that promote microbial growth (also see Chapters 9 and 10). None of this should be taken to indicate that post mortem toxicological analyses are pointless, far from it, but that any findings should be considered with a level of caution appropriate to the circumstances.

Saliva and semen

Characteristics of saliva and semen

Both saliva and semen are products of exocrine glands – that is, they are secretions that are released to the outside of the body through ducts – and both can provide valuable forensic evidence. Saliva is a slightly alkaline secretion that is produced constantly by the salivary glands and released into the mouth. Here it lubricates the membrane surfaces and aids the chewing and swallowing of food. Although it contains the enzyme salivary amylase, that begins the breakdown of starch, saliva does not have an important role in the digestion of carbohydrates. In addition, saliva also contains other enzymes, such as lysozyme, some salts and mucin. The presence of saliva stains can be demonstrated by detecting amylase activity. This is best done with a specific assay method, such as ELISA (enzyme linked immunosorbent assay) that can distinguish salivary amylase from other amylases, such as pancreatic amylase and bacterial amylase (Quarino *et al.*, 2005). Saliva is important forensically because traces may be left at bite marks, and in spit from which it is possible to isolate DNA. Furthermore, a number of drugs (in particular those used illegally in sport) can be detected in saliva; mouth swabs can be used as a source of DNA thereby removing the need for a suspect to provide blood for a DNA test.

Semen is gelatinous fluid produced by the combined secretions of the seminal vesicles, the seminiferous tubules, the prostate gland, and the bulbourethral gland. The typical ejaculate consists of between 1.5 and 5 ml of semen and contains 40–250 million spermatozoa. In addition to the sperm, the ejaculate also contains the sugar fructose that serves as an energy store, citric acid, calcium and a variety of proteins. Semen analysis is usually a feature of cases of rape and sexual assault.

Semen may be released following death. This occurs as a consequence of rigor mortis developing in the *dartos* muscle of the scrotum. The *dartos* muscle consists of smooth muscle fibres that are found in the septum dividing the two testes and also the subcutaneous tissue. Contraction of the *dartos* muscle results in the wrinkling of the skin surrounding the scrotum and the testes are elevated. In addition, rigor may also develop in the muscle fibres found in the seminal vesicles and the prostate and the contraction of all these muscles can result in a discharge of semen. Consequently, one should not assume that the presence of semen denotes recent sexual activity (Shepherd, 2003). For example, if a man is found dead in a public toilet with his genitals exposed and evidence of seminal discharge it should not automatically be assumed that he died or was killed during the course of committing a sexual act.

Demonstrating the presence of saliva and semen

Both saliva and semen fluoresce in UV light. Fluorescence occurs when substances absorb UV radiation and re-emit (some of) it as longer wavelength UV or visible light. Any fluorescing spots observed on the body or clothing are circled with an indelible marker for future extraction and confirmatory analysis. Fluorescence is only indicative since many substances exhibit fluorescence in UV light including urine, certain narcotics, and food colourings such as those used in fruit pop juices.

Commercial semen detection kits are available over the internet for spouses and partners who suspect that their 'significant-other' may be cheating on them. These usually rely on detecting the enzyme acid phosphatase on stained clothing and the assumption that a positive reaction can only be explained by illicit sexual activity. It should be borne in mind that acid phosphatase is a very common enzyme and the test is indicative rather than proof of the presence of semen. Another, albeit more time-consuming, method is to look for the spermatozoa. These may be found within the vagina up to 5 days after sexual intercourse although the frequency of detection declines rapidly with time (Willott & Allard, 1982). Human spermatozoa are extremely small and soon lose their tails, so a sensitive cytological test is required. The most effective method is the 'Christmas tree test' (Allery *et al.*, 2001) that uses the dyes nuclear fast red and picroindigocarmine and stains the sperm red and green. More discriminative tests for semen have been suggested, such as detecting the prostate specific antigen (PSA) that is normally used in the diagnosis of prostate cancer (Maher *et al.*, 2002).

Saliva and semen as forensic indicators

Identifying an individual from saliva or semen samples depends on either serological or DNA profiling. Approximately 75–85% of the population are termed 'secretors', that is, their body fluids include the same or similar profile of antigens, antibodies and enzymes that are found in their serum (although not the blood cells themselves – these are normally confined to the blood vessels). This can therefore, theoretically, enable an assailant to be identified by serological profiling from, say, a bite mark, spit, or semen. However, the serological profile of the blood does not always match that of the body fluids and this reduces the effectiveness of this approach. DNA profiling offers more powerful discrimination but even this technique has its limitations. For example, in the UK, rapists are now aware of the risk of detection by DNA profiling and it is not unusual for them to use spermicidal condoms. It has been suggested that the use of such condoms (if found) might provide an indication of when the assault occurred. In an experimental study, the proportion of viable sperm within spermicidal condoms was found to decline gradually from about 40% to 6% over the course of 3 days (Gosline, 2005). This measurement would be useful if there was a dispute about whether the condom was 'planted' on the suspect or when the event took place. Although these findings are interesting, more work needs to be done on how sperm viability within the condoms is affected by temperature, light and other environmental conditions. Furthermore, some rapists fail to ejaculate or if they had a vasectomy or suffer from certain medical conditions their semen lacks spermatozoa (a condition known as azoospermia). It is estimated that approximately 2% of forensic semen samples lack sperm and although this is a relatively small number it would translate into about 80 reported cases a year in England and Wales (Murray *et al.*, 2007).

Although the presence of an accused man's semen on a victim's clothing is apparently damning evidence of sexual intimacy it is possible that there could be alternative explanations. In January, 2007 Maria Marchese was jailed for 9 years at Southwark crown court, London, following a prolonged hate campaign against Dr Falkowski (a hospital psychiatrist). Ms Marchese had become infatuated with Dr Falkowski and was attempting to break up his relationship with his fiancé. Amongst other acts (which included threats of violence and death) she retrieved a used condom from his dustbin and smeared the semen over her underwear before accusing him of raping her. As a consequence, Dr Falkowski was suspended from his job whilst charged with the offence and was not cleared until 18 months later.

Laser microdissection

In cases of rape and sexual assault, semen samples are usually contaminated with the victim's epithelial cells and/or blood – and hence their DNA. The analysis of mixed sample DNA is difficult, especially if the two are closely related and/or the assailant's DNA is present in very small amounts. The sperm can be separated from other cells on the basis of a preferential lysis technique that causes the destruction of epithelial cells whilst the sperm remain intact. The victim's DNA therefore enters into solution and can be washed away whilst the assailant's DNA remains within the (whole) sperm and can be analysed separately. Unfortunately, the technique is not always fully effective so contamination can remain a problem, especially if large numbers of the victim's epithelial cells are present, whilst there is the risk of a marked reduction in the recovery of the assailant's DNA so it is not suitable where there are few sperm present to begin with. A recently developed way round this problem is to use a technique called laser microdissection separation (e.g. CellCut Plus® and SmartCut Plus®). The sample is mounted on a special membrane covered microscope stage and a glass slide placed on top to prevent contamination. The microscope operative then identifies individual cells or groups of cells of interest via a computer screen and draws around these using the computer controls. A high precision laser with a beam size of $1\,\mu m$ cuts out the shapes and these are automatically transferred to a collection tube and can then be subjected to DNA analysis. Consequently, one has visual confirmation that the appropriate cells are chosen and with good starting material it is even possible to isolate sufficient DNA for STR typing from as few as ten sperm cells – all free from all potential sources of contamination (Di Martino et al., 2004a). Clearly, if many sperm cells are present it is advisable to use more than this (e.g. 150–200), not least because of the limits of the sensitivity of the PCR process and also some of the sperm may contain degraded DNA and these will both limit one's ability to obtain a full DNA profile. It is also possible that more than one man may have committed the assault and therefore sufficient sperm need to be analysed to be sure that the DNA profile is representative. The sample can be stained to facilitate the identification of the cells and it is even possible to extract cells from old slide preparations provided the cover slip can be removed.

Laser microdissection can also be used to extract material for DNA analysis from hair samples (Chapter 4) and it has been suggested that the sticky tapes used to collect forensic material at crime scenes could also be examined for discrete DNA-yielding material. Microbial contamination can compromise DNA analysis so in situations in which, for example, decay is advanced it may be possible to make histological preparations and then excise areas in which microbes are absent or present in smaller numbers.

Spermatozoa are relatively easy to identify even if they lose their tails but, as mentioned above, spermatozoa may be absent or extremely difficult to find in some cases. Fortunately, semen contains cells other than sperm and although it is morphologically impossible to distinguish male and female cells they can be identified using fluorescence *in situ* hybridization with the Vysis X–Y-probe kit (Murray *et al.*, 2007). The kit contains a SpectrumGreen™ labelled Y chromosome probe that hybridizes to the Y chromosome and a SpectrumOrange™ labelled X chromosome probe that hybridizes to the X chromosome. Following hybridization, male cells exhibit one green and one red signal whilst female cells exhibit two red signals. These cells can then be isolated by laser microdissection and their DNA analysed.

Vitreous humor

The term vitreous humor refers to the clear gelatinous matrix located behind the lens within the vitreous chamber. It maintains the shape of the eyeball and so keeps the light receptor cells of the retina flush with the underlying vascular choroid layer. This ensures that the retina provides a smooth light receptive surface. The vitreous humor is formed during embryonic development and, unlike the aqueous humor, is not replaced during adult life. The vitreous humor is usually free of microorganisms and separated from the rest of the body by a series of tough membranes so it tends to decay more slowly than the surrounding tissues. This is not to say that it is metabolically isolated and inert: drugs and metabolites pass from the body into the humor and it contains phagocytic cells to remove debris and invading microorganisms.

Estimation of the post mortem interval

Following death there is a rise in the concentration of potassium within the vitreous humor. This is to be expected because cells have a higher potassium ion concentration than their surrounding extracellular fluid but as long as they are alive, this imbalance, along with those other ions, is maintained by a combination of selective membrane permeability and ion pumps. Once the cells die, the membranes begin to leak and ion pumps stop so ions diffuse down their electrochemical gradients. Consequently, the potassium concentration in the vitreous humor rises. Most workers use linear regression analysis to demonstrate the relationship between the potassium ion concentration and post mortem interval (PMI) with the potassium ion concentration as the dependent variable and the PMI as the independent variable. This makes intuitive sense since the independent variable is the one that is supposed to be 'fixed'. According to Munoz *et al.*, (2001) greater accuracy can be obtained if the potassium ion concentration is used as the independent variable but Madea & Rödig (2006) found that although this did bring statistical improvements these were of little practical value. Madea & Rödig (2006) claim that better accuracy can be obtained by using a form of local regression called the LOESS procedure – this is a procedure in the SAS/STAT® software package and is beneficial where one does not know a suitable parametric form for the regression surface and there are outliers in the data set. Despite this, when they compared the potassium concentrations from 492 cases whose PMI was known with the regression model, only 31% were found to fall within the correct range and the PMI of the majority of the remainder would have been over-estimated. On the whole, the literature indicates that estimates of the PMI from vitreous humor potassium ion concentration are most accurate in the first 24 hours after death. The technique is particularly useful for bodies that have become burnt in fires. Burning interferes with the development of rigor mortis and most other physiological means of determining the PMI during the early post mortem period. Most research indicates that there is little difference between the two eyes in their rates of accumulation of potassium ions etc after death.

There are also changes in the levels of hypoxanthine, xanthine, lactic acid, creatinine and urea in the vitreous humor and by combining assays of potassium with one or more of these substances it might be possible to improve the accuracy of the PMI measurements over longer periods. However, although extracting vitreous humor is straightforward and the ions and metabolites can be analysed quickly and simply there remain many problems to be overcome before these measurements are likely to gain widespread use. To begin with there are relatively few studies on how their levels vary between individuals during life or how they are affected by age and disease. Clearly, if the potassium ion concentration is already at a high or low concentration before death this would impact on the subsequent determination of the PMI. Similarly, there are no standardized protocols and these are important because the storage, handling, and processing of the samples can all impact on the results as can the analytical technique used (Madea & Musshoff, 2007).

Sudden infant death syndrome

Hypoxanthine levels in the vitreous humor are elevated in cases of sudden infant death syndrome (SIDS) – a so far unexplained phenomenon in which an apparently healthy baby dies during its sleep. Deaths under these conditions invariably raise concerns that foul play may have been involved. Hypoxanthine is produced under hypoxic conditions – i.e. when the oxygen supply is limited – and therefore its rise after death is to be expected. Although it is impossible to know exactly when a baby who suffers SIDS dies, the rise in hypoxanthine levels are above those that would be expected from the normal decay process. It is possible that an infectious agent may be involved – the hypoxanthine levels are similar to those of babies dying of infections but significantly higher than those that die suddenly and violently (Vege et al., 1994; Rognum et al., 1991) – but many people believe that SIDS represents the common end point of a variety of conditions and circumstances.

Post mortem toxicology

A number of drugs (e.g. morphine, diazepam, cocaine) can be detected in the vitreous humor and the humor's 'isolation' from the rest of the body means that it is less susceptible to the chemical and microbial decay processes taking place elsewhere (Leikin & Watson, 2003; Skopp, 2004). For example, Pragst *et al.* (1999) were able to detect 6-acetylmorphine (a metabolite of heroin) in the vitreous humor of persons dying of heroin overdose even though it could not be found in their blood. Similarly, the vitreous humor is often sampled for ethanol analysis because it is less susceptible than blood to contamination from microbial ethanol production. However, for most drugs there is limited information on the relationship between their post mortem levels in the vitreous humor and their likely ante-mortem levels in the bloodstream.

Faeces and urine as forensic indicators

Faeces and urine are sometimes found at a crime scene because the culprit wishes to violate further the body or property of his victim. Alternatively, during a violent attack, both the assailant and the victim can experience such stress and fear that defecation or urination is involuntary. Faeces and urine may also be recovered from a toilet that was not flushed and the body of a dead crime victim. Our faeces can reveal a great deal about us. It can indicate our diet, health, where we have been, and, from DNA, who we are. Faecal material is therefore a potentially very useful forensic indicator and should be collected whenever it is present at a case of suspicious death. A further advantage of faeces is that if it becomes dried out or left in an extremely cold environment it can provide evidence hundreds of years after it was voided.

Health

The form, texture, colour and smell of our faeces are all affected by our health and can therefore provide an indication of the condition of the person who produced the faeces. For example, faecal analysis can reveal the presence of the eggs and cysts of parasites and these can provide an indication of the countries a person may have visited. Analysis can also provide an indication of underlying medical conditions (e.g. demonstrating the presence of occult blood).

Diet

We are unable to digest cellulose and therefore a great deal of plant material passes through our guts and can be identified in the faeces. In addition to plant cells, one can often find starch grains, seeds, plant hairs, and xylem, and also pollen that has either been consumed as part of the diet (e.g. honey) or breathed in and swallowed (Fig. 2.16). This can therefore provide an indication of what a person had been eating and therefore, potentially, where they had been eating it and/or the country they may have been living in the recent past. Animal material such as fragments of muscle cells, bits of bone and hair may also be recovered from faeces and used to identify past diet. The animal that was eaten can be identified by analysing for the specific myoglobin (muscle protein). Mineral material (e.g. fragments of quartz or

Figure 2.16 Identifiable vegetable remains present in human faeces. Xylem, plant cell walls and spicules are often found in faeces and can be used to determine the sorts of food consumed. Faeces may also contain pollen and other palynomorphs that were part of the diet (e.g. within honey) or inhaled.

clay minerals) is often found in faeces as a result of matter attached to our food, dust that we have breathed in or a medicine we have taken. This can potentially be used as a geographical indicator.

Case Study: Demonstrating evidence of cannibalism

The Anasazi culture flourished for over a thousand years in the American southwest but then mysteriously died out around the year AD 1200. A wide variety of reasons have been suggested to explain the decline and one of the most controversial of these involves the practice of cannibalism (Graves *et al.*, 2002; Turner & Turner, 1999). The reason the theory is so controversial is that the Anasazi have a reputation for having been peaceful people who survived by hunting and subsistence farming. They were therefore quite distinct from the aggressive Toltecs further south in Mexico who slaughtered thousands as human sacrifices to their gods and who may have indulged in cannibalism. The theory arose from the finding within Anasazi dwellings of disarticulated human bones that bore cut marks indicative of de-fleshing and signs that the bones had been cooked. The finding of defleshed bones is not proof of cannibalism or murderous behaviour because there are records of defleshing and cooking of remains as part of the normal burial procedure in several cultures around the world. The opportunity to provide more conclusive evidence arose when during the course of an archaeological investigation a perfectly preserved sample of human faeces was found in the ashes of a hearth within a dwelling containing the butchered remains of at least four adults and one adolescent. The faeces was unburnt indicating that it was passed after the fire had gone out and the ashes had cooled. Because of the dry conditions, the forensic evidence had been preserved and it was still possible to identify human blood residues on sharp stones that were probably used to cut the flesh from the bones. Immunological tests specific for human myoglobin conducted on the faeces proved positive as did shards from a cooking pot also found in the dwelling (Marlar *et al.*, 2000). Myoglobin is only present in muscle and therefore its presence in faeces indicates the consumption of flesh. Tests on modern faecal samples indicated that human myoglobin is not present in normal faeces even if the donor tests positive for occult blood and therefore exhibits bleeding into the bowel. Further analysis of the archaeological sample revealed an absence of plant remains and this is extremely unusual for people living in that area at that time. The faecal evidence therefore provided strong supporting evidence that not only were the Anasazi being killed but they were also cooked and eaten afterwards.

Today, cannibalism associated with murder arises occasionally and usually involves psychologically deranged individuals. One notorious case is that of Armin Meiwes in Germany. In 2001, Meiwes advertised on the internet for anybody who wished to be killed and eaten to get in touch with him. Surprisingly, several men did, although having met Meiwes they got 'cold feet' and left unharmed. However, one man consented to have his penis cut off before being killed. Both Meiwes and his victim attempted to consume the raw penis but it was considered 'too chewy' so Meiwes tried cooking it. Not being a good cook the penis got burnt and was considered inedible. Meiwes then retired to read a book whilst his victim was left to bleed to death in a bath. However, after a couple of hours the victim was still alive so Meiwes finished him off with a series of stab wounds. All of this is known because Meiwes videotaped the proceedings. Meiwes butchered the body, stored the remains in his fridge and consumed some of them over the following months. Meiwes was arrested in 2002 when he started advertising for new victims to come forward. Initially he was convicted of manslaughter on the grounds that his victim was willing to be killed but at a subsequent retrial he was found guilty of murder and sentenced to life. While in prison, Meiwes is alleged to have become a vegetarian.

This information is included to illustrate that one should never be surprised at what humans are willing to do to one another or have done to them. One can also use this case as an exercise to identify the various stages at which different forms of forensic evidence could have been used to recreate the crime scene had Meiwes not thoughtfully provided a video recording.

Extracting DNA from faeces

The retrieval of human DNA from faeces is difficult because it contains substances that interfere with standard DNA extraction and analytical techniques (Hopwood et al., 1996). For example, the human DNA needs to be differentiated from the large amounts of bacterial DNA and the DNA amplification process can be inhibited by the presence of bile salts and plant polysaccharides (Lantz et al., 1997). However, a commercially available kit for the extraction of DNA from faeces, the QIAGEN QIAamp® DNA Stool Mini Kit, is now available and is reportedly extremely effective (Vandenberg & van Oorschot, 2002; Johnson et al., 2005).

In 1995, the murderer of Monica Jepson, a 66-year-old widow, left few clues behind at the nursing home in Birmingham where she was staying. Just about all the police had to go on were a partial fingerprint and some faeces left on the nursing home's fire escape. At the time, DNA profiling techniques were still in their infancy although it was possible to eliminate one suspect. By 2001, techniques had improved and a full profile was obtained. This was then compared to the National DNA Database© and a match obtained with John Cook. His DNA was present on the database as a consequence of a previous offence and when brought in for questioning, his fingerprints were found to match those left at the crime scene. He was subsequently charged with murder and sentenced to life in prison.

Unless a person has a kidney or urinary tract infection, urine does not normally contain many cells although it is possible to extract DNA from urine and urine stains. It is easier to obtain complete DNA profiles from women's urine than that of men because they shed more epithelial cells (Prinz et al., 1993; Nakazono et al., 2005) and also, presumably, because during the menstrual cycle, their urine will contain blood cells.

Drugs and poisons

Although some poisons as well as drugs and their metabolites can be detected in faeces (e.g. metabolites of cocaine) it would not normally be analysed if alternative fluids or tissues were available. This is because faeces is chemically and physically highly complex and contains a large number of cells. Furthermore, microbial decomposition/alteration of the drugs etc. would be highly likely. By contrast, urine generally exhibits a lack of cells coupled with low protein and lipid levels and this simplifies the detection of drugs and their metabolites. Urine analysis is therefore commonly performed where it is thought possible that drug use or poisoning might have been a factor in a person's death. However, it is difficult to relate post mortem

Table 2.3 Summary of forensic information that can be obtained from body fluids and waste products

Sample	Forensic information	Test
Blood, serological markers	Identification	Blood typing Rhesus factor Enzyme polymorphisms
Blood, molecular markers	Identification	DNA profiling
Blood, chemistry	Poisoning	Specific tests (e.g. carboxyhaemoglobin)
	Drug use	Specific tests (e.g. heroin metabolites)
Blood, infections	Previous travels	Specific tests (e.g. malaria, leishmaniasis)
	Source of infection (if being spread recklessly or deliberately)	Specific tests (e.g. HIV, hepatitis)
Bloodstains	Sequence of events during a violent assault	Bloodstain pattern analysis
	Linking an assailant to a victim and/or location	Bloodstain pattern analysis + DNA profiling
Saliva and semen	Identification	DNA profiling
	Linking an assailant to a victim and/or location	DNA profiling
Faeces and urine	Identification	DNA profiling
	Linking an assailant to a victim and/or location	DNA profiling
	Previous diet (and hence linking the individual to a location)	Undigested food residues in faeces
	Poisoning	Specific tests for the poison or its metabolites
	Drug use	Specific tests for the drug or its metabolites (e.g. morphine, EPO)

levels in the urine to ante mortem levels in the blood and even high levels may not be synonymous with a fatal overdose, whilst if a person dies rapidly the drug/poison may not have had time to reach the bladder. It should also be noted that high levels of protein, lactic acid and the enzyme lactate dehydrogenase in the urine either as a consequence of ante mortem metabolic disorders or post mortem decay can interfere with the immunoassays of drugs such as benzodiazepines and morphine and thereby result in false positives. It is therefore important to check for the presence of substances such as these that might interfere with the test.

An interesting piece of research conducted by Zuccato et al., (2005) demonstrated the feasibility of detecting cocaine and its main metabolites in sewage water. They then scaled up their results to estimate the amount of the drugs flowing through the sewage water systems of four Italian cities. Their findings indicated a level of drug use far above that which the authorities believed to occur. For example, they estimated that the 5 million people living along the river Po were consuming about 200 000 lines of cocaine every day. It is possible that a similar approach might be used to covertly monitor levels of drug use in a house, nightclub or prison by analysing the drug levels in the sewage pipes.

Future directions

There is still a need to develop a reliable means of estimating the age of bloodstains. It might be achieved through estimating the rates of DNA degradation although this is likely to be affected by a range of biological and environmental factors.

The analysis and interpretation of blood spatter patterns will probably be speeded up and, hopefully, improved with increased use of automated scanning and computer-based techniques. If the blood spatter pattern at a crime scene can be reliably recorded as computer files it will be easier to send these to experts around the country or between countries for their opinion. Although this can already be done through the use of photographic evidence, computer technology allows one to record the scene in three dimensions and therefore to gain a more accurate impression of the scene. This evidence could also be presented to the jury when a case comes to court.

Quick quiz

(1) Briefly explain the uses and limitations of red blood cell antigens for identifying an individual.

(2) What is meant the term a 'presumptive test' for blood?

(3) State three presumptive tests for blood and the circumstances in which you might use them.

(4) In relation to bloodstains, distinguish between the appearance and causes of cast off stains, impact spatter and transfer stains.

(5) How can an analysis of the vitreous humor help determine the PMI?

(6) What is laser microdissection and why is it a useful technique in forensic science?

(7) What is azoospermia and what relevance does it have to an investigation into alleged rape?

(8) Why does the inability to detect drug metabolites in a sample of urine not exclude the possibility that a person died of a drug overdose?

(9) Why is it difficult to extract human DNA from faeces?

(10) How would it be possible to determine whether someone had been eating whale meat?

Project work

Title

Insect modification of bloodstain patterns.

Rationale

Insects can modify bloodstain patterns by crawling though stains and thereby smearing them or transferring blood that becomes attached to their legs or bodies. Flies often feed on blood and due to their habit of regurgitation can transfer blood spots to distant parts of a room. Insect-modified stains or transfer stains need to be distinguished from those caused by the loss of blood from a wound or falling from a weapon.

Method

Blood would be provided to different species of insect within a caged arena. The arena could be covered with different substrates (e.g. white card of different textures). One could compare the regurgitation patterns of different species of fly (e.g. blowflies [*Calliphora* spp.] and houseflies [*Musca* spp.]). One could compare the time taken for regurgitation stains to be produced, the influence of fly density, number of regurgitation patterns produced over a given time period, whether there was any difference between male and female flies. Male and female flies could be provided with both sugar solution and blood. Female flies are said to require a protein meal to produce their eggs and might therefore be expected to produce more blood regurgitation spots.

Title

The influence of soil type on the ability to detect blood.

Rationale

There are reports that it is possible to detect blood on soil several months after it was spilt (see previous text). This ability is likely to be affected by chemical and biological (e.g. microbial abundance and diversity) properties of the underlying soil.

Method

Known volumes of blood obtained from a butcher or slaughterhouse would be placed on the surface of soils (e.g. peaty, sandy, clay loam etc) or hard surfaces (e.g. tarmac or concrete). The use of a stencil or similar means to produce a set pattern will help to monitor any change in detectable area. At set periods of time the presence of blood would be monitored using luminol or other presumptive test. The ability to produce transfer stains from the original stain could also be tested by walking across the test area after varying periods of time. Changes in microbial abundance and diversity in the stained region might also be monitored (see Chapter 10) and compared with the surrounding unstained region. The attractiveness of the stained region to flies could be monitored by placing a collecting trap above it.

3 Molecular biology

Chapter outline

The Structure of DNA
DNA Sampling
DNA Profiling
Polymerase Chain Reaction
Short Tandem Repeat Markers
Single Nucleotide Polymorphism Markers
Determination of Ethnicity
Determination of Physical Appearance
Determination of Personality Traits
Mobile Element Insertion Polymorphisms
Mitochondrial DNA
RNA
DNA Databases
Future Developments
Quick Quiz
Project Work

Objectives

Discuss what is meant by the term 'DNA profiling' and evaluate the different methods currently available by which it can be conducted.

Evaluate the advantages and limitations of the different DNA-based methods for determining gender and personality traits.

Discuss how the National DNA database should be developed so as to enhance its effectiveness without adversely affecting civil liberties.

Essential Forensic Biology, Second Edition   Alan Gunn
© 2009 John Wiley & Sons, Ltd

The structure of DNA

DNA is quite literally 'the stuff of life' as it contains all the information that makes us who we are. Indeed, some biologists suggest that the sole reason for any organism's existence is to ensure that its DNA is replicated and survives into the next generation. It is found in the nucleus of cells and also the mitochondria. Therefore, with the exception of certain specialized cells that lack these organelles, such as mature red blood cells, every cell in the body contains DNA. Furthermore, unless heteroplasmy occurs (see later), this DNA is identical in every cell, does not change during a person's lifetime and is unique (identical twins excepted) to that individual. DNA is composed of four nucleotides (adenine, thymine, cytosine and guanine), and phosphate and sugar molecules. Nuclear DNA takes the form of a ladder twisted into the shape of a double helix in which the rails are composed of alternating sugar and phosphate molecules whilst the nucleotides act as rungs joining the two rails together. Adenine is always joined to thymine and cytosine is always joined to guanine.

Within the nucleus, DNA is found in structures called chromosomes. Human cells contain 23 pairs of chromosomes and these vary in shape and size. Twenty-two of these pairs are referred to as the 'autosomal chromosomes' and these contain the information that directs the development of the body (body shape, hair colour etc.). The remaining pair of chromosomes are the X and Y 'sex chromosomes' that control the development of the internal and external reproductive organs. Each chromosome contains a strand of tightly coiled DNA. The DNA strand is divided into small units called genes and each gene occupies a particular site on the strand called its 'locus' (plural 'loci'). The total genetic information within a cell is referred to as its 'genome'. There are about 35 000–45 000 genes and, on average, they each comprise about 3000 nucleotides although there is a great deal of variation. These genes code for proteins that determine our hair and eye colour, the enzymes that digest our food and every heritable characteristic. Surprisingly, only a small proportion of the genome actually codes for anything and between these coding regions lies long stretches of repetitive non-coding regions that exhibit a great deal of variability. Each gene exists in two alternative forms, called 'alleles', one of which is found in each of the pair of chromosomes. If DNA profiling detects only one allele, this is usually interpreted as a consequence of a person inheriting the same allele from both parents. If three or more alleles are detected then this is an indication that the sample contains DNA from more than one individual. Mitochondrial DNA is arranged slightly differently to nuclear DNA and will be dealt with separately.

Because the sequence of nucleotides along the nuclear DNA chain is unique to every individual and is the same in every cell in his or her body, it is often stated to be similar to a 'bar-code' for identification. Although the entire human genome has been sequenced, it is not necessary to go to these lengths in order to identify an individual. Indeed, the majority of the DNA in every one of us is virtually the same, so sequencing all of it would be a pointless exercise. Instead, forensic scientists concentrate on regions of the genome that exhibit a high degree of variability. The sources of variation that occur in human nuclear and mitochondrial DNA are summarized in Fig. 3.1 and a comprehensive guide to the analysis of DNA in a forensic context is provided by Butler (2005).

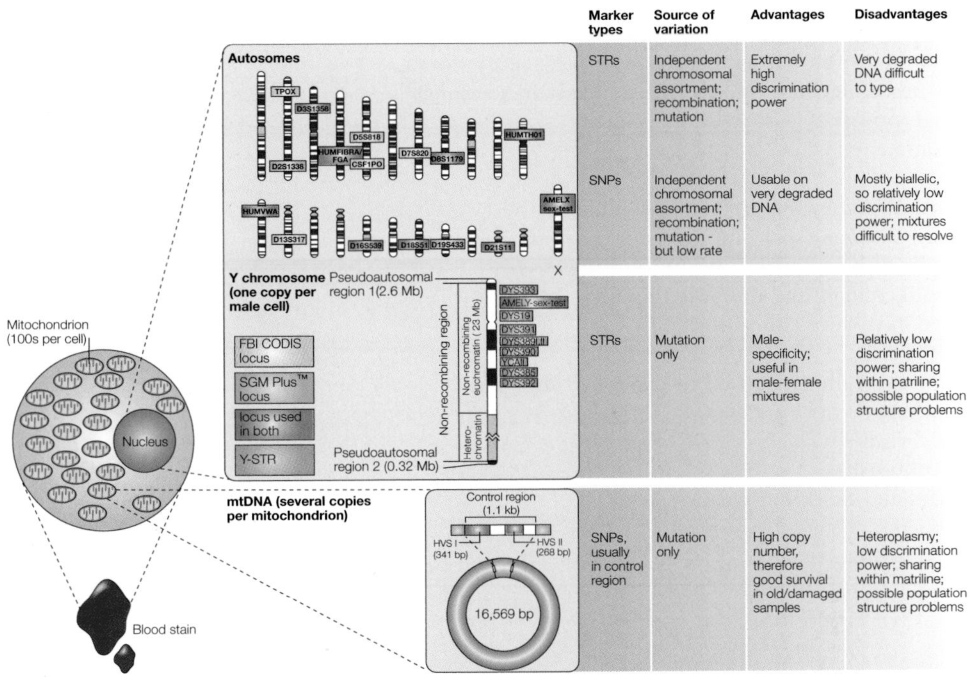
Marker types
Source of variation
Advantages
Disadvantages
Autosomes
TPOX
D3S1358
D5S818
HUMFIBRA/FGA
CSF1PO
D7S820
D8S1179
HUMTH01
D2S1338
HUMVWA
D13S317
D16S539
D18S51
D19S433
D21S11
AMELX sex-test
X
Y chromosome (one copy per male cell)
Pseudoautosomal region 1(2.6 Mb)
Non-recombining region
Non-recombining euchromatin (23 Mb)
Hetero-chromatin
DYS393
AMELY-sex-test
DYS19
DYS391
DYS389I/II
DYS390
YCAII
DYS385
DYS392
FBI CODIS locus
SGM Plus™ locus
locus used in both
Y-STR
Pseudoautosomal region 2 (0.32 Mb)
Mitochondrion (100s per cell)
Nucleus
Blood stain
mtDNA (several copies per mitochondrion)
Control region (1.1 kb)
HVS I (341 bp)
HVS II (268 bp)
16,569 bp
STRs
Independent chromosomal assortment; recombination; mutation
Extremely high discrimination power
Very degraded DNA difficult to type
SNPs
Independent chromosomal assortment; recombination; mutation - but low rate
Usable on very degraded DNA
Mostly biallelic, so relatively low discrimination power; mixtures difficult to resolve
STRs
Mutation only
Male-specificity; useful in male-female mixtures
Relatively low discrimination power; sharing within patriline; possible population structure problems
SNPs, usually in control region
Mutation only
High copy number, therefore good survival in old/damaged samples
Heteroplasmy; low discrimination power; sharing within matriline; possible population structure problems

Figure 3.1 Sources of human nuclear and mitochondrial DNA. Further details are provided in the text. Abbreviations STR = short tandem repeat; Y-STR = short tandem repeat occurring on the Y chromosome; SNP = single nucleotide polymorphism; FBI CODIS = US Federal Bureau of Investigation Combined DNA Index System; HVS = hypervariable site; Mb = megabase; mtDNA = mitochondrial DNA; SGM = second generation multiplex. (Reprinted by permission from Macmillan Publishers Ltd Jobling, M.A. and Gill, P. (2004) *Encoded Evidence DNA in Forensic Analysis*, copyright © 2004.)

DNA sampling

Because virtually all of our cells contain DNA and we lose cells all the time, for example whenever we blow our nose, brush our teeth or comb our hair, it is possible to isolate DNA from a wide variety sources (Table 3.1). Indeed, it is so easy to leave a trail of DNA that crime scene investigators must wear masks, and disposable over-suits and over-shoes to avoid contaminating the location. DNA contamination can also occur via mortuary instruments (Rutty *et al.*, 2000) so it is preferable that samples are taken from a body before it is moved. Similarly, all DNA samples need to be kept apart from the moment they are collected to avoid the possibility of cross contamination. For example, if samples from a victim and a suspect are transported in the same container (even if they are in separate bags) or processed at the same time, there is a risk of cross contamination. As the sensitivity of DNA analysis increases, particularly with regard to techniques such as low copy number STR analysis the risks of contamination during the collection, storage and processing increase and the results need to be interpreted with care (see later). For example, DNA can be recovered from bed sheets even if a person slept on them for only one

Table 3.1   Potential sources of human DNA for forensic analysis

Body fluids: blood, semen, saliva, urine, faeces, vomit
Tissues: skin, bone, hair, organs, fingernail scrapings
Fingerprints
Weapons
Bites
Discarded chewing gum
Drug packages spat out after storage in the mouth
Cigarette butts
Handkerchiefs and discarded tissues
Used envelopes and stamps
Cutlery
Used cups, mugs, bottled or canned drinks
Clothing
Hairbrushes
Toothbrushes
Shoes and other footwear
Plasters
Used syringes

night (Petricevic *et al.*, 2006). This means that one could prove that a man and a woman shared a bed but not that they shared it at the same time.

The collection, transport and/or storage of liquid and tissue samples is sometimes problematic but this can be overcome by using commercially available FTA® cards that are produced by Whatman International Ltd. These cards contain chemicals that lyse any cells in the sample and immobilize and stabilize the DNA and RNA that is released. The cards also contain chemicals that preserve the DNA and RNA and thereby allow the samples to be stored at room temperature for long periods. The cards can be pressed against liquid samples (e.g. saliva or semen) or the liquid can be dropped onto them. Tissue samples or blood clots can be squashed onto the cards. When required for analysis the DNA or RNA can be readily extracted (Smith & Burgoyne, 2004).

It should be noted that according to the UK 2004 Human Tissues Act it is illegal to take a sample of a person's DNA without their consent except under certain conditions (e.g. to prevent or detect a crime or to facilitate a medical diagnosis). It is therefore illegal for a man to test DNA samples surreptitiously from his offspring to determine whether he really is the father.

DNA profiling

The sequencing of an individual's DNA is known as 'DNA profiling' or 'DNA typing' and although the term 'DNA fingerprinting' is sometimes used, it is not really appropriate. This is because the courts and the scientific community accept fingerprints as unique identifying features. For example, identical twins do not have the same fingerprints even though they share the same DNA profile. Consequently, experts presenting DNA-based evidence in court, talk about probabilities of a match between two samples rather than stating 'yes, they match' or 'no, they do not match'. The possibility that 'an evil twin brother' committed the crime does not belong solely in the realms of fiction. Several cases have arisen in the UK and America in which DNA recovered from a crime scene could have been derived from either of two brothers. Often, these can be resolved by other evidence, such as fingerprints or one of the twins being in prison at the time of the offence. However, in the absence of such evidence and if the twins fail to cooperate with the police, it is an extremely difficult situation for the prosecution to resolve. It also indicates that DNA databases need to be updated so as to keep track of where twin matches could occur.

If the DNA profile of a suspect does not match that of the evidence, it is said to be an 'exclusion'. This is a useful result because, in conjunction with other evidence, it enables police to rule a suspect out of their enquiries or a convicted person to prove his or her innocence. If the interpretation of the DNA profile fails for some reason, such as owing to contamination or DNA degradation, the results are said to be 'inconclusive'. This means that further evidence must be sought before a suspect can be either excluded or confirmed as being relevant to the enquiries. If the profiles are deemed to match then the suspect's profile is called 'an inclusion' and the significance of the match has to be calculated by quantifying the 'random match probability (Pm)'. This is a statistical test that estimates the chance of two unrelated people sharing the same DNA profile. Where the markers are not linked (e.g. autosomal STRs – see later), and are therefore inherited independently, the Pm can be estimated by multiplying the individual allele frequencies in a sample population (Jobling & Gill, 2004). Consequently, the more loci that are included in the analysis and the greater the heterozygosity of each of these loci, then the smaller will be the value of the Pm and the greater will be the probability that the two profiles originated from the same person.

Despite the high discriminating power of DNA-based evidence, in the absence of other corroborating evidence it is unlikely to be accepted by a court as proof that a person was (or was not) responsible for a particular crime. The Pm value can be compromised by several factors that need to be taken into account. Firstly, if the DNA has begun to degrade, it may not be possible to obtain a full profile that accounts for all the loci. Secondly, both the victim of a crime and the suspect may be closely related (many serious crimes are committed by people related to their victim) and therefore share many alleles by descent. Similarly, they may both originate from the same sub-population some of which are characterized by high levels of inter-marriage. Discriminating between the remains of closely related people who die in natural or man-made disasters can sometimes be problematic for the same reasons (Kračun, et al., 2007).

The prosecutor's fallacy and the defence attorney's fallacy

A fallacy occurs when one's reasoning relies upon a misunderstanding or misrepresentation of the facts and this therefore results in the argument one is presenting becoming invalid. The terms 'prosecutor's fallacy' and 'defence attorney's fallacy' were originally coined by Thompson & Schumann (1987) and relate to the inappropriate presentation of statistical evidence in court proceedings. The terms are often used in American literature and although they are relevant to the presentation of any type of evidence in which numerical data are involved they are especially appropriate to the presentation of molecular evidence. The prosecutor's fallacy occurs when the prosecution attempts to prove a suspect's guilt solely on the basis of associative evidence. Associative evidence is that which involves matching material from two or more sources. For example, it might be fibres retrieved from a suspect and those found on his victim, DNA profiles from a semen sample and a suspect or soil samples found on a shoe and those found at a crime scene. For example, let us assume that a match is found between the DNA profile of a suspect and that recovered from a bloodstain found on a windowsill following a burglary. It would be incorrect for the prosecution to claim that 'the DNA profile recovered from the windowsill indicates there is only a 1 in 10 million chance that the suspect is innocent'. This is wrong because the prosecutor is assuming the guilt of the suspect and ignoring all other evidence. For example, the suspect might have been in hospital undergoing heart surgery at the time of the offence. Even assuming that the suspect had the opportunity to commit the offence, the prosecutor is assuming that because the probability of finding matching DNA profiles if the suspect is innocent is very small then it follows that the probability of the suspect being innocent is also very small.

The defence attorney's fallacy is the mirror image of the prosecutor's fallacy and assumes that associative evidence is not relevant. For example it can arise when the defence attempts to argue along the lines that if the frequency of particular DNA profile is one-in-a-million and the UK has a population of 60 million therefore there are at least 59 other people than the suspect who might have committed the crime. This assumes that all 60 had an equal likelihood of leaving the DNA profile found at the crime scene. This is clearly impossible because of those 59 other people, some will be babies, others might be disabled or in nursing homes and most would be many miles from the crime scene.

Current STR DNA profiling techniques (see later) can result in extremely low *Pm* values and therefore the statistical probability that a person chosen at random might have a particular profile becomes vanishingly small – possibly as low as 10^{-10}. This can make the prosecutor's fallacy an extremely easy trap to fall into. In addition, contrary to what one might expect, matching or nearly matching DNA profiles have been reported from relatively small DNA databases. The reasons how this might occur along with other aspects of the statistical analysis of DNA profiles can be found in Weir (2007).

Forensic applications of DNA profiling

Although the use of DNA profiling to identify both victims and assailants in serious crimes such as rape and homicide gains most attention in the media the technique is routinely used in minor crimes such as burglary, car theft and to identify the senders of hate mail and poison pen letters. Indeed, one of the principal purposes of the setting up of the National DNA Database (NDNAD) (see later) was to facilitate the solving of so-called volume crimes. DNA profiling is also used for paternity testing and in particular it is sometimes used to establish family relationships among people claiming visas and/or asylum – although in the latter instance its use needs to be done with extreme care (e.g. Karlsson et al., 2007). DNA profiling is also used in insurance company fraud investigations where there is a dispute over who the driver was in a traffic accident. For example, when a car is in an accident and the airbags are activated, they deliver a considerable blow and this, coupled with the forces from the accident means that the bags become contaminated with saliva, nasal mucus, vomit and, not infrequently, blood. Consequently, DNA profiles obtained from the bags can be used to demonstrate who was sat where at the time of the accident. Similarly, where there are claims that food was contaminated with blood or other body fluids it is not unusual for these to prove to have originated from the complainant. The disaffected employee responsible for intentionally contaminating food with spit, urine and other body fluids during processing in the factory or kitchen can also be identified by DNA profiling.

DNA profiling is not restricted to solving crimes involving human victims and Lorenzini (2005) relates an interesting case in which it was used to solve a case of wildlife poaching. A poacher had snared a wild boar in an Italian National Park and then killed it with a knife, leaving the corpse under a bush so that he could collect it after nightfall. Whilst he was away, conservation officers found the body and attempted to arrest the poacher when he returned. However, the poacher claimed that the boar was already dead when he found it and they had to let him go – though they confiscated the boar's body. A post mortem on the boar indicated that the knife wound was inflicted when the animal was still alive and a search of the poacher's home yielded a bloodstained home-made knife. DNA analysis indicated that the blood originated from a wild boar rather than a domestic pig and that the DNA profile matched that of the dead boar. As a result, the suspect was successfully prosecuted for animal cruelty and poaching.

The popularity of TV dramas and documentaries featuring forensic science has lead to concerns that the public will develop unreal expectations of forensic investigations. In particular, in the real world, investigations are seldom quick and they are also fallible. Furthermore, if the DNA is badly degraded it may be impossible to obtain a full profile, or any profile at all. In addition, there are concerns that criminals will be alerted and learn how to avoid detection. However, we constantly shed skin cells and hair and short of dressing in a full-enclosure body suit and mask (which is likely to draw attention to oneself and doesn't facilitate rapid movement) it is difficult to prevent leaving DNA evidence wherever we go. Furthermore, indulging in vigorous or violent activities inevitably results in more of one's DNA being shed into the locality. Maintaining public confidence in the use of molecular evidence is a more serious issue and it is essential that both proponents and critics of the techniques retain a sense of proportion. In the following pages, some of the more commonly used molecular techniques are briefly discussed and their advantages and limitations are summarized in Table 3.5.

Case Study: The identification of Colin Pitchfork

The first and most famous case of DNA profiling in a forensic investigation involves that of Colin Pitchfork, who in 1988 was sentenced to life in prison for the rape and murder of two young girls in the town of Narborough, Leicestershire. The first girl was raped and murdered in 1983 and a semen sample recovered from her body indicated that the culprit had blood group A and an enzyme profile shared by only 10% of the male population. However, with no further leads, the investigation never progressed any further. In 1986, another girl was raped and murdered in the same area and the semen sample showed the same characteristics. These findings led the police to believe that both murders were the work of the same man and that he lived in the vicinity. A local man, Richard Buckland was subsequently arrested and whilst he confessed to the second girl's murder he denied having anything to do with the first girl's death. The police were convinced that they had arrested the correct person and asked Dr Alec Jeffries of Leicester University for help as he had developed a means of creating DNA profiles. Dr Jeffries' findings indicated that, whilst both girls were raped and murdered by the same man, this could not have been Richard Buckland. This emphasizes the need for caution even when a suspect confesses to a crime.

With their chief suspect exonerated (another first for DNA profiling), the police undertook the world's first DNA mass intelligence screening in which all the men in the area, 5000 in all, were asked to provide DNA either as a blood sample or a saliva swab. Of these samples, only those exhibiting the same blood group and enzyme pattern as the murderer were subjected to DNA profiling. This was a major operation, not least because the profiling techniques were much more time consuming than those in use today, and took 6 months to complete. At the end of this time, the operation drew a blank – none of the profiles matched those of the murderer. There followed a lull of about a year until a woman reported that she had overheard a man saying that he had provided a DNA sample in place of his friend Colin Pitchfork. Colin Pitchfork, who was a local baker, was therefore arrested and his DNA profile was found to match those of the semen samples recovered from the two murdered girls.

Polymerase chain reaction

Kary Mullins invented the polymerase chain reaction (PCR) in 1983 and it has since become one of the most powerful techniques in molecular biology. It is an enzymatic process that enables a particular sequence of the DNA molecule to be isolated and amplified (copied) without affecting the surrounding regions. This makes it very useful in forensic casework in which DNA samples are frequently limited in both quantity and quality. For instance, PCR has been applied to the identification of DNA from saliva residues on envelopes, stamps, drink cans, and cigarette butts (Withrow *et al.*, 2003). It also has the advantages of being sensitive and rapid.

Table 3.2 Summary of reactions involved in the polymerase chain reaction (PCR)

(a) Incubate the sample at 94–97 °C to separate the DNA helix into two separate strands and denature the DNA.

(b) Reduce the temperature to 50–60 °C to allow the primers to 'anneal' to the DNA.

(c) Raise the temperature to 70–72 °C to initiate the 'polymerization' stage in which *Taq* DNA polymerase enzyme uses the DNA template identified by the primers and the nucleotides adenine, guanine, cytosine and thymine as building blocks to reproduce a complimentary copy of the template (Figs 3.2 and 3.3).

(d) Repeat the procedure in successive cycles of denaturing, annealing and polymerization so that in a very short time the original sequence is 'amplified' thousands, or even millions of times.

(e) After the amplification step, the PCR products (sometimes called 'amplicons') are separated on the basis of their length. In the past, this was done using flat bed gel electrophoresis but this is being superseded by capillary electrophoresis as it is faster and can be automated. It is common practice to investigate several different sequences at the same time in a process referred to as 'multiplexing'. This is achieved by designing primers that produce allele size ranges that do not overlap (Fig. 3.4).

However, PCR is not suitable for the analysis of long strands of DNA, so it cannot be used in the older Restriction Fragment Length Polymorphism (RFLP) analyses in which the strands often contain thousands of bases.

Once a region of the DNA molecule has been identified as worthy of investigation, the flanking sequences are ascertained so that PCR primers can be designed to identify the beginning and end of the sequence. The primers consist of short sequences of DNA that bind or hybridize onto their complimentary sequences on the test DNA sample. Once the primers have been designed, the PCR process is carried out as outlined in Table 3.2.

Taq DNA polymerase is so-called because it was discovered in the thermophilic bacterium *Thermus aquaticus*. This enzyme has replaced the *E. coli*-derived DNA polymerase that was used in early PCR reactions because it is more heat stable. It is not harmed by the denaturation part of the PCR cycle therefore removing the need to add fresh enzyme after each denaturation. Consequently, an excess of *Taq* DNA polymerase, primers and nucleotides are added at the start of the process and adjusting the annealing temperature controls the specificity of the reaction.

The primers are labelled with differently coloured fluorescent dyes, and therefore, after electrophoresis, the PCR products can be detected by exposure to a laser beam that induces fluorescence at specific emission wavelengths that are then detected with a recording CCD camera. The results are printed out as a trace referred to as 'an electropherogram' (Fig. 3.5). Sometimes the machine may misinterpret a colour (e.g. it mistakes blue for yellow) and this gives rise to a false peaks – this phenomenon is called 'bleed through' or 'pull up'. This can be recognized by careful analysis of the electropherograms across the colour spectrum. Other potential sources of error are 'stutter peaks' that occur immediately in front of (commonly) or after (less commonly) a real peak. Stutter peaks are easy to identify and exclude from the interpretation when the sample is derived from a single person but if it contains

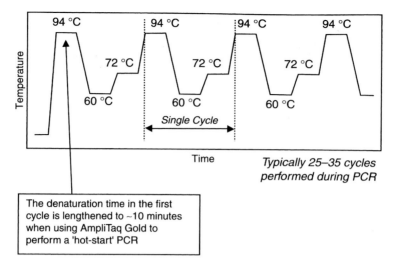

Figure 3.2 Diagrammatic representation of the PCR thermal cycling process. Each cycle takes about 5 minutes and the whole process lasts about 3 hours. (Reproduced by permission of Elsevier from Butler, J.M. (2005) *Forensic DNA Typing*, 2nd edn. Copyright Elsevier (2005).)

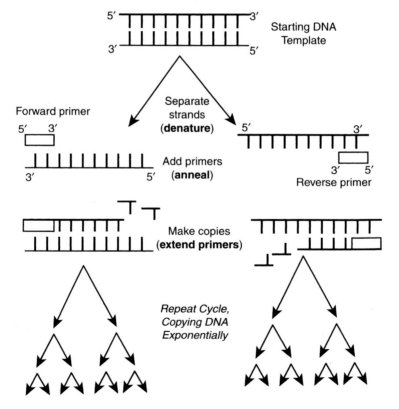

Figure 3.3 Diagrammatic representation of the amplification step of the PCR process. (Reproduced by permission of Elsevier from Butler, J.M. (2005) *Forensic DNA Typing*, 2nd edn. Copyright Elsevier (2005).)

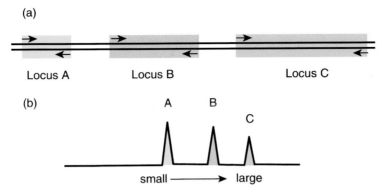

Figure 3.4 Diagrammatic representation of the multiplex PCR process. (a) The arrows represent three sets of primers that have been designed to amplify three different loci Locus A, Locus B, and Locus C. (b) The three loci are different sizes and can therefore be resolved easily on the basis of size separation. (Reproduced by permission of Elsevier from Butler, J.M. (2005) *Forensic DNA Typing*, 2nd edn. Copyright Elsevier (2005).)

mixed DNA it can be difficult to discern a stutter peak from a real one. In addition, random flashes may occur owing to air bubbles, contaminants and other interferences thereby resulting in background 'noise' that may mask small peaks or even be mistaken for peaks themselves. Re-running the sample will usually identify 'false peaks' because they are unlikely to occur in exactly the same place twice. Surprisingly, there appear to be no universally accepted guidelines concerning the lower accepted limits that distinguish a 'true peak' from 'background'.

Quantitative (real time) PCR

This technique is based on the PCR process and is designed to both quantify and amplify the targeted DNA. Two of the commonest means of quantification are the inclusion of a fluorescent dye (e.g. SYBR® Green) in the PCR reaction that intercalates with the DNA as it is produced and the TaqMan® assay (Fig. 3.6). In the intercalation assay, the binding of SYBR® Green to double stranded DNA results in an increase in fluorescence, therefore, as more amplicons are produced the greater the fluorescence detected. Because the dye binds to any double stranded DNA molecule the intercalation method is non-specific (i.e. it will not distinguish between DNA molecules). In the TaqMan® assay oligonucleotide probes (TaqMan® probes) are added that bind to a specific internal region of DNA between the forward and reverse PCR primers. The probes have a 'reporter' dye attached to their 5′ end and a 'quencher' dye attached to their 3′ end. When a high energy dye (reporter dye) is in close proximity to a low energy dye (quencher dye) there is a transfer of energy from the high energy dye to the low energy dye – this is what happens in the intact probe and it results in the fluorescence of the reporter dye being very low or not detectable. When, during the PCR process the Taq DNA polymerase replicates a template on which a TaqMan® probe is bound the enzyme (which has 5′-nuclease

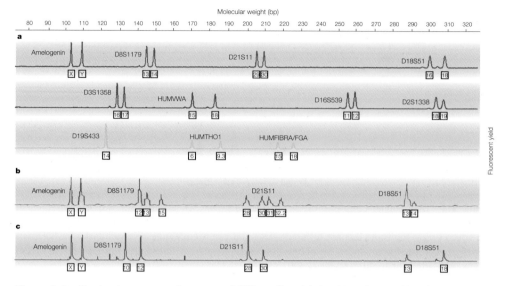

Figure 3.5 Electropherograms of autosomal STR profiles (a) An SGM Plus profile of a man (note the two peaks for the amelogenin locus). The profile is displayed in green, blue, and yellow channels of a four-colour fluorescent system. The red channel was used for size marker and is not shown here. Most of the STR loci are heterozygous (i.e. there are two peaks) and the alleles are evenly matched (i.e. the peaks are the same size). The number beneath each peak indicates the size of each allele in repeat units. (b) Part of an SGM Plus profile of a mixed sample. Only the green channel is shown. The sample is obviously a mixture as three alleles are present at D8S1179 and four at D21S11. The minor component of a mixture can only be identified if it is present above the level of 'background noise'. The amelogenin peaks are of approximately the same size indicating that this is a mixture of DNA from two men. If the mixture was from a man and a woman the X amelogenin peak would have been higher than the Y peak. (c) Part of an STR profile following low copy number testing. This process can lead to heterozygote imbalance at some loci. For example, note the differences in peak height at D21S11 and D18S51. Owing to stochastic effects (see text), no two amplifications of the same sample yield behave identically and therefore duplicate LCN PCR amplifications should be undertaken and only those alleles present in both electropherograms should be recorded. (Reprinted by permission from Macmillan Publishers Ltd. Jobling, M.A. and Gill, P. (2004) *Encoded Evidence DNA in Forensic Analysis*, copyright © 2004.)

activity) splits the probe thereby separating the reporter dye and the quencher dye – hence fluorescence of the reporter dye is increased whilst the fluorescence of the quencher dye decreases. The method is extremely sensitive and can detect as little as a twofold increase in the level of a DNA sequence. Because custom-designed primers are used this method is more specific than the intercalation method but in both cases, with each cycle of the PCR process more DNA is produced and this is measured as an increase in fluorescence. The DNA product is therefore 'quantified' as it accumulates in 'real time' and hence the terms 'real time' and 'quantitative' PCR (various abbreviations are used including qPCR, RT-PCR, and QRT-PCR). Once sufficient DNA is produced it can be sequenced or used for Southern Blotting.

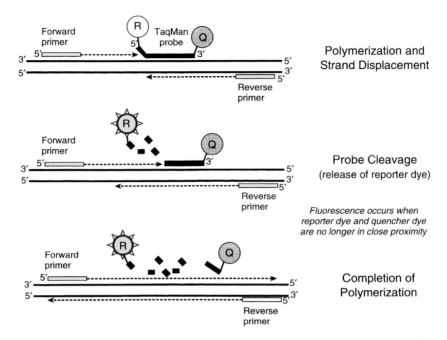

Figure 3.6 Diagrammatic representation of the TaqMan® assay. (Reproduced by permission of Elsevier from Butler, J.M. (2005) *Forensic DNA Typing*, 2nd edn. Copyright Elsevier (2005).)

One of the most important features of qPCR is that by choosing appropriate TaqMan® probes the analyst can determine whether or not there are sufficient amounts of specific genomes present to undertake STR analysis (see below). STR analysis is currently the cornerstone of DNA profiling but many forensic samples do not produce profiles owing to low levels of DNA in the sample or partial profiles owing to the DNA being degraded. To avoid wasting material and expensive reagents it is useful to know the suitability of the sample for STR analysis before starting the assay. Swango *et al.* (2006) have described a relatively simple means of addressing these problems. They did this by undertaking simultaneous qPCR on a relatively short (67 base pair [bp]) and a long (170–190 bp) STR amplification targets and incorporating an internal control to enable the normalization of sample data to account for tube-to-tube variation, PCR inhibition etc. As would be expected, the yield of the longer target provided the best indication of whether the sample contained sufficient amplifiable DNA to produce a good STR profile whilst the ratio of the three targets indicated the quality of the DNA in the sample.

Short tandem repeat markers

Short tandem repeats (STRs), also referred to as 'microsatellites' or 'simple sequence repeats (SSRs)', are brief lengths of the non-coding region of the human genome consisting of less than 400 base pairs (hence 'short') in which there are 3–15 repeated units, each of 3–7 base pairs (hence 'tandem repeats'). These STR sequences, or 'markers', can be divided into three categories: 'simple', 'compound' and 'complex'. Simple STRs are those in which repeats are of identical length and sequence units. Compound STRs consist of two or more adjacent simple repeats whilst complex STRs have several repeat blocks of different unit length and variable intervening sequences. STRs occur on all 22 pairs of autosomal chromosomes and the X and Y sex chromosomes. STRs can vary greatly between individuals, this diversity resulting from the effects of mutation, independent chromosomal variation and recombination. However, STRs found on the Y chromosome exhibit less diversity than those on other chromosomes because they do not undergo recombination. Consequently, STR diversity on Y chromosomes results solely from mutation.

There are over 2000 STR markers suitable for genetic mapping but only a few of them are used routinely for forensic DNA profiling. In the UK, the Forensic Science Service (FSS) currently uses the SGMplus™ system (SGM+) that utilizes10 autosomal STR markers whilst in America the FBI uses 13 markers. The DNA profiles are then stored on computer databases – in the UK this is the NDNAD whilst in America it is the Combined DNA Index System (CODIS). The use of a standard set of markers and computerized systems facilitates comparisons between the DNA profiles of suspects, convicted offenders, unsolved crimes and missing persons. If all 10 (or 13) STR loci in two DNA samples are found to have identical lengths then this is compelling evidence that they originated from the same person. Commercial testing kits are available for these loci and the kits also include a marker at the amelogenin locus to enable sex determination. Amelogenin is a substance involved in the organization and biomineralization of enamel in developing teeth. In humans, the gene is expressed on both sex chromosomes but that on the X chromosome is six base pairs shorter than that on the Y chromosome. Consequently, following PCR and electrophoresis, males being heterozygous (XY) express two peaks (or bands) (Fig. 3.5a) whilst females, being homozygous (XX) express a single peak. The test is not foolproof and problems can arise if there is a deletion of the amelogenin gene on the Y chromosome – an important consideration in some ethnic groups, such as Malay and Indian populations (Chang *et al.*, 2003).

A DNA profile report will often take the form of a table of alleles and a hypothetical example obtained using the ProfilerPlus™ system is illustrated in Table 3.3. The numbers relate to the position of the alleles at each gene locus. For example, in suspect 1, the gene D3S1358 is heterozygous and expresses alleles 15 and 16 whilst in suspect 2 the gene is homozygous and allele16 is expressed on both chromosomes. Suspects 1 and 2 have different profiles from that found in the semen stain and are therefore classed as 'exclusions' whilst the profile of suspect 3 is the same as that found in the stain and is therefore 'an inclusion'. The statistical frequency of that combination of alleles occurring in the population is then calculated by reference to a sample population. For example, for the locus VWA, if 8% of Englishmen expressed allele 15 and 21.6% expressed allele 16, the frequency of this pair of alleles would therefore be $2 \times 0.08 \times 0.216 = 0.0346$, or 3.46% of the male English population. If the frequencies at all the loci are added together, then the frequency estimate for the whole DNA profile will be extremely small – perhaps one in a hundred million or more. Obviously, a great deal depends on the sample population used to generate the allele frequencies and corrections may need to be

Table 3.3 Table of alleles illustrating the hypothetical DNA profile of a semen stain and that of three suspects. Amel = amelogenin test for gender; Sus = suspect number

					Allele Loci					
	D3S3158	VWA	FGA	D8S1179	D21S11	D18S51	D5S818	D13S317	D7S820	Amel
Stain	16,18	16,16	19,25	13,14	29,30	17,17	11,11	10,11	9,10	XY
Sus 1	15,16	16,16	19,25	13,14	29,30	14,17	11,11	10,11	9,10	XY
Sus 2	16,16	15,16	21,23	14,14	27,28	17,17	10,11	8,9	8,9	XY
Sus 3	16,18	16,16	19,25	13,14	29,30	17,17	11,11	10,11	9,10	XY

made to allow for this. For example, certain allele combinations will be common among close relatives, sub-populations or ethnic groups but might be rare in the population at large. Consequently, the frequency of certain DNA profile characteristics may be extremely low as a national average but common among family members or an ethnic group.

Because STR analysis relies on the identification of sequences that are much shorter than those required for RFLP analysis, it is less vulnerable to problems associated with DNA degradation. STR analysis can therefore be effective on older body fluid stains or corpses at a later stage of decomposition than would be the case with RFLP analysis. However, DNA degradation can still present difficulties. For example, peak heights may be reduced thereby making it difficult to distinguish them from background 'noise'. Indeed, some peaks may disappear entirely whilst others remain visible thereby resulting in an inaccurate profile. Evidence of DNA degradation is often exhibited by a progressive decline in peak height with increasing sequence length. This is because longer sequences are more vulnerable to the effects of degradation. If the profile contains DNA from more than one individual, this problem can be exacerbated because the two (or more) DNA samples may not degrade at the same rate or in an identical manner. If one of a pair of alleles fails to be recorded, this is referred to as 'allele dropout'. Consequently, a heterozygous individual may appear homozygous at one or more gene loci. Similarly, an additional allele may be observed – this is referred to as 'allele drop-in'. Allele drop-in results from contamination and becomes obvious when the allele does not appear in repeated independent PCR reactions.

The shortness of the STR markers means that only small amounts of DNA are required (although if the amounts are extremely small, problems can arise – see above). The marker sequences can be easily amplified using PCR and their shortness also reduces the risk of differential amplification. Because PCR amplification occurs in a non-linear manner, reproducibility is affected by stray impurities and the shorter the sequence, the less risk there is of this occurring. Moving the forward and reverse primers in as close as possible to the STR sequence can further reduce the size of the STR amplicons. This procedure is known as miniSTR analysis. Butler *et al.*, (2003) have produced a set of miniSTR primers that allows the maximum reduction in size for all 13 CODIS STR loci and also several of those used in commercial STR kits. This approach is useful where there are problems with conventional STR analysis through allele dropout and reduced sensitivity of larger STR alleles (Schumm *et al.*, 2004).

Case Study: Determination of paternity from aborted chorionic villi

Paternity testing is seldom a problem after a child is born or even before birth if the baby is well-developed; however, if only aborted material from an early stage of pregnancy is available it is more difficult. This is because the maternal DNA is so much more abundant than the foetal DNA that it is preferentially replicated in the PCR analysis. Robino *et al.* (2006) have described an interesting case in which they successfully overcame this problem. The case began when a severely mentally and physically disabled 21-year-old woman was brought to the emergency room of a hospital suffering from vaginal bleeding. Whilst there she passed a large blood clot and subsequent histological examination of this revealed the presence of chorionic villi – in short, she had suffered an abortion at an early stage of pregnancy. The chorion is an extra-embryonic tissue layer that is formed by the embryo and ultimately becomes the main embryonic part of the placenta. As the chorion develops it forms finger-like projections called chorionic villi that grow into the mother's *decidua basalis* – a part of the endometrium of the uterus that is shed after birth. The woman was mentally incapable of giving consent to sex and therefore the pregnancy must have resulted from sexual assault. A criminal investigation was therefore instigated. Her condition meant that she was also unable to indicate who had assaulted her so the culprit could only be identified from DNA analysis although standard protocols could not be employed for the reasons outlined above. The authors used laser microdissection (see Chapter 2) to isolate tissue from the foetal chorionic villi without including the surrounding maternal tissue and took blood from the woman to act as a reference sample. All the samples were then subjected to STR analysis for 15 loci and the Amelogenin locus for sex determination. The results (Table 3.4) demonstrated that the foetus was female and that its profile was sufficiently different from the woman's to indicate that the foetal tissue was successfully isolated whilst sharing one allele in common with the woman at each locus thereby indicating that she was the mother. Furthermore, eight loci in the foetus were homozygous and this strongly indicated that the woman was assaulted by someone genetically closely related to her. Consequently, blood samples were taken from the woman's brother, father and maternal and paternal grandfathers and subjected to the same STR analysis. The results also indicated that there were sufficient incompatibilities between the alleles of the woman's father and grandfathers for them to be excluded as potential culprits. However, the woman's brother provided an appropriate match for each of 15 loci on the foetus's STR profile and the probability that he was responsible was judged to be extremely high.

Y-short tandem repeat markers

Over 200 STR markers have been identified on the human Y chromosome. Between nine and eleven of these markers are used routinely in forensic science and commercial kits are currently available for at least six of them (Sinha *et al.*, 2003). They are particularly useful in cases of rape and sexual assault where there are mixed male and female DNA profiles and therefore separating the two is a major challenge. Unlike conventional STR analysis, there is typically only one peak or band for each STR type in Y-STR analysis and these can only originate from DNA from a male. In the case of multiple sexual assaults more peaks will be found, depending upon the number of men involved. The simultaneous detection of multiple Y-STR loci produces additional genetic information without consuming additional DNA (Tun *et al.*, 1999). A further advantage of Y-STR analysis is that it enables DNA profiles to be made in cases of sexual assault in which the man did not produce sperm owing

Table 3.4 Case study of paternity determination from aborted chorionic villi: STR profiles of the woman, the foetus and her male relatives. Note the eight homozygous loci in the foetus – this is unusually high and is suggestive of incest. Table redrawn from Robino *et al.*, 2006.

Locus	Woman	Foetus	Brother	Father	Paternal grandfather	Maternal grandfather
D8S1179	15	15	15	13,15	13,15	10,13
D21S11	28,29	29,30	30	29,30	27,29	28,31.2
D7S820	10	10	10,12	8,10	8	10,12
CSF1PO	11,12	11	10,11	11	11	10,12
D3S1358	15	15	15	15	15,16	16,18
TH01	8,9	8	8,9	9	6,9	6,8
D13S317	11,12	12	12,13	11,13	9,11	12
D16S539	9,11	9,14	9,14	11,14	11	9,12
D2S1338	17,20	17,18	18,20	20	20,24	18,20
D19S433	16,16.2	14,16.2	14,16.2	14,16	12,14	13,15
vWA	14,17	17	17,18	14,18	14,17	17,19
TPOX	8,9	8,11	8,11	8,9	9,10	8,11
D18S51	10,22	10,15	10,15	15,22	14,15	13,14
Amelogenin	X,X	X,X	X,Y	X,Y	X,Y	X,Y
D5S818	13	13	13	13	11,13	13
FGA	20,24	20,21	20,21	20,21	21	21,23

to a medical condition or being vasectomized. In these circumstances, the absence of sperm would mean that only a very small amount of male autosomal DNA would be present and the female's autosomal DNA would swamp this. By specifically targeting the Y chromosome STR markers it would be possible to target the minute amount of male DNA that would be present.

Because Y-STR markers exhibit lower variability than autosomal STR markers, their discriminatory power is much less and unless a mutation occurs, all male relatives – sons, fathers, brothers etc. – will share the same profile. This needs to be taken into account when assessing the strength of the evidence and could be a major problem when the suspect comes from an inbred population or a criminal family. However, like other DNA profiling techniques, their value can be as great in excluding suspects as in identifying a culprit.

Low copy number STR analysis

Conventional STR multiplex analysis works best where there is at least 1 ng (nanogramme = 10^{-9} g) of good quality DNA present and fewer than 28 PCR cycles are required to generate sufficient material for a full PCR profile. However, many forensic samples contain much lower levels of DNA than this and/or the DNA is degraded and in these circumstances a different approach is required (e.g. Irwin *et al.*, 2007). Low copy number STR analysis (LCN-STR) is employed where there is less than 100 pg (i.e. <0.1 ng; picogramme = 10^{-12}g) DNA (Gill, 2001a) and despite the miniscule amounts of DNA present a full STR profile can be generated. Some workers consider that defining LCN-STR in terms of the amount of DNA present in the sample is not appropriate and prefer to consider it to be an approach adopted for the analysis of results that occur below the stochastic threshold (i.e. the point below which their interpretation of peaks or bands would be considered unreliable) using normal techniques. Using LCN-STR, it is possible to obtain a full profile from the DNA of a single cell. LCN-STR may employ up to 60 PCR cycles and although it is extremely sensitive the results need to be interpreted with care, especially where the DNA of two or more people is present. For example, where there is only a very small amount of DNA available to begin with, there may be insufficient material to enable a second laboratory to undertake confirmatory testing. Further problems can arise through 'stochastic fluctuation' – i.e. random events that, in this case, become magnified owing to the small amounts of starting material and the number of replications. For example, allele dropout may occur from the unequal replication of the two alleles found at a particular heterozygous locus (Fig. 3.5c) thereby making it appear homozygous. Similarly, the low amount of starting material means that an amplification that occurs by random chance or a tiny contamination become magnified thereby resulting in allele drop-in. In addition, large stutter peaks can be formed and these may be mistaken for allele peaks. Gill (2001a) and Gill *et al.*, (2000) provide good discussions of the problems associated with LCN-STR and how the technique can be utilized to provide reliable results.

The meaningfulness of DNA detected at extremely low levels is liable to robust questioning during court proceedings. For example, a newspaper or package found at a crime scene could reveal the profiles of several people who handled the item during its production, transport, storage and sale. Similarly some people shed DNA more than others (Phipps & Petricevic, 2007; Lowe *et al.*, 2002) and their DNA can be transferred to items they themselves never handled (e.g. via a handshake) – this is known as secondary transfer. Indeed, the last person to handle an object may not be the one to yield the dominant DNA profile. This not only complicates the analysis but could lead to arrest of someone with a criminal record (or just on a DNA database) but who had nothing to do with the crime scene. Alternatively, the person responsible for a crime could argue that the extremely low levels of their DNA recovered from a crime scene resulted from contamination or might even have been planted by the police or somebody wishing to frame them. A successful prosecution would normally, therefore, rely on additional evidence other than the finding of very low levels of DNA by a process such as LCN-STR.

DNAboost™

DNAboost™ is a computer software technique developed by the Forensic Science Service (FSS) that is used to analyse mixed or highly degraded DNA profiles that are too difficult to interpret using existing procedures. Following DNA sequencing, the software generates all feasible profiles and compares these to the profiles of named individuals on the NDNAD. At the time of writing the technique was still being trialled and was somewhat controversial.

Familial searching

Despite the ever increasing size of the NDNAD, many DNA profiles recovered from crime scenes fail to yield a match. However, people who are genetically related share similar DNA profiles and the degree of similarity is a reflection of the closeness of relatedness (e.g. Table 3.4). Therefore, by narrowing the search for similarity from 'compared with all records on the database' to 'compared with a particular subset based on profile characteristics, ethnicity, and residence' it is sometimes possible to identify a family link. This is based on the following broad social assumptions:

(1)   Most crimes, whether they are petty burglary or murder are committed by people who live locally and/or know the area. Therefore, if a match is not found on the NDNAD the search should initially be focused on those of the right age etc. living nearby.

(2)   Criminal behaviour often runs in families and a child raised among people who commit crimes is more likely to offend than one who was raised by a non-offending family. Therefore, even though person X is not on the NDNAD their brother might be.

(3)   Most recorded crimes are committed by (and on) persons living in poor socio-economic conditions and they tend to remain close to the place where they were born. For example, many criminal gangs are highly territorial, their territory being based on where they grew up and some of their members may be related.

Needless to say, these are sweeping generalizations but in the absence of other leads they can provide an initial line of enquiry. Once a potential family link is established those members not on the NDNAD can be requested to donate DNA and this may lead to the identification of the culprit. Inevitably, this approach can be controversial because families/communities who have little trust in the police can feel that they are being unfairly picked on or it is a subterfuge to get their DNA onto the NDNAD. It can also cause serious family problems through revealing instances of previously unknown paternity or non-paternity and who is on the NDNAD. The alternative of requesting DNA samples from everyone living in the vicinity of a crime could be even more controversial as well as being a large, costly and time-consuming exercise.

Case Study: The murder of Lynette White

In the early hours of February 14th 1988, Lynette White, a 20-year-old prostitute was brutally murdered in her flat in the Cardiff docklands red light district. Her throat was cut, her wrists and face slashed, and she was stabbed over 50 times – particularly around her breasts. The crime scene had all the hall marks of a frenzied sex attack and resulted in a great deal of media coverage and a major police operation (the two are not uncommonly connected). The location and time of Lynette's death were not in doubt – the bloodstains indicated that she died where she was found and her watch, which was damaged during the attack, indicated that this occurred between 1.45 and 1.50 am (concurrence evidence). Furthermore, the stains on the walls and Lynette's clothing included blood from someone other than Lynette – it is not unusual for attackers to injure themselves (see Chapter 5). DNA profiling was still in its infancy at the time and most of the analysis was therefore based on serology. Together they indicated that the blood originated from a man with a rare combination of five blood group characteristics. Witnesses in the dockland area at the time of the attack stated that they had seen a white man with bloodstained clothes acting in a distressed manner. From their descriptions a known paedophile referred to in the published accounts only as Mr X was identified. He remained the main suspect until DNA profiling excluded him. With the main suspect exonerated, the police became desperate for a result. They proceeded to bully a confession out of Lynette's pimp, Stephen Miller, and got him to implicate others. Miller was 26 years old at the time but with an IQ of 75 and the mental age of 11 he was described as being 'vulnerable'. Unknown to him, he had an alibi that would have made it physically impossible for him to have killed Lynette – the police were aware of this but chose not to take it into account. The police also charged Yusef Abdullahi (who was at work on a ship at the time of the murder), Tony Paris and two other men. All five men were black and known to the police. The main evidence against them was Miller's 'confession' and the statement of two prostitute friends of Lynette that they had seen the five men murder her – although it allegedly took 48 hours in police custody for them to volunteer this information. The trial took place two years after the crime was committed and both it and the presentation of some of the forensic evidence were flawed. Miller, Abdullahi and Paris were found guilty of murder and jailed for life whilst the other two men were acquitted. Subsequently, the convictions of the 'Cardiff Three' as they came to be called were quashed at appeal and the police were left to start their investigation again. Over the following years new methods of DNA analysis were developed but re-analysis of the stored material failed to yield usable profiles.

In June 1999, the case was re-opened and passed to an independent forensic science laboratory – the Forensic Alliance – headed by Dr Angela Gallop. By this time few blood stains were left for analysis. Fortunately, they found that a scrap of bloodstained cellophane cigarette wrapping that was retrieved from beside Lynnette's body included one small spot with a different profile to hers. This was not sufficient to prove that the blood belonged to the murderer – for that they would need to find the same profile elsewhere in the flat in circumstances that would indicate an association with the murder. Although the murderer had left a trail of bloody handprints along the hallway as he left the building the strips of stored stained wallpaper failed to yield a usable DNA profile – probably because the use of fingerprint sprays on them had degraded the DNA (see Chapter 2). It was now over ten years since Lynette was murdered and her flat had been cleaned and redecorated twice so the chances of recovering evidence from the crime scene appeared remote. Nevertheless, in 2003 the flat was examined again and layers of paint carefully removed from the front door and the skirting board close to where her body was discovered. Tiny traces of blood were found and these included one that had DNA profiles of both Lynette and the man whose blood was found on the cellophane. Lynette had only used the flat for a week before she was murdered so there was a low probability of the profiles being deposited on top of one another at two different dates. Photographic records and knowledge of how bloodstain patterns are formed (see Chapter 2) helped the investigators to focus their search. For example, many of the wounds were inflicted whilst Lynette was lying on the floor and the frenzied nature of the attack meant that numerous cast-off stains would be projected onto the lower part of the walls and leak beneath the skirting board. In addition, the attacker would be covered in blood and, being wounded himself, would leave transfer stains of his own and Lynette's blood on the walls and doors as he fled the scene. Lynette's clothing was also re-examined and bloodspots with the same male DNA profile as that found elsewhere were found. A link was therefore made – the suspect's blood had been recovered from Lynette's clothing, mingled with her blood, and from near to her body: he must have been bleeding on and around Lynette at or around the same time that she was killed.

All of those originally charged and also the 'witnesses' provided DNA samples and were quickly excluded from the investigation. The suspect's DNA profile was also entered onto the NDNAD but no match was found. The comparison was therefore restricted to men living in the Cardiff area who expressed particular features of the suspect's DNA profile. By concentrating on the rarest component of the profile, the list of candidates was reduced to 600 names. The list was then reduced to 70 by comparing common components within profiles and ultimately to one individual whose profile was very similar to that of the suspect. This individual was a teenager who wasn't even born at the time of murder but there was a good chance that he was related to the murderer. The police therefore obtained DNA samples from all the boy's relatives and ultimately a cousin of his, Jeffrey Gafoor was found to have an identical profile to the suspect. He was arrested and admitted that he had initially paid Lynette £30 for sex but after arriving at the flat he changed his mind and demanded his money back – this resulted in an argument, Gafoor then drew a knife and in his words 'lost it'. He was jailed for life in 2003. This case indicates how easily miscarriages of justice can arise, how DNA evidence can be recovered many years after the event and how a person can be identified using the NDNAD even if their DNA isn't on it.

Case Study: The manslaughter of Michael Little

Michael Little was a lorry driver who worked for the Ford Motor Company and at about 12.30 am on 21st March 2003 he was driving his truck along the M3 motorway when his windscreen was shattered by a brick hurled from a footbridge across the road. The brick struck Mr Little in the chest but he was able to steer his lorry to the side of the road and switch off the engine before he died of heart failure.

The police believed that whoever threw the brick was probably local since the footbridge connected Camberley and Frimley (in Surrey, UK), and was often used by residents after a night out. Forensic analysis of the brick revealed a bloodstain that yielded a mixed DNA profile for Mr Little and an unknown other man. Because of the small amount of DNA available from the unknown man it was analysed using low copy number DNA analysis and this produced a partial DNA profile. This partial profile was then shown to match the full profile of a DNA sample recovered in a vandalized car in Frimley. Earlier in the evening, someone had attempted to steal the car by smashing one of its windows and hotwiring it. The attempted theft failed so the car was pushed into a hedgerow and abandoned. However, in breaking the window the would-be thief hurt his hand and left blood behind in the vehicle. It subsequently transpired that the thief had a partner and they had attempted to steal the car on their way home after a night out drinking. Frustrated in their attempts to steal the car they proceeded to walk home and along the way they each picked up a house brick from a driveway and these were subsequently thrown from the footbridge. The DNA profiles recovered from the brick that struck Mr Little and the car did not match any of those on the NDNAD but from the nature of the crime and the DNA profile characteristics the police believed they were looking for a white male who was probably under the age of 35 who lived locally. They therefore requested men who fitted this description (and who were willing) to donate DNA but after analysing 350 samples they were no closer to an answer. A familial search was therefore made on the NDNAD with the criteria limited to persons living in the Surrey/Hampshire regions who shared common profile characteristics with the unknown man's DNA. This yielded a list of 25 names but one of these stood out because it matched the profile of the unknown man in 16 out of 20 areas, thereby suggesting that they could be related. Consequently, DNA samples were requested from his relatives – this led to the identification of 20-year-old Craig Harman and he was arrested on 30 October 2003. Harman was initially charged with murder, attempted theft of a car and the theft of two house bricks. He was ultimately jailed for 6 years for manslaughter. This case was somewhat unusual because there was no direct contact between Little and Harman and also indicates the speed with which an identification can be made even in difficult cases provided a good DNA profile is obtained.

Single nucleotide polymorphism markers

Single nucleotide polymorphisms (SNPs) arise from differences in a single base unit (Fig. 3.7) and are the commonest form of genetic variation. They are found throughout the genome, including the X and Y sex chromosomes. Everyone has their own distinctive pattern of SNPs and this, therefore, provides a means of identification. Using a technique called mini-sequencing, the base at a given SNP can be determined and once the bases at several sites at different loci are known one can produce a profile similar to that of an STR profile. Using allele frequencies for each SNP, the likelihood of two persons sharing the same SNP profile can be estimated. Because

T T G A C G T
A A C T G C A

T T A A C G T
A A T T G C A

Figure 3.7 Diagrammatic representation of a single nucleotide polymorphism.

the maximum number of alleles at each site is only 4 (A, C, G or T) 50–100 SNPs need to be examined to achieve the same discriminatory power as STR-based profiling (Gill *et al.*, 2004; Gill, 2001b). However, a process called microarray hybridization allows numerous SNP loci to be examined simultaneously, so it does not take long. The stability of SNPs, compared with STRs, means that they are less likely to be lost between generations and they are sometimes used in paternity cases. However, a statistical simulation study by Amorim & Periera (2005) indicated that relying exclusively on SNP analysis would result in more inconclusive results than STR analysis.

Because SNP analysis requires minute quantities of sample and the segment size can be even smaller than that needed for STR analysis, the technique can provide information even when the DNA is severely degraded. However, its effectiveness is compromised in mixed DNA samples because it could be difficult to distinguish which SNP belonged to which person. Furthermore, a quantitative test would be required in this context and this is not possible with some of the current SNP assays. There would be less of a problem if the mixture were composed of DNA from a single male and a single female because Y-linked SNPs could only originate from the male. However, if more than one male could have contributed to the DNA sample or the sexes were the same (e.g. male rape), their separation could be difficult. A possible solution to this problem would be to identify tri-allelic SNPs (Phillips *et al.*, 2004) although according to Brookes (1999) these are 'rare almost to the point of non-existence'.

Determination of ethnicity

All ethnic groups share the same alleles used in the current STR Plus profile although in some groups certain alleles tend to be more or less frequent than in others. Although it is possible to make tentative inferences about a person's ancestry from their STRplus™ profile (Lowe *et al.*, 2001) it is not an especially reliable technique for this purpose.

SNPs show much greater potential than STRs for the determination of ethnic origin (Kidd *et al.*, 2006) and a commercially available test DNAWitness™ is already available. DNAWitness™ uses 176 SNPs to determine the percentage of European, East Asian, Native American and sub-Saharan African BioGeographical Ancestry (BGA) in a DNA sample (www.dnaprint.com). The test uses reference DNA samples collected from these geographical regions and an unknown sample is screened against them using a statistical procedure called 'maximum likelihood estimation'. The apparently odd (to Europeans) inclusion of 'Native American' in the BGA list is because this test was developed with an eye to the American genealogy market and many Americans would dearly like to have 'proof' of Native American ancestry. The technique has its limitations and is complicated by the increasing mobility of our societies and the frequency of intermarriage. However, it can be helpful when attempting to put a 'face' to a culprit because even eyewitness statements can be unreliable. Indeed, it is often stated in forensic literature that about 50% of eyewitness identifications are incorrect. A classic example of this was the case of Derek Todd Lee, who was found guilty of the rape and murder of seven women in Louisiana USA in 2003. Although the police had a record of the culprit's DNA from the crime scenes, this did not match any of those stored on the FBI's CODIS database. In addition, the DNA tests used for the CODIS database do not provide any information on racial origin. The police were therefore reliant on eyewitness descriptions – and these indicated that the culprit was a white male. It therefore came as a surprise when subsequent tests using DNAWitness™ indicated that the culprit's profile was 85% sub-Saharan and 15% Native American BGA. This information allowed the police to broaden the investigation and follow new leads. After about 2 months, the police arrested Derek Todd Lee whose genetic heritage matched that indicated by the SNP profile and whose STR profile proved to match that from the crime scenes (www.dnaprint.com).

Determination of physical appearance

Being able to narrow down the list of suspects is a crucial part of any criminal investigation. Eyewitness statements are obviously valuable in this regard but they may not be available and are notoriously unreliable. If a full profile DNA sample is obtained from a crime scene and provides a perfect match to one already on the NDNAD then the police know exactly who they are looking for but if there is no match then the only information may be whether the suspect is a man or a woman. Our physical characteristics, sometimes referred to as our phenotypic expression, (e.g. skin and hair colour, iris pigmentation, height, and shape of the jaw) result from the interaction of many genes with one another and are sometimes also affected by age, environment, disease and diet. Because of the complexity of the interactions there are relatively few reliable tests available for most aspects of physical appearance and although skin coloration, can be implied from ethnicity this needs to be done with caution (Parra *et al.*, (2004).

Most forensic studies on phenotypic expression focus on pigmentation (Tully, 2007) although this is a complex phenomenon and influenced by over 120 genes that interact with one another and have varying effects. The melanocortin 1 receptor (MC1R) gene exhibits considerable variability among Caucasian individuals and has been studied in detail. MC1R is mainly found on the surface of melanocytes and controls their production of the pigments eumelanin (black–brown) and pheomelanin (red–yellow). People with red hair express MC1R polymorphisms that reduce the ability of the melanocytes to synthesize eumelanin/overproduce pheomelanin. Consequently, the melanocytes produce mainly pheomelanin and the individual tends to have red hair and pale, freckled skin. Branicki *et al.* (2007) identified forensically useful SNPs in the MC1R gene that could be used to predict the possession of red hair although not its shade and neither were they able to use analysis of the MC1R gene to predict iris colour. A commercial test (Retinome™) for iris coloration based on SNP analysis is available from DNAPrint™ Genomics that is reportedly accurate in over 90% of cases in which the individual is of predominantly European descent. On its own, information such as this probably has limited usefulness since a suspect can reduce the likelihood of detection by colouring their hair or putting on a wig and wearing contact lenses to disguise iris coloration.

In addition to providing a reliable indication of a particular physical character trait, for any method to be useful in a typical crime scenario it should require a minimum amount of sample, provide a result quickly and be cost effective. Most current methods are performed sequentially (i.e. one test is done at a time) and they are therefore slow. Traits such as height and weight are likely to be particularly recalcitrant to prediction owing to complicating environmental factors, such as diet. For example, between 1957 and 1977 the average height of Japanese men increased by 4.3 cm and that of women by 2.7 cm. Similar increases are seen in other populations that have experienced improvements in diet and health care provision. To be of practical use a predictive test would also need to predict a character within a relatively narrow range. For example, predicting that the suspect was a male between 1.65 and 1.85 metres tall and weighed 65–90 kg would simply indicate that he was within the range of most adult men in England and Wales.

Determination of personality traits

The factors that determine whether or not a person develops criminal tendencies are exceedingly complex and beyond the scope of this book. However, it is becoming increasingly apparent that certain genetic characteristics expressed in combination with life experiences can predispose a person to particular behaviours. For example, men expressing a particular allele combination that results in low brain monoamine oxidase A activity and who are mistreated as children are more likely to be violent as adolescents and/or adults than those who do not (Bernet *et al.*, 2007). Similarly, men with a particular allele combination coding for the brain 5-hydroxytryptamine (serotonin) transporter and who experience stressful events are more likely to become depressive or suicidal than those with a different allele combination. Note that in both cases it is the combination of genotype and life experience that make a particular behaviour more likely to occur and they do not mean that the behaviour *will* occur.

Mobile element insertion polymorphisms

An alternative to SNPs for the identification of racial characteristics from DNA is the analysis of mobile element insertion polymorphisms based on short interspersed elements (SINEs). The commonest class of these are the so-called *Alu* elements that are about 300 nucleotides long. Most *Alu* elements are fixed at a particular locus but a few subfamilies are polymorphic for insertion presence/absence and can be used to determine genetic relationships between populations (Watkins *et al.*, 2003). Ray *et al.* (2005) undertook a blind study using 100 *Alu* insertion polymorphisms and was able to use them to correctly infer 18 individuals as being of African, Asian, Indian or European descent. SINE analysis requires standard PCR laboratory equipment but suffers from the drawback of being time consuming if manual systems are used. On the positive side, unlike SNPs, *Alu* insertions are not at risk of forward and backward mutations and it is possible that fewer insertions will need to be analysed in future once the most reliable ones are identified. In addition, gender can be determined by analysing *Alu* insertions (Hedges *et al.*, 2003) and this is therefore a useful method to employ if it is thought that amelogenin gene deletion has occurred on the Y chromosome.

Table 3.5 Advantages and disadvantages of the most commonly used methods of forensic DNA profiling

Genetic marker	Advantages	Disadvantages
Autosomal STRs	Small sample size. Slight DNA degradation not a problem. Excellent discrimination.	Discrimination seriously reduced if DNA badly degraded.
Autosomal SNPs	Extremely small sample size. Discrimination possible with badly degraded DNA. Tests for ancestry available.	Discrimination power lower than for STRs. Ability to distinguish mixed DNA profiles lower than STRs.
Y-linked STRs	Male specific therefore useful for mixed gender DNA samples. Small sample size. Slight DNA degradation not a problem.	Low discrimination especially among male relatives. Discrimination seriously reduced if DNA badly degraded.
Y-linked SNPs	Male specific therefore useful for mixed gender DNA samples. Very small sample size. Slight DNA degradation not a problem.	Low discrimination. Difficult to distinguish individuals if DNA from more than 2 males present.
SINEs	Gender determination not affected by allele deletion. Tests for racial origin possible.	Limited number of studies to confirm effectiveness.
Mitochondrial DNA	Very small sample size. Effective with even badly degraded DNA. Heteroplasmy may facilitate identification.	Lower discrimination than STRs. Limited discrimination if individuals are maternally related. Heteroplasmy may prevent accurate identification.

Mitochondrial DNA

Mitochondria are intracellular organelles that generate about 90% of the energy that cells need to survive. The numbers of mitochondria found in a human cell depend upon its energy needs and vary from zero in the mature red blood cell to over 1000 in a muscle cell. They are thought to descend from bacteria that evolved a symbiotic relationship with pre- or early eukaryotic cells many hundreds of millions of years ago. With time, the symbiotic relationship became permanent but the legacy is reflected by present day mitochondria retaining their own bacterial type ribosomes and their own DNA (referred to as mtDNA) that is distinct from that found in the cell nucleus. Each mitochondrion contains between 2 and 10 copies of the mtDNA genome. The inheritance of mtDNA also differs from nuclear DNA in that it is exclusively generated from the maternal side. This is because the sperm head is the only bit of a spermatozoon that enters the egg at the time of fertilization. Usually, the spermatozoon's tail and the mid piece (which is the only bit containing mitochondria) shear off as the head enters the egg's perivitelline space. Occasionally, a few mid piece mitochondria are incorporated at fusion but these are subsequently destroyed by the egg (Manfredi *et al.*, 1997). Consequently, the only mtDNA present in the developing embryo is that derived from the egg and therefore as usual (it is alleged) the workforce is exclusively female.

Human mtDNA is a circular DNA molecule that contains 16 569 base pairs that code for 37 genes that in turn code for the synthesis of two ribosomal RNAs, 22 transfer RNAs and 13 proteins. Unlike nuclear DNA, the mitochondrial genome is extremely compact and about 93% of the DNA represents coding sequences. The remaining, non-coding region is called the control region or displacement loop (D-loop) (Fig. 3.1). The D-loop region consists of about 1100 base pairs and it exhibits a higher mutation rate than the coding region and about 5–10 times the rate of mutation within nuclear DNA. The mutations occur as substitutions in which one nucleotide is replaced by another one: the length of the loop region is not changed. The mutations result from mtDNA being exposed to high levels of mutagenic free oxygen radicals that are generated during the mitochondrion's energy generating oxidative phosphorylation process. The substitutions persist because mtDNA lacks the DNA repair mechanisms that are found in nuclear DNA. These mutations result in sequence differences between even closely related individuals and makes analyses of the D-loop region an effective means of identification (Budowle *et al.*, 1999). Because the mtDNA is inherited only from the mother, it also allows tracing of a direct genetic line. Furthermore, unlike the inheritance of nuclear DNA, there are no complications owing to recombination.

The D-loop is divided into two regions, each consisting of about 610 base pairs, known as the hypervariable region 1 (HV1) and hypervariable region 2 (HV2). It is these two regions that are normally examined in mtDNA analysis by PCR amplification using specific primers designed to base pair to their ends. This is then followed by DNA sequence analysis. Because of the high rate of substitutions, it is possible to analyse just these short regions and still differentiate between closely related sequences. It has been estimated that mtDNA may vary about 1–2.3% between unrelated individuals (Inman & Rudin, 1997) and although mtDNA

(a)

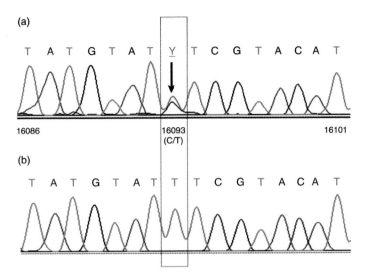

(b)

Figure 3.8 Diagrammatic representation of (a) heteroplasmy and (b) homoplasmy at position 16093. In (a), the nucleotides cytosine (C) and thymine (T) are present at position 16093 whilst in (b) only thymine is found. (Reproduced by permission of Elsevier from Butler, J.M. (2005) *Forensic DNA Typing*, 2nd edn. Copyright Elsevier (2005).)

sequencing does not have the discriminating power of STR DNA profiling, it can prove effective where STR DNA analysis fails.

The mtDNA sequence of all the mitochondria in any one individual is usually identical – this condition is referred to as 'homoplasmy'. However, in some people, differences in base sequences are found at one or more locations (Fig. 3.8). These differences arise from them containing two or more genetically distinct types of mitochondria. This condition is known as 'heteroplasmy' and it can have a significant impact in forensic investigations (Lo *et al.*, 2005). Heteroplasmy used to be considered relatively rare but it is now believed to occur in 10–20% of the population (Gibbons, 1998). To make matters worse, it is now apparent that heteroplasmy is not necessarily expressed to the same extent in all the tissues of the body. For example, two hairs from a single person might have different proportions of the base pairs contributing to the heteroplasmy and this might result in an exclusion rather than a match (Linch *et al.*, 2001). This is because heteroplasmy may result from the high mutation rate or from either inheritance at the germ line level or the level of somatic cell mitosis and mtDNA replication.

In forensic science, mtDNA analysis is most frequently used where the samples do not contain much nuclear DNA – for example, a fingerprint or a hair shaft – or where the DNA has become degraded through the decomposition process or burning (Bender *et al.*, 2000). Because there are numerous mitochondria in a single cell and each mitochondrion contains multiple copies of the mitochondrial genome, it is possible to extract far more mtDNA than nuclear DNA. Epithelial cells, which are the commonest cell type used in forensic casework, contain an average of 5000 molecules of mtDNA (Bogenhagen & Clayton, 1974). Mitochondrial DNA analysis does, however, suffer from a number of problems. For example, all maternally related individuals are likely to have the same mtDNA sequences, so the discriminating powers are limited compared with autosomal STR analysis. Heteroplasmy can be considered either a problem or a useful trait depending on the circumstances. It can create problems because the mixed sequence is also what would be expected if there were more than one individual contributing to the DNA profile. A difference of only one base pair between the mtDNA profile of the sample and the suspect is considered insufficient to prove either a match or exclusion whilst a difference in two or more base pairs is grounds for exclusion. By contrast, heteroplasmy can provide an identifying characteristic where the suspect expresses the same heteroplasmy characteristics as the sample.

Other common problems associated with mtDNA analysis are that detecting differences in sequences is more time consuming and costly than determining differences in lengths – as is accomplished using STR analysis. In addition, the rarity of mtDNA sequences has to be determined by empirical studies and the results are not as statistically reliable as those for other types of analysis. Finally, owing to the high copy number per cell there is always a risk of contamination and cross-contamination associated with mtDNA sequencing.

Case Study: Identification of the remains of Tsar Nicholas II

Towards the end of the Russian Revolution, on the night of either 16th or 17th July 1918, Tsar Nicholas II and his family were shot and their bodies disposed of. For many years, it was believed that after they were shot, their bodies were butchered, covered in sulphuric acid, burnt and thrown down a nearby mineshaft. However, there were also rumours about various members of the family escaping and their fate was shrouded in mystery. Following the collapse of Soviet Russia, a previously secret report became public knowledge and it transpired that although the bodies were thrown down a mineshaft, they were subsequently retrieved and buried in a concealed pit about 12 miles north of the town of Yekaterinberg. In 1991, the pit was located and the skeletons of five females and four males were retrieved from it. These were believed to include those of the Tsar, his wife Tsarina Alexandra, and three of their daughters. The morphology of the skeletons indicated that those of Prince Alexei and Princess Maria were missing. This agreed with the contents of the secret report that stated that after the family was shot, two of the bodies were burned. However, it still needed to be determined whether the remains included all the other members of the royal family.

STR and mtDNA testing confirmed that a family group was present in the grave: the other bodies belonged to three servants and the family doctor. Further mtDNA analysis indicated a match between all 740 tested nucleotides from the mitochondrial D-loop region for the remains presumed to be those of the Tsarina and her children and those of Prince Philip, Duke of Edinburgh. Because Prince Philip's maternal grandmother was the Tsarina's sister he inherited the same mtDNA profile as the Tsarina and the Tsarina's children inevitably also had the same profile. Mitochondrial DNA analysis of the bones thought to belong to the Tsar proved more complex but also even more interesting. Profile matches were found between these remains and those of two distant maternal relatives except for position 16169. At this point, the bones expressed a cytosine (C) nucleotide but that of the relatives expressed a thymine (T). Cloning experiments on the bones demonstrated the presence a mixture of mtDNAs that differed only at this single position, the C form accounting for 70% of the mtDNA. This suggested that the Tsar exhibited heteroplasmy (Gill *et al.*, 1994).

Although the results indicated that the bones were 98.5% certain to belong to the Tsar, this was not sufficient for some commentators who suggested that there were problems with sequence background effects or contamination during the recovery, storage or analysis of the samples. The Russian Orthodox Church had canonized the Tsar, so a great deal depended upon identifying the remains with absolute certainty because they would become objects of veneration and pilgrimage. Eventually, the mystery was solved when permission was gained to open the coffin of the Tsar's brother, Grand Duke Georgij Romanov, who died of tuberculosis in 1899. Mitochondrial DNA analysis of the Duke's bones, demonstrated an identical mtDNA profile to those presumed to belong to the Tsar – including the same C and T heteroplasmy at position 16169. This was the first instance in which heteroplasmy was used to facilitate human identification. The results, together with morphological studies on the bones and all the other DNA studies meant that the likelihood ratio for the remains' authenticity was in excess of 100 million (Ivanov *et al.*, 1996).

RNA

Ribonucleic acid (RNA) is chemically similar to DNA but is a single stranded molecule, has the sugar ribose (as opposed to deoxyribose) in its backbone and incorporates the base uracil (as opposed to thymine). Humans, like all eukaryotic organisms contain three types of RNA each of which is produced in a process called transcription from information coded for in DNA. Each type of RNA has a specific function:

- Messenger RNA (mRNA) carries the information that is necessary to make a particular protein from a gene on the DNA molecule to ribosomal RNA.

- Ribosomal RNA (rRNA), as its name suggests, is found in the organelles known as ribosomes that are located in the cytoplasm. During translation, rRNA binds to transfer RNA and catalyses the formation of peptide bonds between amino acids in order to make proteins.

- Transfer RNA (tRNA) molecules transfer specific amino acids to rRNA during protein synthesis and place them in the correct orientation on the mRNA.

Although of fundamental significance to normal cell function, RNA is seldom employed as a source of forensic information (Bauer, 2007). This is partly because

RNA is considered to be an unstable molecule that is rapidly broken down after death by a combination of the body's own enzymes and those of microbes. However, the rate of RNA degradation depends partly upon its location and it is possible that the extent of the degradation could be used as an indicated of the PMI during the first few hours or days after death (Bauer *et al.*, 2003). So far, this work remains at the preliminary stages and it is uncertain how complicating factors such as cause of death, underlying disease and post mortem environmental conditions would affect the accuracy. Similarly, it is possible that RNA degradation could be used to age blood and other stains but more work needs to be done to confirm how effective this would be in real-life situations. RNA could also be used to identify the provenance of body fluids. This is particularly relevant where the investigator wishes to distinguish between menstrual blood and that shed from a wound: this could arise in cases of alleged sexual abuse/rape and bloodstains are found on clothing or bedding. These cases are particularly problematic when the alleged abuse occurs within families since the presence of a father's DNA on a daughter's duvet or underwear would have little significance. Although all cells contain the same DNA, they produce different proteins depending upon their function and therefore express different mRNA profiles. For menstrual blood, the enzymes matrix metalloproteinase 7 and matrix metalloproteinase 10 have proved effective markers (Juusola & Ballantyne, 2007). These enzymes are expressed during menstruation when they are responsible for the breakdown of components of the extracellular matrix. This breakdown occurs as part of normal endometrial remodelling and the enzymes are absent during the early and mid-secretory phases. It is worth remembering that individual differences and health factors can affect tests such as these so any results should always be treated with care. For example, matrix metalloproteinases, including matrix metalloproteinase 7, are over-expressed in certain cancer cells.

DNA databases

The National DNA Database® (NDNAD) was established in April 1995 following a recommendation from the Royal Commission on Criminal Justice in 1993. Scotland and Northern Ireland have their own DNA databases but export the profiles of all persons they arrest to the NDNAD. The NDNAD is governed by a combination of the Home Office, the Association of Chief Police Officers and the Association of Police Authorities and with invited representatives from the Human Genetics Commission. It was the first national DNA database to be established in the world and in 2007 it contained over 4 million profiles (Table 3.6). This represents about 6% of the UK population although about 10–13% of the profiles are thought to be replicates (e.g. through people using aliases). The FBI's DNA database, CODIS, contains numerically more DNA profiles but represents only about 0.5% of the US population. The NDNAD is run by the Forensic Science Service (FSS) under contract from the Home Office and the FSS is the main organization that loads profiles onto the database, undertakes profile searches and matches and reports back to the police authorities.

The size of the NDNAD and the way in which it is used and managed raises many important ethical questions that are comprehensively debated in a report

Table 3.6 Information stored on individuals on the NDNAD in 2008

---

(1) Name, gender, date of birth and ethnic appearance as described by the arresting officer – this latter information can be of dubious authenticity.

(2) Type of DNA sample used (e.g. mouth swab, blood, hair, semen)

(3) Type of DNA test employed (e.g. STRplus™)

(4) DNA profile: in the case of STRplus™ this would consist of a string of 20, two digit numbers and the amelogenin sex indicator

(5) Data on the police force that collected the sample

(6) Arrest summons number: this provides a link to the Police National Computer which is stores criminal record and police intelligence information

(7) A unique bar code that identifies the record and provides a link to the stored DNA sample

---

produced by Nuffield Council on Bioethics under the chairmanship of Sir Bob Hepple QC (Nuffield Council on Bioethics, 2007) but remain largely unresolved. Some of the questions are as follows:

(1) Should the NDNAD retain the DNA data of everybody regardless of age or gender or be restricted to specific individuals?

(2) Should the DNA data be retained for ever or for a specific time period and if the latter, how long should this be?

(3) Should the DNA data be restricted to, for example, STR profiles stored as a computer file or should a DNA sample be retained to permit further analysis at a later date?

(4) Who should have access to the DNA data?

Who should be on the NDNAD?

In England and Wales the police routinely take DNA samples from anyone arrested on suspicion of having committed a recordable offence. (Recordable offences are those for which a custodial sentence could be imposed if found guilty although the Home Office has suggested that the distinction between recordable and non-recordable offences may be removed in the future.) This occurs regardless of whether or not the person is subsequently charged or whether the person's DNA is relevant to the alleged offence. For example, urinating against a lamppost or organizing a one-man demonstration against fly fishing outside the House of Commons without informing the police beforehand would both be sufficient reason for someone to be arrested, fingerprinted and have their DNA taken. Similarly, people may be arrested *en masse* (for example, after a football match or during a demonstration) simply because they are in the vicinity of a single individual causing a nuisance. Although not charged, their DNA may be taken and entered on the NDNAD. Victims of crime and persons who voluntarily give a DNA sample during a screening exercise may also find their DNA entered on the NDNAD. Although volunteer donors have the right to ask that their DNA is used only for a specific investigation many are confused by the consent forms and end up agreeing to their DNA being entered on the NDNAD and once this happens they cannot have the information withdrawn at a later date: a DNA profile is deemed to be the property of the police force that collected it for analysis.

Owing to biases inherent to any system, certain social groups, in particular young Black males, are over-represented on the NDNAD and this fuels claims of racial discrimination. Some estimates suggest that if current trends continue by 2010 50% of Black males will have their details stored on the NDNAD compared with less than 15% of White males. Some commentators (e.g. Sedley, 2005) suggest that the best way of avoiding these problems is to make it compulsory for everyone to have their DNA on the NDNAD – presumably as part of a national identity scheme. This is usually accompanied by the suggestion that 'if you haven't done anything wrong you have nothing to fear'. This indicates a rather naïve understanding of the justice system since innocent people are sometimes convicted of crimes they did not commit whilst being the subject of a police investigation because your DNA profile is similar to the partial profile recovered from, say, a terrorist bomb factory can be an extremely stressful experience and lead to family breakdown and financial loss even if never charged. In addition, what counts as 'wrong' varies with time and we cannot predict how society will change. For example, in the UK it is not long ago that homosexuality was a criminal offence, it was virtually impossible for a wife to prosecute her husband for rape, fox hunting was a popular country pursuit and smoking was permitted on aeroplanes.

Another argument commonly used against extending the coverage of the NDNAD is that it alters the fundamental relationship between the individual and the state. At its inception, the purpose of the NDNAD was intended to be the monitoring of the active criminal population but the majority of those now on the database have either never been found guilty of an offence or committed some minor misdemeanour for which they were punished and are highly unlikely to offend again. Therefore, DNA samples are now being collected on the basis that an individual might have done something in the past and/or they (or indeed someone related to them) might possibly do something in the future. The individual has therefore changed from 'an ordinary member of the public' to a 'potential criminal'. Some would dismiss this as paranoia and point to the several murder investigations that have been solved with the aid of data on the NDNAD placed there as a consequence of individuals committing minor offences.

That the NDNAD has contributed to the solving of many criminal offences is without doubt and it has proved invaluable in solving some atrocious crimes. However, although the size of the NDNAD continues to increase, the number of cases in which DNA profiling has contributed to the solving of crime has remained more or less the same. Indeed, Home Office figures indicate that it is the number of DNA profiles obtained from crime scenes that is important to the solving of crimes rather than the number of people on the NDNAD. This goes back to the well-known fact that the majority of crimes are committed by a minority of people – and these are often already well-known to the police.

How long should information be stored on the NDNAD?

It makes sense that biological information from a crime scene should be stored for a long time to permit future analysis. However, the storage of an individual's data is a different matter. Currently, in England and Wales, the records of a 13-year-old girl caught stealing a bag of sweets and a 42-year-old serial rapist would be stored on the NDNAD for the same amount of time – that is, for life. This lack of distinction between offences is a cause for concern because it is not unusual for young people to do things that they later regret and it is a sad society that retains the records of every individual's every minor transgression throughout the whole of their life. If everyone was on the NDNAD then it would automatically have to be for life and this would therefore remove the 'stigma' of membership. Indeed, it would be absence from the database that would arouse suspicions. Furthermore, it is argued that persons who have committed minor crimes may go on to commit worse ones and that presence on the NDNAD can help in the identification of remains. It should, however, be borne in mind that DNA samples are stored by commercial companies and therefore the annual cost associated with the storage of millions of samples could be substantial. The European Court of Human Rights has ruled (4th December, 2008) that it is unlawful for the UK to keep DNA samples belonging to innocent individuals but the police remain loathe to lose what they see as a vital resource is in the 'fight against crime'.

What form of DNA information should be retained?

Those opposed to DNA databases argue that the information could be used to identify personal characteristics such as risk of contracting cancer or a degenerative disease and that these could then be made available to third parties such as employers or insurance companies. The FSS is expected to function as a commercial organization and sells its services to both the police and commercial customers. It would therefore be under pressure to make the most of one its prime assets although the Human Tissues Act is supposed to guard against non-consensual disclosure of DNA data except under specific circumstances. The current STRplus™ DNA profiles cannot provide meaningful information on personal characteristics. However, if the whole genome is stored such that it could be subjected to analysis in the future, for example, using SNP profiling, it is perfectly feasible that within a few years it will be possible to determine a great deal about an individual's appearance, character and susceptibility to mental illness and disease. It could therefore be argued that once an STRplus™ profile is generated for an individual their DNA sample should be destroyed because they could always be re-sampled at a later date should an instance arise in which the police required additional DNA evidence.

Who should have access to the data stored on the NDNAD?

Access to information on the NDNAD would be of benefit to many official organizations both in the UK and overseas and, at one level, it could be argued that making the information available in any situation in which a crime was committed or could be committed in the future should be facilitated. For example, people trafficking, drug smuggling and international terrorism could all benefit from greater cooperation between countries and security organizations. However, the more people who have access to information the greater the opportunity for both mistakes and misuse to occur. Currently, the number of people who have access to the NDNAD is carefully controlled but this cannot be guaranteed for the future. In particular, an obvious development aim is to design hand held or mobile devices that would enable police to generate DNA profiles from a crime scene or an individual and upload them onto the NDNAD for direct comparison. This would inevitably require more people to gain access and this in turn would facilitate the planting of evidence and the illegal downloading or tampering with stored information. For example, persons on witness protection schemes are sometimes provided with a new name and identity but they cannot be provided with a new DNA profile.

Future developments

As our understanding of the human genome evolves and laboratory techniques are refined we will undoubtedly continue to extract more and more information from increasingly small amounts of DNA. Indeed, it is now possible to obtain information from such small amounts of DNA that the problem of contamination and secondary transfer are serious practical issues. For at least 10 years, the arrival of DNA chip technology has been promised that will revolutionize forensic science. The method utilizes large arrays of DNA probes bound to a silicon, glass or polypropylene surface. Sometimes referred to as a 'Lab-on-a-chip', these would be the size of a credit card and be able to extract the DNA from a sample, amplify specific sequences and analyse them to provide a DNA profile (Fedrigo & Naylor, 2004; Wang, 2000). Theoretically, such devices could be used at the scene of a crime and the profile compared with those held on a national database to provide an instant identification. However, this in itself would present serious practical difficulties.

The expansion of the NDNAD is raising ethical and legal issues that must be addressed if the general public is to retain confidence in the authorities responsible for law and order. Collecting DNA samples from the public in England and Wales appears to have become an end in itself rather than the collecting of samples from crime scenes (some estimates suggest that fewer than 20% of crime scenes are forensically examined). In addition the presentation of forensic science information, particularly that relating to molecular biology needs to be improved. Molecular science is a complex subject and overburdened with abbreviations and abstruse terminology. Consequently, it can be hard for a jury to truly understand the strengths and weaknesses of the 'evidence' put before it. The same can also be said of many prosecution and defence lawyers who become blinded by science and seduced by the statements such as there being a 'one in 16 billion chance of the DNA sample coming from another person' without questioning the nature of profiles or the collection and processing of the samples and whether it is appropriate to calculate the probability on the basis that anyone in the world could have committed the crime. There is therefore a need to develop a mechanism that would enable DNA-based evidence to be presented in court in a standardized and clear way so that both its strengths and weaknesses are apparent. The public also needs to be educated to accept that the portrayal of forensic science on TV and film is seldom realistic and it is not always possible to retrieve a perfect DNA profile from a crime scene.

Whilst DNA databases have led to the arrest of several rapists and murderers, it is also worth remembering that most of the victims of violence know their assailant and the crimes are a consequence of sudden, unplanned anger or opportunism and fuelled by drink or drugs or both. The highly intelligent, charismatic serial killer who is a staple feature of crime fiction and film plots is, mercifully, an exceedingly rare individual. Most violent crimes are committed by poorly educated people living in the less affluent parts of our society and their victims come from the same background. These crimes are usually solved by standard police procedures. Indeed, despite all the technical advances, pitifully few cases of rape and sexual assault result in successful prosecutions. The solving of many crimes does not therefore depend on ever more sophisticated technologies but on the appropriate use of existing ones coupled with effective police work.

Quick quiz

(1) List ten common sources of DNA that you might find at a crime scene.

(2) In relation to DNA profiling results, distinguish between 'an inclusion', 'an exclusion' and 'an inconclusive'.

(3) Explain the purpose of the amelogenin gene test and also the limitations of this test.

(4) Explain the value and limitations of Y-STR markers for identification.

(5) Distinguish between STRs and SNPs and their advantages and limitations for identification purposes.

(6) In relation to mitochondrial DNA, explain what is meant by the term 'heteroplasmy' and how it can be both a problem and an advantage for identification purposes.

(7) What is the genetic basis of 'familial searching'?

(8) Explain how ethnicity might be inferred from a DNA profile.

(9) What is the value of low copy number STR profiling and why do results gained from it have to be treated with caution?

(10) Give two arguments for and two arguments against the development of a universal DNA database.

Project work

Title

Environmental factors affecting the rate of degradation of DNA in bloodstains.

Rationale

It has been suggested that the degree of DNA degradation may reflect the length of time since death in human tissues. Owing to their size and complexity, tissues may not be the ideal candidates to investigate this relationship. Bloodstains, by comparison, are simpler, easier to work with and, whilst there are a variety of methods available to determine how long a person has been dead, there are no reliable means of ageing bloodstains.

Method

Fresh blood would be collected from an abattoir and used to prepare bloodstains of varying sizes on different surfaces. The stained items would then be left under different environmental conditions (e.g. low and high temperature, dark and light, high and low humidity) and at varying time intervals the amount of DNA degradation that had occurred would be determined using different marker systems such as autosomal STRs, Y-linked STRs, mtDNA, SNPs and mRNA.

Title

The influence of age and gender on DNA secondary transfer.

Rationale

Age and gender are known to affect the formation of fingerprints and it would be useful to know whether this is also reflected in the extent to which DNA is transferred between people and objects during normal contact.

Method

This project would require ethical approval. Male and female volunteers, preferably not genetically related and aged from children to elderly would be asked to shake hands and then after varying times to grasp a uniform inert object such as a plastic rod (e.g. a sterilized mop or brush handle). Swabs would be taken from each individual at the start of the experiment, after shaking hands and from the plastic rod and then subjected to DNA profiling. The experiment can be adapted to include variables such as hand washing, physical activity (e.g. rubbing a separate plastic rod at the start to loosen the skin surface) and the nature of the material onto which DNA is transferred.

4 Human tissues

Chapter outline

The Outer Body Surface
Hair
Bones
Teeth
Future Developments
Quick Quiz
Project Work

Objectives

Describe the different ways in which fingerprints are formed and the methods for their detection.

Explain how hair can provide information on a person's identity and their exposure to chemicals such as methodone and explosives.

Discuss the strengths and limitations of retinal/iris scans, tattoos and scar tissue as unique identifying features.

Compare the use of bones and teeth as means of determining a person's personal characteristics.

The outer body surface

Skin

Skin covers the whole surface of our body and is the largest organ in terms of both weight and surface area. It varies in thickness from about 0.5 mm on the eyelids to 4 mm or more on the heels. Repetitive abrasion of the skin surface results in the formation of calluses, the distribution of which, along with the condition of the fingernails, can provide an indication of a person's trade or activities. For example, manual labourers develop thick calluses over their palms and fingers and their fin-

Essential Forensic Biology, Second Edition   Alan Gunn
© 2009 John Wiley & Sons, Ltd

gernails become chipped and ragged. At the opposite extreme, musicians develop calluses on highly localized regions and string instrument players may allow certain fingernails to grow long to facilitate plucking. Violinists typically develop calluses on the dorsal surface of their left second and third fingers over the proximal inter-phalangeal joints – these calluses have their own name – 'Garrod's pads'. Similarly, wind players develop calluses on the mid portion of the upper lip.

Structurally, skin can be divided into two regions, the outer epidermis and the thicker inner dermis. The epidermis is composed of several layers of cells that are often described as being a keratinized stratified squamous epithelium. Roughly translated, this means that the cells contain the fibrous protein keratin, have the appearance of a wall of bricks (stratified), are flat (squamous), and they form the outer surrounding layer (epithelium). Embedded within the epidermis are melano-cytes: round cells with long, slender projections that contain the pigment melanin and are responsible for giving skin its coloration. Where there is an overgrowth of melanocytes a round, flat or raised area called a nevus or mole is formed. The dis-tribution of these can be useful for identification. Beneath the epidermal layers lies the dermis that is largely composed of connective tissue but also contains blood vessels, nerve endings, glands and hair follicles.

Tattoos and scars

Men, and to a lesser extent women, have adorned themselves with tattoos for thou-sands of years. However, in modern western societies, their use until recently has tended to be restricted to certain male groups or professions – such as sailors (Sperry, 1991). Nowadays, men and women of all ages and degrees of affluence wear tattoos. Until the skin is lost during the decay process, these tattoos remain visible and provide identifying characteristics (Fig. 4.1). Tattoos can remain visible even after a person has suffered severe burns and they have proved useful in the identification of people who died during explosions and fires. Should they become obscured during the decomposition process tattoos can be rendered visible by treat-ment with 3% hydrogen peroxide (Haglund & Sperry, 1993). Even if the tattoos are irretrievably lost, their former presence can be predicted from finding the pig-ments within nearby lymph nodes (Hellerich, 1992). Although they are normally found on the outer body surfaces, some people are tattooed inside the lips or on the gums. Tattoos can indicate a relationship, lifestyle preference, mental state, service in the armed forces, a means of covering up intravenous drug use, or gang membership (Mallon & Russell, 1999). For example, there is a probability that a man with a tattoo depicting foxes racing down his spine and disappearing into his anus may be a homosexual (and given to extrovert behaviour) whilst a woman with four dots on each finger may be a lesbian. Drug addicts' tattoos often include prophesies of doom such as 'born to lose', whilst those of female prostitutes are more light hearted, such as 'to hell with housework' and 'pay as you enter' inscribed in an appropriate location. Names are not a good indication of a relationship because the tattoo may remain long after the object of affection has become a distant memory. Criminal gang members, especially the Japanese Yakuza, Russian *mafiya* and South African 'numbers gangs' often have extensive and elaborate tattoos. Riley

Figure 4.1 Tattoos can be found in the most unlikely places. The more unusual the tattoo, the more useful it can prove in identification. (Reproduced from Shepherd, R. (2003) *Simpson's Forensic Medicine*, 12th edn. Copyright 2003, Hodder Arnold, London.)

(2006) provides an interesting guide to the tattoos employed by American gangs and how one can distinguish street gang tattoos from those employed by prison gangs. The Russian *mafiya*, and in particular the *vorovskoi mir* (thieves of the world) have a reputation for using tattoos as a code to indicate amongst other things the wearer's position within the criminal hierarchy, their past crimes and which prisons they were held in. However, accoding to Galeotti (2008) the 'old style' criminal bosses are being increasingly superceded by *avtoritety* ('authorities') who are dispensing with the rituals and tattoos.

Owing to the current popularity of tattoos, it is important to keep an open mind when attempting to interpret their significance because the tattoos could represent a person's fantasy about themself rather than the reality of their existence or be the consequences of a 'good idea' at a time of excessive inebriation. However, unless one is actually a member of a criminal gang it is very foolhardy to sport its tattoo because it risks attracting the attention of the police and the violent indignation of the gang.

Permanent scar tissue is formed when there is irreversible injury to a tissue and is characterized by the deposition of collagen fibres, localized remodelling and reduced blood supply. Scar tissue forms in our internal organs and tissues as well as on the surface of our skin and whether the scars result from accidents or medical operations, their presence can be useful for identifying an individual and deducing past events. However, scars are seldom so characteristic that they can provide definitive identification. The effectiveness of the healing process is affected by numerous factors, such as the nature of the wound, the part of the body, general health and age. Interestingly, wounds inflicted during early gestation will heal rapidly and perfectly. A patchwork of healed and partially healed cuts is a common feature of self-harm. These wounds can be anywhere on the body although they are usually in places that can be hidden by clothing and are always restricted to sites that the victim can reach.

Fingerprints

The deepest layer of the epidermis is known as the basal layer. In the 1920's, Kristine Bonnevie (the first female professor in Norway) suggested that when a developing human embryo is about 10 weeks old, the basal layer at the tips of the fingers grows faster than the surrounding upper epidermal layers and the lower dermal layers. This would result in the basal layer being stressed and thrown inwards into a series of folds that are manifested on the surface as 'friction ridge skin'. Similar skin is also found on the palms of the hands, the soles of the feet and the lower surface of the toes. The patterns of loops, whorls and arches that are formed are unique to every person. This hypothesis of friction skin formation has never been conclusively proven but computer modelling studies lend it strong support (Kücken, 2007).

Regardless of the precise mechanisms by which they are formed, once we are born we already have our own unique fingerprint pattern that remains the same for the rest of our life. As we grow bigger the fingerprint ridges do not change in shape but become further apart – indeed, there is a relationship between height and ridge width. Minor cuts, burns or bruises result in temporary defects to the ridge pattern and once they have healed, the normal fingerprint pattern is restored. Deeper injuries, such as serious burns, result in the formation of permanent scar tissue – and this is also a good identifying characteristic. Cosmetic surgery cannot erase the fingerprint pattern because to be effective it would require the removal of so much tissue that a person would be unable to use their hands effectively. However, bricklayers, cement workers and those who routinely work with highly abrasive substances can wear away the surface of their friction ridges and therefore leave poorly defined prints. It has been suggested that criminals could attempt to reduce the risk of leaving prints by coating the surface of their fingers with nail varnish before committing a crime. A number of studies have attempted to link personal characteristics to fingerprints but their reliability is limited. For example, Acree (1999) and Gungadin (2007) both found that men tend to have a lower fingerprint ridge density than women but their 'cut off' point differed: Acree found that a pattern density of ≤ 11 ridges per $25\,mm^2$ was most likely to belong to a man whilst a density of ≥ 12 ridges per $25\,mm^2$ was most likely to belong to a woman; by contrast, Gungadin's figures were ≤ 13 ridges per $25\,mm^2$ for men and ≥ 14 ridges per $25\,mm^2$ for women. Whether this is in any way related to height differences between men and women is not known.

Types of fingerprint

The study of fingerprints is known as dactyloscopy. It has been used as a forensic tool since the 19th century and as early as 1906 a conviction was upheld in England solely on the basis of fingerprint evidence. Even identical twins do not have the same fingerprints and courts therefore accept them as unique identifying characteristics. This makes fingerprints an ideal forensic test: fingerprints are often found at a crime scene (unless careful efforts were made to remove them or not to make them in the first place) and they can be quickly, easily and cheaply collected, processed and analysed. And they provide an extremely reliable result.

Fingerprints are normally divided into three classes: plastic, visible and latent (Fig. 4.2). Plastic fingerprints are those that are formed when a person touches a soft or semi-solid substance such as soap or unset putty. This results in a shallow three-dimensional record of the friction ridges that is either transient or long-lasting depending on the substance and the circumstances. For example, plastic fingerprints can be identified from 2500-year-old clay seals recovered from the ruins of the Mesopotamian city of Ur. Visible fingerprints, as their name suggests, are those that can be seen without further enhancement. They result from a combination of sweat and the oily secretions from the skin glands being deposited on a contrasting surface thereby making the print visible. Alternatively, they may be formed when a hand covered with substances such as blood, ink or engine oil is pressed against a contrasting surface. Latent fingerprints are those that need to be enhanced in some way before they become visible. This is because sweat and body oils are colourless, so unless the prints are on a contrasting surface, they cannot be seen easily. Washing one's hands before handling an object reduces the quality of latent fingerprints because it removes the sweat and body oils. Similarly, young children tend to leave poor quality fingerprints that degrade quickly because they do not produce as much body oil as adults. Latent prints may also result from the removal of a layer of material, such as fine dust, thereby leaving a 'negative impression'.

Characteristics of fingerprints

Fingerprints can be classified according to their distinctive patterns of loops, arches and whorls (Fig. 4.3). However, identifying a fingerprint is not easy and relies on comparing two sets of prints in terms of their characteristic features and the spatial relationship of these to one another in a three stage process: (i) matching the overall pattern of friction ridges, (ii) matching the characteristics features of those ridges, and (iii) matching unique features such as the number and distribution of sweat pores and the three-dimensional shape of individual ridges. There is little consensus as to how much similarity between two sets of prints is necessary for them to be considered a 'match'. In the UK, a 16-point minimum was required for many years whilst in France it was 17 and in the USA each state set its own standards. With time, it was accepted that there is little scientific or statistical basis for any numerical standard and in 2001 the UK adopted a non-numerical approach that incorporates 'objective criteria'. Computer programmes speed up the matching process and police forces in England and Wales use a biometric database called IDENT1 – this supersedes the National Automated Fingerprint Identification System (NAFIS). In 2007 IDENT1 contained sets of ten-prints for over 7 million individuals, 2.86 million palm prints and over 1.5 million unidentified marks from crime scenes. In America, the FBI operates a similar system called the Integrated Automated Fingerprint Identification System (IAFIS) and has access to over 75 million fingerprints.

Fingerprinting the living and the dead

Changes to the law introduced in the Criminal Justice Act 2003 allow police officers in England and Wales to take both DNA and fingerprint evidence from a person

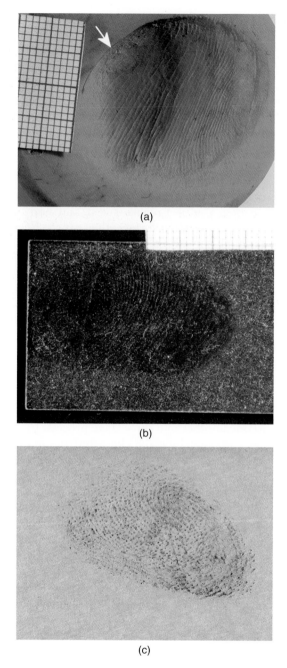

(a)

(b)

(c)

Figure 4.2 Different types of fingerprints. (a) Plastic fingerprint left in Plasticene. Note the fingernail impression – arrow. Each square = 1 mm², (b) Visible fingerprint left on a dusty glass slide. Each square = 1 mm². (c) Latent fingerprint on white paper rendered visible using ninhydrin spray.

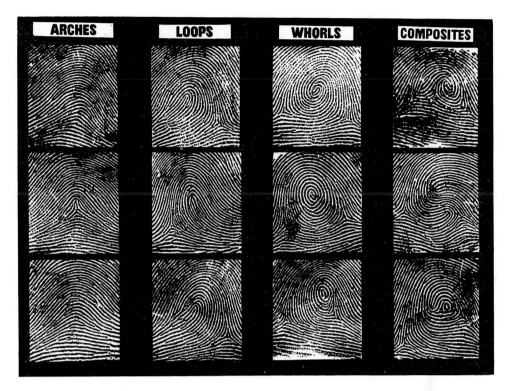

Figure 4.3 Fingerprint characteristics showing the distinction between arches, loops, and whorls. Composites, as the name suggests are where two or more features are present. (From Howe, 1950)

after they are arrested and detained at a police station. Previously, this could only be done after a person was charged with a criminal offence. Fingerprints were traditionally taken by asking a person to dip their fingers in a thin layer of printer's ink and then press their fingers onto fingerprint card. Each finger was recorded separately and the procedure could be time consuming. Nowadays, fingerprints are usually taken electronically using 'LIVESCAN' digital fingerprinting consoles – the 'donor' places their hand on a glass platon and the fingerprints are recorded electronically. The digital images are then entered instantly into IDENT1. Consequently, within a short period of time, it is possible to check whether a person is lying about their identity or that prints recovered from a crime scene resemble those of someone already on the database. However, the computer programme's main purpose is to speed up a search: final confirmation of a match always relies on physical observation of the prints by a fingerprint officer.

Fingerprints provide an excellent means for the identification of an unknown body provided that it is fresh – in which case it can be printed in the same manner as a living person. However, once the body starts to decay then the operation becomes more difficult (Kahana *et al.*, 2001). For example, with time the skin

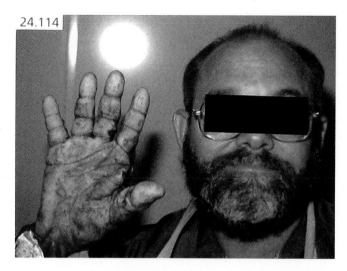

Figure 4.4 During decomposition it is not unusual for the outer layer of skin to slough off. When this occurs on the hands it is called 'degloving'. It is possible to insert a finger or even the whole hand around the cast skin in order to make prints from it. (Reproduced from Dolinak, D. *et al.*, (2005) *Forensic Pathology Theory and Practice*. Copyright © 2005, Elsevier Academic Press.)

becomes wrinkled, especially if it has been exposed to water or damp whilst fluid exudates and fungal hyphae can prevent ink from binding and interfere with scanning. If the epidermal layer sloughs off from the dermis (a common occurrence in bodies found in water) it is possible to insert ones own finger or a mould inside the cast skin and make a print (Fig. 4.4). Mummification and burning results in the skin shrinking, wrinkling and drying and in extreme cases this can make fingerprinting impossible (Fig. 1.1). In the latter cases, the fingers can be cleaned of grease, dust and carbon particles and then coated with liquid latex to produce a cast that is then placed onto a support in the shape of a human finger – the cast can then be manually printed or scanned (Porta *et al.*, 2007). This approach has the added advantage of providing a permanent record that could even be examined by scanning electron microscopy if desired. There still appears to be a need for a comparison study of latex prints against traditional prints to determine the extent to which detail is lost or artefacts introduced and it would be interesting to know whether latex prints could be obtained from bloated and decaying corpses. The latter could be initially tested by fingerprinting volunteers who have swum for an hour or taken a long hot bath to ensure that their fingers exhibit pronounced wrinkling.

Detection of fingerprints

Searching for fingerprints is a specialist skill and should be undertaken with extreme care owing to the ease with which prints may be damaged or lost if the incorrect method is chosen. All fingerprints, whether visible or latent, need to be given a unique code and their location recorded. To begin with, all the visible prints must be located by passing a strong beam of light over the scene or object. The angle of the beam of light should be altered to make the prints 'stand out'. Sometimes latent fingerprints can be rendered visible simply by viewing them under different lighting conditions – oblique lighting is often effective. The use of lasers tuned to an ultra-violet wavelength has also been suggested (Akiba *et al.*, 2007). However, the use of special fingerprint powders or sprays is usually required, the choice of which depends on the nature of the print, the substance it occurs on and the personal preferences of the investigator. Powders stick to the moist or oily parts of the print and are employed on non-absorbent surfaces such as plastic or glass and they tend to be brushed on (Sodhi & Kaur, 2001). They usually consist of two components: an adhesive such as starch and a colorant such as an inorganic salt or an organic dye that adsorbs onto the adhesive. Ideally, 'once only' brushes should be used owing to the potential for fingerprint brushes to pick up and transfer DNA (van Oorschot *et al.*, 2005). 'Magnetic' powders are useful on textured surfaces, such as leather. These powders are not themselves magnetic but they contain iron and are applied with a magnetic wand. Magnetic attraction therefore sticks the powder to the wand and this is dragged slowly across the surface of an object. The components of fingerprints exert a stronger attractive force than the magnet, so the particles remain behind and thereby render the prints visible.

There are various means of chemically identifying fingerprints and the choice depends on the surface and whether the objects are required for other forensic examinations – some tests can interfere with one another. One of the most commonly used reagents is ninhydrin – this reacts with the amino acids found in sweat (mainly alanine, aspartic acid, glycine, lysine, ornithine and serine) to form a pink to reddish purple coloured compound. The ninhydrin reagent may be sprayed or brushed on, or even used as a dip. This method is especially effective for prints found on paper, wood and other porous surfaces. It is possible to speed up the process of development of the prints by conducting the procedure at 50–70 °C and 60–80% relative humidity. This is known as the accelerated ninhydrin process of fingerprint development.

Silver nitrate is useful for identifying latent prints on brass objects such as ornaments and ammunition cartridges, but also porous surfaces such as wood. The silver nitrate reacts with sodium chloride in sweat to form silver chloride: this is light sensitive and under the influence of UV light decomposes to silver and chlorine. Consequently, the print becomes visible as a greyish or brownish stain – although this fades with time. Unfortunately, silver nitrate will react with any source of sodium chloride and this may obscure the print. Consequently, it would not be a good method for detecting prints on the brass fittings found on a sea-going yacht. Iodine fumes react with the oily components of latent fingerprints thereby making them visible as a brownish colour. However, iodine prints are unstable and must be photographed immediately or 'fixed' in some way to preserve them.

Exposing latent fingerprints to ethyl or methyl cyanoacrylate (superglue) vapour renders the prints visible through the formation of a white deposit. The prints can then be enhanced by the application of fluorescent dyes or by observing the prints in UV light. This method has the advantage of being simple and cheap and works best for prints on glass, china and other non-porous surfaces. It is also useful for detecting latent prints on gun cartridge cases. It should be done at room temperature and in a closed container, partly to protect the operator from the fumes but also to ensure that a high humidity of about 80% is maintained.

Although prints produced by a bloody hand (or foot) may be visible, it is often necessary to treat or enhance them before an accurate record can be made. First the print is fixed with 2% w/v sulphosalicylic acid and then covered with a 0.5% w/v solution of 2,2′-azino-di-[3-ethylbenzizolinesulphonate] diammonium salt (ABTS). Immediately before use, hydrogen peroxide is added to the ABTS solution to act as an oxidant. The haemoglobin in the blood catalyses the breakdown of the hydrogen peroxide to produce reactive oxygen radicals (Chapter 2) and these bring about the conversion of ABTS to a green coloured compound. The print is then washed, dried and photographed. In the past, 3,3′-diaminobenzidine (DAB) was commonly used for enhancing bloody fingerprints but it is carcinogenic and therefore its continued use cannot be recommended. ABTS, by contrast, is equally sensitive and safe.

In addition to these methods, there are many other chemicals that can be used to render fingerprints visible. Sometimes a print may be subjected to more than one detection method in order to obtain as accurate a record as possible. Fourier transform infrared (FTIR) chemical imaging is showing potential as a means of enhancing the quality of the fingerprint image captured from either untreated prints or those revealed using techniques such as cyanoacrylate fuming (Tahtouh *et al.*, 2007). Images are collected digitally as an array of pixels (just as in a digital camera) and FTIR imaging identifies the spectra of each pixel at all spectral frequencies and thereby identifies the chemicals present. FTIR imaging therefore combines data on spatial information (i.e. images) with data on chemical composition with a high degree of resolution. The process can provide images of fingerprints on surfaces on which it would otherwise prove difficult to resolve the print ridges, such as highly patterned banknotes and also identify the presence of chemicals of forensic interest such as drugs or explosives within the print pattern.

Once a print is located and rendered visible, it should be photographed. The print may then be left *in situ* or lifted and stored separately from the object it was found on. If it is to be preserved on the object, the print should be protected by clear tape, varnish or lacquer. There are several means of lifting a fingerprint but it is best to use special fingerprint tape as this has a very smooth adhesive surface. The tape is pressed against the print and then lifted, after which it is pressed into a 'lift card'. Lift cards have a smooth surface and are usually transparent. The prints may then be scanned or photographed again and stored.

Problems with fingerprint analysis

Although it is generally accepted that everyone has their own unique fingerprints (the chances of two persons having identical fingerprints is usually stated as 1 in 64 billion), correlating the prints found at a crime scene with those of a suspect is not a simple matter. For example, the prints may be smeared, poorly formed or 'partial prints' such as half a thumbprint and the prints of several people may overlap. Furthermore, prints decay with time at a highly variable rate that depends upon what they are composed of, the matrix they are formed on and the environmental conditions. For example, the lipid components of latent fingerprints can change dramatically with time (Archer *et al.*, 2005) and this should be borne in mind when deciding which enhancing agent should be used to render the prints visible. Exposure to very high temperature can also destroy fingerprints or render them difficult to interpret. This is a problem when investigating the remains of bombs used in terrorist attacks.

There is currently no accepted method for ageing fingerprints. Consequently, finding a suspect's fingerprints on a gun would indicate that they had handled it but not when they had handled it. This can be important when attempting to prove that a suspect committed a murder or was present in a room but who claims to have picked up the weapon or visited the room before or after the crime was committed.

Although the use of digital technology has increased the speed with which fingerprints can be recorded, stored and compared it is not without its problems and these must be appreciated in order to avoid mistakes (Cherry & Imwinkelried, 2006). For example, there could be problems with the resolution of the scanned image – narrow fingers tend to produce poorer images than wide fingers because they touch fewer sensors and therefore leave fewer dots on a 500 pixels per inch (ppi) scan. By comparison, 35 mm black and white forensic film has the equivalent of 6000 ppi. Imagine trying to distinguish between the numbers one and seven using dots: the fewer the dots the poorer the resolution and with only three dots it would be impossible. Clearly, if detail is not captured potential points of similarity or difference could be missed. It is therefore essential to use the highest sensitivity setting when capturing images and to know the resolution of those stored on a databank. Similarly, if images are compared on a computer screen the screen's resolution and the settings can influence one's ability to spot differences. It is also worth noting whether the image was processed in any way after collection and consider whether it was indexed correctly.

The introduction of biometric passports and identity cards in some countries has lead to concerns from civil liberties groups and worries over whether the information could be misused. In Germany in 2008 the Chaos Computer Club obtained what it claimed was the fingerprint of the German Interior Minister, Wolfgang Schaube from a water glass. To illustrate what they saw as the potential for biometric data being misused they printed images of the fingerprint and gave these away with issues of the magazine *Die Datenschleuder*. The images were reproduced on a film of flexible rubber with a glue backing that enabled the 'print' to be attached to the user's own fingers. This technique has been used successfully to 'fool' biometric readers in the past. This calls into question the reliability of biometric readers that use fingerprint identification; we cannot avoid leaving our prints on objects and these could easily be acquired by third parties. It also means that it is possible to frame someone for an offence by taking surreptitious copies of their prints and then transferring these to an object that would link them to a crime. It should be borne in mind that taking someone's fingerprints without their consent (except in the course of a police investigation) is a criminal offence.

Despite these difficulties and unlike most other forms of forensic evidence fingerprinting is seldom presented with a numerical indication of the likelihood of a match between two prints. It is believed that either they match or they do not. Indeed, a 1979 resolution of the International Association for Identification – which in 2007 had 6800 members – stated that an expert giving 'testimony of possible, probable or likely (fingerprint) identification shall be deemed to be engaged in conduct unbecoming'. A further problem is that although those undertaking the final analysis are invariably highly experienced experts they are also subject to the same subconscious influences as anyone else. For example, in a fascinating series of experiments Itiel Dror and co-workers demonstrated how experts are swayed in their interpretation of fingerprints by the context in which they are presented (Dror & Charlton, 2006; Dror et al., 2005, 2006). For example, they would change a decision regarding a set of fingerprints that they had previously pronounced upon in court when presented (unknowingly) with the same prints in a different context.

The extent to which mismatches are made is not known but is probably very small. However, when a mismatch does occur the consequences for the victim are catastrophic. Two case studies are presented here of instances where mistakes were made. This is not to cast doubt on the technique but because they illustrate why care needs to be taken in interpreting forensic evidence and that it is important to be willing to admit that one might have made a mistake.

Case Study: The misidentification of Brandon Mayfield as being involved in the Madrid train bombings

On March 11 2004 several bombs exploded on trains at Madrid, Spain. The bombs resulted in the deaths of over 190 people and many more were seriously injured. Although the Spanish government initially blamed the 'home-grown' Basque terrorist organization ETA it quickly became apparent that it was the work of Muslim extremists. The Spanish police found fingerprint evidence on a plastic bag that had contained detonator caps and, via Interpol Madrid, sent these electronically to the FBI so that a search could be made using their database. Three FBI investigators and an external examiner all agreed that the prints matched those of Brandon Mayfield, an American lawyer, and he was promptly arrested and spent two weeks in prison as a 'material witness'. It didn't help that Mayfield was a convert to Islam, had an Egyptian wife and represented a man (albeit regarding child custody) who pleaded guilty to conspiring to help al Qaida and the Taliban fight the US forces in Afghanistan. Shortly after Mayfield's arrest, the Spanish police contested the identification but the response of the FBI was defensive and a Unit chief was dispatched to Spain to argue the FBI's opinion that there was a '100% positive' match between Mayfield's prints and those found on the bag. The Spanish police remained unconvinced and subsequently arrested an Algerian man whose prints were a much better match. Despite this Mayfield remained in prison and was only released once the story was about to go public. Initially it was claimed that the mistake arose owing to the poor quality of the computer images but it was later admitted that this was not a factor. Instead, the problem arose from the decisions being made in the context of the case, an unwillingness to accept error, and the examiners also knowing one another's findings before looking at the prints themselves (Stacey, 2004).

Case Study: The accusation of perjury against detective pc Shirley McKie

A similar case to that of Brandon Mayfield occurred in Scotland in 1997 when detective pc Shirley McKie was accused of illegally entering the home of a murder victim after a print similar to hers was found on a wooden doorframe above the body. She denied entering the home. There was never any suggestion that McKie was involved in the crime but she was charged with perjury for lying on oath that she had not visited the house. She was subsequently cleared of the charge but her career was destroyed and it took another nine years for her to clear her name. Although experts from the Scottish Criminal Records Office (SCRO) stated that they had identified 45 matches between the prints on the doorframe and those of McKie independent experts were extremely critical of their findings. For example, the independent experts stated that the SCRO were misconstruing contours in the wood for ridge patterns and some points were double marked so as to increase the number of matches. Despite this, the SCRO rigidly held to its original decisions. As in the Mayfield case, problems arose not only through the actual examination process but also an unwillingness of experts to admit to mistakes. In addition, like DNA-based evidence, once an expert has proclaimed a match there is a tendency to assume guilt rather than question the validity of the conclusions – although that is changing.

Lip prints and ear prints

The use of lip prints for identification purposes is occasionally alluded to in forensic literature but there have been no published studies that have verified their individuality, there appears to be no accepted standards or methods for their identification and no databases. It would therefore be an interesting topic for students to investigate. Lip prints are currently most useful as a potential source of DNA from which a much more accurate identification may be obtained. There is considerably more information on the use of ear prints – much of it controversial. Scientists can seldom resist burdening the English language with ugly and unnecessary new words so it should come as no surprise to learn that the study of ear prints is sometimes referred to as 'earology'. Ear prints are typically left at a crime scene when a person listens at a door or window before committing an offence. For example, in May 1996, a frail elderly woman was smothered to death during a burglary in Fartown, Huddersfield. At the subsequent trial, in 1998, Mark Dallagher, known locally as a petty burglar, was charged with the woman's murder. Despite protesting his innocence, claiming that an ankle injury would have made it difficult to commit the crime and supplying an alibi, he was found guilty. This was largely on the testimony of expert witnesses for the prosecution who stated that his ears yielded prints exactly the same as those found on a windowpane at the crime scene. It did not help that Dallagher's alibi said that she was asleep and 'on medication' at the time of the offence. Furthermore, a fellow prisoner claimed that Mark Dallagher had confessed to him while he was being held awaiting trial. However, at a retrial that ended in January 2004, he was exonerated when it transpired that DNA isolated from the ear prints at the crime scene did not belong to him and there were serious concerns about the reliability of the identification of the prints themselves. For example, when he first compared Mark Dallagher's ear prints with those found on the windowpane, one of the expert witnesses had written that the two were 'definitely not' the same (www.forensic-evidence.com/site/ID/DNAdisputesEarlID.html). The prison 'confession' was also dismissed as being unreliable. The use of prison informants is always controversial as prisoners often attempt to gain favours or remission by telling the authorities tales they wish to hear. Alternatively, the informant may be pressured into concocting a story or be simply dishonest or malicious.

The use of ear prints for identification is difficult because when a person presses their ear against a substrate, it will be deformed to an extent that depends upon the pressure exerted and the substrate. Consequently, even prints from the same person will be, to an extent, variable. When gathering prints from willing donors this can be overcome (to some extent) by requiring them to listen for a sound emanating from behind a glass plate. The prints can then be recovered in a similar manner to fingerprints. This approach is therefore similar to the situations in which ear prints would be left at a crime scene.

There are insufficient studies or databases to know whether or not an ear print is truly a unique identifying characteristic of an individual. However, this does not negate the value of the ear print analysis because a numerical indication of the closeness of the match between two sets of prints may be enough to establish whether it is worth continuing to associate an individual with a criminal investigation. Work has begun on computerized systems for analysing ears and ear prints and the EU has financed the melodramatically titled FearID project – Forensic Ear Identification project (Alberink & Ruifrok, 2007). Some police forces already take ear prints as well as fingerprints and these are entered onto a central computer database. If the ear is pierced, then this provides a further identifying characteristic – although it may also result in the formation of only a partial ear print (Abbas & Rutty, 2005; Swift & Rutty, 2003). Ear piercings can also be helpful features to note when identifying a dead body. Even if the ear jewellery has been lost or stolen, scar formation may remain where the ear was pierced. The inexplicable popularity among young persons for piercing other bits of their anatomy with metal rings and studs also provides identifying features should they indulge in criminal acts or their body be found under suspicious circumstances. The possibility of identifying an individual from closed circuit TV recordings (CCTV) by observing the characteristics of their ears has been suggested but this could be thwarted by wearing a hat, hoodie, turban, hijab or a wig of long hair so its effectiveness would be limited.

The retina and iris

Retinal and iris scanning are being used increasingly for personal identification: for example for high security access at airports and in military establishments. Retinal scanning involves mapping the distribution of blood vessels on the retina by shining a low intensity infrared light through the eye and picking up the reflected light on

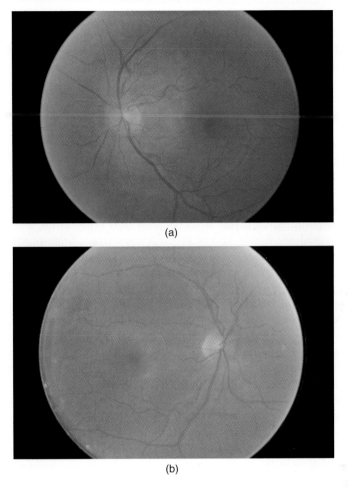

(a)

(b)

Figure 4.5 Retinal surface of (a) left and (b) right eyes illustrating the complex network of blood vessels.

a video camera. The blood vessels in the retina absorb more of the infrared light and therefore stand out from the surrounding tissues (Fig. 4.5). The iris is said to contain over 400 distinguishable characteristics although only a proportion of these are used in current scanning technologies. Apart from overall form and colour, these characters are formed by random events during the tissues' formation and growth. The distribution of these characters is determined using algorithms and it has been estimated that the chances of two persons sharing more than 70% of their iris characteristics is approximately one in 7 billion (Daugman, 2004; Daugman & Downing, 2001). Even identical twins have different iris characteristics. Iris scanning technologies are currently faster than those for retinal scanning and have the added advantage of permitting the user to be up to a metre away from the camera. There appear to be no published studies on how retinal and iris scan characteristics change after death but because the retina tends to detach and fold shortly after death this technique is unlikely to be much help where there is an unidentified dead body. It has been suggested that the degree of retinal detachment and folding could be used as an estimate of the post mortem interval but there are few studies on this and it is affected by many variables, such as temperature and manner of death.

Hair

Our body is normally covered with hair except for the surfaces of the palms, the palmar (lower) surface of the fingers, soles of the feet and the lower surface of the toes. Although hair comes in various lengths and thickness depending upon where it grows, its basic structure remains the same. A hair consists of two portions: a shaft projecting from the skin and a root that is embedded within the skin. The hair shaft is formed from three layers of dead cells that contain large amounts of the protein keratin and are firmly bound together by extracellular proteins. Keratin is a robust molecule that is resistant to decay and there are few animals capable of digesting it. Consequently, hair remains long after the soft body tissues have decomposed. The outermost layer of a hair is called the cuticle, this region is the most heavily keratinized and it has a scale-like appearance (Fig. 4.6). The middle layer contains pigment granules in the case of darkly coloured hair but these are reduced in number or absent in grey or white hair. The inner layer, which may be absent in thin hair, also contains pigment granules. The hair root extends down into the dermis of the skin where it is surrounded by the hair follicle and at the base of this is the hair bulb. Within the hair bulb are living germinal cells that are responsible for the continued growth of an existing hair or its replacement should the hair fall out. A hair with its follicle attached therefore provides a good source of DNA. Unfortunately, hairs with well developed follicles are those that are actively growing (i.e. in the anagen phase) and they are firmly embedded in the body and unlikely to be detached unless pulled with some force. It is only the old hairs that are referred to as telogenic hairs that fall out easily and these have small bulbs that contain relatively little DNA or it is badly degraded. Furthermore, keratin interferes with the PCR process and therefore it is important to separate the cells containing DNA from the rest of the hair. One way of doing this is to use laser microdissection (Chapter 2) to separate away the few useful cells from the rest of the hair (Di Martino *et al.*, 2004).

All the hairs on our body are constantly growing and then being lost and replaced. Loss may be passive through it falling out naturally or through gentle pulling – as when we comb our hair. Clothing, pillows and combs are therefore good places to look for hairs. Hairs may be transferred between people during vigorous bodily contact, such as sex, and the more vigorous the encounter, the more likelihood that transfer will take place. In cases of sexual assault, one may therefore find scalp, facial and pubic hair from the attacker on the body of the victim. In the latter situation, and also during fights, it is also common for both victim and the attacker to pull out one another's hair. In the case of an assault or homicide it is therefore important to examine the body and clothing of the victim and also the locality for hair that is different from that of the victim. Microscopic analysis of scalp hair can provide some identifying features. For example, a transverse section through a hair

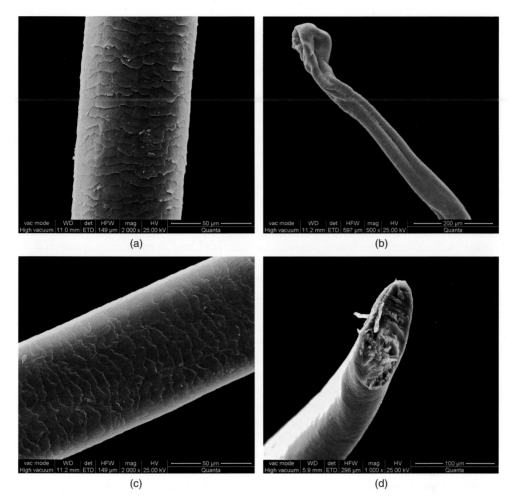

vac mode | WD | det | HFW | mag | HV | ———— 50 µm ————
High vacuum | 11.0 mm | ETD | 149 µm | 2 000 x | 25.00 kV | Quanta

(a)

vac mode | WD | det | HFW | mag | HV | ———— 200 µm ————
High vacuum | 11.2 mm | ETD | 597 µm | 500 x | 25.00 kV | Quanta

(b)

vac mode | WD | det | HFW | mag | HV | ———— 50 µm ————
High vacuum | 11.2 mm | ETD | 149 µm | 2 000 x | 25.00 kV | Quanta

(c)

vac mode | WD | det | HFW | mag | HV | ———— 100 µm ————
High vacuum | 5.9 mm | ETD | 298 µm | 1 000 x | 25.00 kV | Quanta

(d)

Figure 4.6 False colour scanning electron micrographs of human hair: (a) scalp hair showing the scaling pattern, (b) scalp hair pulled out by the root showing the bulb (a good source of DNA), (c) pubic hair showing scaling pattern, (d) cut end of a pubic hair showing its flattened elliptical cross-section.

shaft can indicate its shape: straight hair appears round, curly hair is kidney-shaped, and wavy hair is oval. Its pigmentation can also provide an indication of coloration. However, coloration can be altered through the use of hair dyes whilst a skilled hairdresser can soon alter the length and appearance of person's hair and a criminal could wear a wig during or after committing a crime. The physical appearances of hairs recovered at a crime scene might therefore provide some evidence but DNA extracted from the hair provides a more reliable means of identification.

Ehleringer *et al.* (2008) found that drinking water from around the USA had different oxygen and hydrogen isotopic ratios and these 'signatures' were replicated in samples of hair gathered from barber shops. The technique was not sensitive enough to identify a city but could enable a region to be predicted. This region would presumably be the one in which a person was living whilst their hair was growing. Further work will need to be done to determine how quickly the ratio changes once one moves to a region with a very different isotopic signature.

Hair, especially if it is greasy or contaminated with blood, acts as a trap for small particles. Consequently, it picks up and retains pollen and other palynomorphs (Chapter 9) as well as soil and inorganic particles. Obviously, if a person is struggling on the ground or a body is dragged across a surface hair will pick up even more of these substances and they can act as excellent forensic indicators of place and, sometimes, time of year. This is discussed in detail by Wiltshire (2006a) who also provides two case examples.

Assimilation of drugs and poisons

Drugs such as methadone and poisons such as lead and arsenic are sequestered in hair and can be detected long after the last dose was administered and even after death. By contrast, most drugs and poisons are eliminated rapidly from the body when we are alive and they are lost with the soft tissues during the decay process. Hair analysis is therefore valuable in determining whether a person was taking drugs either voluntarily or unsuspectingly (Musshoff & Madea, 2007). The latter scenario is typical in cases of 'date rape' in which the victim is rendered incapable by their drink being laced with a hypnotic such as Rohypnol ® (flunitrezapam) or zolpidem: these drugs have the added problems of causing memory impairment (Villain *et al.*, 2004). The breakdown product of Rohypnol, 7-aminoflunitrezepam, is easier to detect in hair than flunitrezepam itself because it is more basic and therefore binds better to the melanin in hair and can be detected up to a month after ingesting a single dose of Rohypnol (Negrusz *et al.*, 2001). However, it is worth remembering that a negative result does not mean that Rohypnol was not taken because there are there are big differences between individuals which makes it impossible to determine how much Rohypnol was ingested or when. Similarly, some persons use Rohypnol as a recreational drug and therefore its presence in their hair may not be an indication that they unintentionally drank a spiked drink.

For many drugs there is uncertainty concerning how the results of hair analysis relate to circulating drug levels at the time of death and also the potential for environmental contamination to occur. For example, cannabinoid residues may be found as a result of illicit drug use or exposure from the atmosphere. Consequently, it is important that hair samples be decontaminated before analysis – although there is no consensus as to the best procedure. The presence of specific metabolites in a particular ratio to the parent drug is usually taken as a good indication that the drug was taken into the body. Coloration is also an important factor in the effectiveness with which hairs assimilate chemicals. For example, in experimental rats, white hairs assimilate less methadone than black hairs (Green & Wilson, 1996). Needless to say, hair treatments such as colouring and bleaching and post mortem events such as colonization by fungi, and exposure to sunlight or damp all affect the recovery of chemicals to varying extents. Because hair grows at a fairly constant rate (depending upon the area of the body and our age) the amount of the chemical along the length of a hair can indicate when it was administered.

Case Study: Did arsenic cause the madness of King George III?

The analysis of a few strands of hair taken from the body of King George III has shed new light on the possible reasons for his fits of madness (Cox *et al.*, 2005). King George III died in 1820 and at this time it was a popular practice to keep lockets of hair from famous people. In this case, the hair was placed in an envelope that ultimately found its way to the London Science Museum. When analysed, in 2004, the hair was found to contain arsenic at levels over 300 times the point at which it causes toxic effects. Since the 1970s it has commonly been believed that King George's madness was a consequence of him suffering from the medical condition porphyria. Porphyria is the term used to cover at least seven related blood disorders that arise through the deficiency of enzymes involved in the production of the haem part of haemoglobin. This leads to the accumulation of chemicals called porphyrins and their precursors in the blood stream. Most sufferers do not experience severe illness but in some the symptoms include seizures, anxiety and mental confusion. Effects such as these are much more common in women than in men but as arsenic is capable of triggering acute attacks of porphyria, this could explain why King George suffered so unexpectedly badly. It does, however, beg the question of how he came to ingest such remarkably high quantities of arsenic. The even distribution of arsenic along the length of the hairs indicated that it was not surface contamination or the result of a single poisoning but had been acquired throughout the time the hair was growing. King George's medical records mention the use of skin creams containing arsenic but this would have been insufficient to explain the amounts found in his hair. Similarly, the common practice of wearing wigs that were powdered with dust containing arsenic is also thought to be an unlikely explanation – although neither of these would have helped. The answer probably lies in him being treated with potions such as emetic tartar – antimony potassium tartrate – to alleviate his mental confusion. At the time, doctors believed passionately in the powers of bleeding, purging and vomiting for the treatment of virtually all ailments and, as its name suggests, emetic tartar was a used to induce vomiting. Unfortunately, antimony and arsenic often occur together in minerals and it was then impossible to completely separate the two. Consequently, in an attempt to cure King George of his 'madness', his doctors were actually making his condition worse.

Assimilation of explosives

It is possible to detect traces of explosives in hair, even after it is washed (Oxley *et al.*, 2005). This includes explosives such as TATP (triacetone triperoxide – also known as acetone peroxide and 'Mother of Satan') that was used in the London Tube Train bombings in July 2005. Many explosives have low vapour pressures but even these may become either adsorbed or absorbed, or both, onto hair. However, traces of explosives may also be assimilated onto hair by contamination from hands or sharing a contaminated towel. Consequently, detecting traces of explosives on a person's hair is not necessarily proof that they were involved in terrorist activities.

Bones

The study of skeletal remains associated with events that are likely to lead to criminal proceedings is called forensic anthropology or forensic osteology whilst the study of the location and recovery of those remains is called forensic archaeology. Needless to say, there is a lot of overlap between the two and both build upon the considerable body of knowledge built up over the years in traditional anthropological and archaeological research. Only a brief overview can be provided here and those wishing for more detail are advised to consult specialist textbooks such as Klepinger (2006).

The identification of a complete human skull or the larger bones is easy but distinguishing the small bones and sesamoid bones from those of other animals requires more skill. This can be especially difficult where the bones originated from a young child as many of them exhibit different morphological features to those of an adult. Identification can also be problematic where skeletons are disarticulated and only partial remains are found. Consequently, it is not unusual for the police to be alerted to the presence of bones on a beach or moor that subsequently prove to be those of a sheep or pig. Similarly, a person digging their garden or out walking in the countryside may dismiss human bones as those of a dead pet or wild animal and therefore of no consequence. Even oddly shaped pieces of wood and stones are occasionally confused with bones (Fig. 4.7). If the provenance of a bone is uncertain it may be possible to extract DNA to confirm whether or not it is of human origin

Figure 4.7   It can be easy to mistake both natural and man-made objects for bones. This piece of wood bears a superficial resemblance to a large femur.

but failing that histological analysis of the bone microstructure may be required (Hillier & Bell, 2007). Anatomical measurements of bones can provide an indication not only of whether or not they are of human origin but also of gender, ethnicity, and age at the time of death. The bones may also provide clues to the time since death, the presence of underlying diseases, the cause of death, past movements, and, if DNA is extracted, identity (Byers & Myster, 2005).

The human skeleton

The skeleton of an adult human comprises 206 named bones but there are even more in a child. The reason for this discrepancy is that some bones fuse together as we get older. The skeleton of an adult can be divided into two compartments: the axial skeleton and the appendicular skeleton. The axial skeleton comprises the skull, hyoid bone, and the bones of the ears, the vertebral column and the thorax. The appendicular skeleton comprises the bones of the shoulder blades, the upper and lower limbs and of the hip girdle. Most of the bones are paired on the left and right hand sides of the body. The bones can be classified into five types based upon their shape: long, short, flat, irregular and sesamoid. The long bones are those such as the femur in the thigh. These bones are longer than they are broad and are usually slightly curved because this shape allows weight to be more evenly distributed along their length. Short bones are approximately as broad as they are long and typical examples are the majority of the carpal bones found in the wrist. As their name suggests, the flat bones are flat and 'plate-like'. They consist of two almost parallel layers of compact bone and examples include the bones of the skull, the breastbone and the ribs. These bones provide protection for the underlying soft tissues and an extensive area for the attachment of muscles. The irregular bones are those of the vertebral column and some of the facial bones. The sesamoid bones gain their name from their shape resembling that of a sesame seed. They are found within those tendons that bear heavy stresses and strains and they protect them from excessive wear. Most sesamoid bones are only a few millimetres in size, such as those in the palms of the hand, and their number can vary between individuals. An exception to this generalization is the two patellae or kneecaps. In addition, small sutural bones are found between the sutures of some of the cranial bones although their number varies considerably between individuals and they may be absent entirely.

The ability to identify individual bones only comes with a great deal of practical experience and is not aided by most of the anatomical features having long complicated names. However, it essential if a skeleton is disarticulated and/or the remains are commingled (mixed together – as in a mass grave). For example, the presence of three tibiae would indicate the presence of at least two people. If there was one right tibia and two left tibia then there could be two or three people. If both the left tibiae were noticeably shorter than the right tibia then it is still possible that there might be only two people since one of the victims might have had a deformed leg – closer analysis of wear patterns would confirm or refute this.

Male   Female

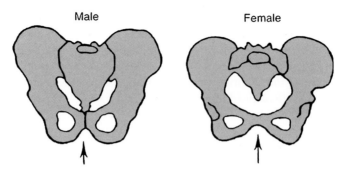

Figure 4.8 Diagrammatic representation of the adult male and female pelvis. There are several morphological differences but one of the most obvious is that the subpubic arch (arrowed) is noticeably narrower ('V-shaped') in men than in women ('U-shaped'). (Reproduced from Saukko, P. and Knight, B. (2004) *Knight's Forensic Pathology*, 3rd edn, copyright 2004, Hodder Arnold.)

Determination of gender

Gender determination is seldom a problem whilst a body is in the early stages of decay because the genitalia and secondary sexual characteristics (e.g. facial hair) are present. Individuals who have undergone a sex change operation or were born hermaphrodites can present more difficulties but a DNA test (Chapter 3) can usually solve the matter. If only skeletal evidence is available, it is still possible to determine the gender with a high degree of accuracy in adults but it is much more difficult in children (Duric *et al.*, 2005). The most helpful indicator is the pelvis, which exhibits numerous differences between men and women (Fig. 4.8). For example, in a woman the pelvis is smoother, lighter and more spacious than in a man. Similarly, the area of the pelvis called the sub pubic arch has an inverted 'U shape' in women but an inverted 'V shape' in men and there are several other differences. There are also gender differences in the dimensions of the bones of the skull (especially the mandibles), the long bones and the sternum. Some of these bones cannot by themselves provide an accurate determination of gender but when used in combination with two or more other bones (and if there is a known marked sexual dimorphism within the population) the accuracy can be increased to more than 95%.

Determination of race

The terms 'race' and 'racial characteristics' often generate emotionally-charged controversy (Bamshad *et al.*, 2004) and some workers prefer to use the term 'ethnicity' because it includes variables such as language, religion, and social factors. There is no such thing as a 'pure human race' and many anthropologists, such as Brace (1995), consider that it would be more helpful to consider human variation in terms of clines of traits rather than discrete populations. Despite this 'race' is commonly referred to in forensic and medical literature and it can be a useful concept in the identification process provided that its limitations are recognized. For example, within the forensic literature the terms 'white' and 'black' usually relate to differences between Americans of European and African descent respectively. However, beyond this they have no meaning whatsoever since they are context dependent: nobody is pure white or pure black and a person who would be considered white in one country/community might be considered black in another and vice versa. There is no consensus on how many racial categories there are but the terms 'Caucasian', 'Negro' and 'Mongoloid' are in common usage. (There is a similar lack of agreement in assigning ethnic criteria.) The term 'Caucasian' was originally coined in 1775 by the German scholar Johan Blumenbach. He recognized five distinct human races: Caucasian, Mongolian, Ethiopian, Native (North and South) American, and Malay. He chose the term Caucasian because he thought that the Caucasus region produced 'the most beautiful race of men'. He also thought that humans originated in the Caucasus region and all subsequent racial types were derived from them. Although most scientists believe that humans originated in Africa, Blumenbach's thoughts on human origins are not entirely fanciful as shown by the discovery in 2007 of early hominin remains provisionally dated at 1.7–1.8 million years old near the town of Dminisi in Georgia.

Although it is possible to assign a skull to one of three broad groupings of Caucasian, Negro and Mongoloid, further discrimination is seldom possible (Fig. 4.9). Among the morphological differences are the shape of the mandibles and the size of the nasal openings – the latter tends to be narrower and higher in Caucasians than Negroes. Krogman & Iscan (1986) have disputed the value of mandible meas-

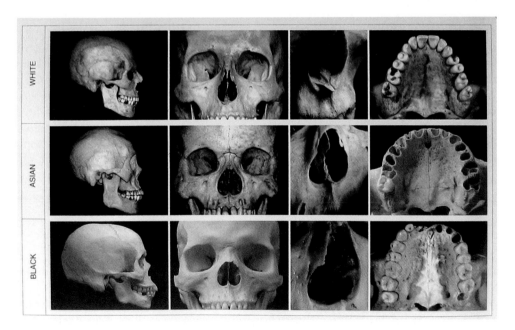

Figure 4.9 The skulls of adult humans can usually be ascribed to one of three racial categories: Caucasian (or 'white'), Negro (or 'black'), or Mongoloid (or 'Asian'). See text for details of discriminating features. (Reproduced from Dolinak, D. *et al.*, (2005) *Forensic Pathology: Theory and Practice*. Copyright © 2005, Elsevier Academic Press.)

urements for racial discrimination but Buck and Vidarsdottir (2004) used a computer-based method of geometric morphometric analysis of mandibles to identify unknown sub-adult individuals with an accuracy of over 70%. This is noteworthy because up until adulthood many anatomical characteristics change in proportion to one another because growth does not occur uniformly. Consequently, most of the work on race has been done on adult skeletons. However, it should also be noted that an accuracy of 70% suggests that three out of ten skeletons would be incorrectly identified. Furthermore, much of the work on morphological distinctions between Caucasian/Negroid individuals is based on the US studies and therefore relate to the major populations found in that country and are not necessarily applicable elsewhere.

Determination of stature

Stature refers to a person's natural height when standing in an upright position and, clearly, it is an important consideration in the identification process. Although it would appear to be a simple measurement to make it presents difficulties even whilst a person is alive – once they are dead, things only get worse (Bidmos, 2005). Stature varies naturally throughout the day owing to differential loading on the spinal vertebrae and errors are commonly made when making the measurements. If the only records available are those made by the individual concerned, they should be treated with some suspicion. Men, especially, tend to overestimate their height. Furthermore, as we enter old age, our bones tend to shrink and our posture changes. Consequently, during adulthood, stature is not a constant but changes gradually with time. Once a body is skeletonized, determining stature is not simply a matter of laying the bones out on a bench and measuring the length of those that contribute to height. This would not be the way the bones would be arranged in relation to one another during life and the cartilage at the joints would be missing. A further problem that can arise is that some of the bones might be missing. Nevertheless, provided one has at least one of the long bones – such as the femur, tibia, or humerus – it is possible to estimate stature with reasonable accuracy. The length of the long bones is proportional to height so by reference to a table or regression equation one can arrive at an estimated height (Trotter, 1970). Obviously, the accuracy of this approach increases with the number of long bones available for measurement and allowances have to be made for gender and population and it is important that any estimates are accompanied by an indication of the 95% confidence intervals.

Determination of age

If all the soft body parts are decayed, one could use a combination of both dental and skeletal characteristics to estimate how old a person was when they died. These characteristics are not especially accurate and become less so as a person ages. In terms of skeletal development, remains can usually be categorized into one of eight groups: perinatal (foetal), neonatal, infant, young childhood, late childhood, adolescence, young adult, and older adult. The term 'perinatal' refers to unborn babies and their age can be determined with some accuracy from the measurement of the cervical, thoracic and lumbar vertebrae (Kosa & Castellana, 2005). The development of an unborn child proceeds at a regular and predictable rate because the foetus is protected from the outside environment and if food is lacking, its growth will be at the expense of the mother. Neonates are babies that have been born but whose teeth are not yet emerged. At this time, the baby has very small bones and many of these, such as the pelvis and those of the skull, are not yet fused together. However, there is a lot of individual variation in the speed with which these events take place. The teeth begin to emerge during infancy and these provide a fairly accurate indication of age in children. In addition, the bones start to ossify (harden) although, again, there is a great deal of variation between individuals in when and how rapidly this process develops. In late childhood, more of the bones begin to ossify and the permanent teeth make their appearance.

During adolescence, the long bones grow rapidly in length. This is brought about by the activity of the chondrocyte cells in the regions of the bones called the epiphyseal plates. The chondrocytes multiply and form a layer of cartilage that causes the epiphyseal plate to become wider and hence the bone elongates. As the cartilage forms, the chondrocytes on either side of it die off and their place is taken by another cell type, the osteoblasts. The osteoblasts convert the cartilage into bone and so the bone shaft grows. By late adolescence, the cartilage of the epiphyseal plates becomes completely replaced with bone tissue in a process known as epiphyseal plate closure. Once closure is complete further lengthening of the bone is not possible although changes may still take place in circumference. The number of epiphyseal plates differs between bones and the timing of their closure varies both within and between the different types of bones (Fig. 4.10). It is therefore possible to estimate age with reasonable accuracy on the basis of the extent of epiphyseal plate closure within the skeleton. For example, in the clavicles (collarbones) of men epiphyseal closure at the acromial end of the bone (i.e. that next to the scapula) occurs at 19–20 years of age. However, at the opposite sternal end (i.e. that next to the breastbone) epiphyseal closure does not occur until 25–28 years of age (Krogman & Iscan, 1986). The levels of the sex hormones heavily influence the timing of plate closure during puberty. Consequently, gender has to be considered when attempting to estimate age from skeletal characteristics – closure usually occurs at a younger age in women than in men owing to their generally earlier maturation.

After adolescence, age determination from skeletal remains becomes more problematic. One method is to observe the degree of closure of the cranial sutures in the skull – as the plates fuse together with time, the sutures become less distinct. Many forensic texts mention the fusion of the basilar suture (sphenoid–occipital synchondrosis) within the skull as an indicator of early adulthood (Fig. 4.11). Fusion is said to be completed between the ages of 18 and 25 but its position makes it extremely difficult to observe in both the living and the dead and some workers believe that fusion may well occur at an earlier age in many individuals. An alternative approach is to observe the shape of the rib bones, and the degree of pitting of the cartilage that connects the ribs to the breastbone (Kunos et al., 1999; Iscan et al., 1984). To begin with, the ends of the ribs are flat and the cartilage is smooth but with increasing age the rib ends become ragged and the cartilage becomes pitted. The reliability of this approach has been questioned (Schmitt & Murail, 2004) and the rib shafts

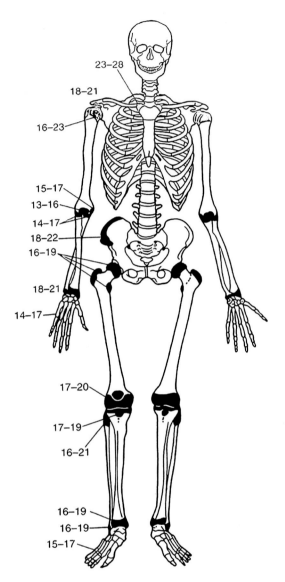

Figure 4.10 Diagrammatic representation of the human skeleton to show the ages at which epiphyseal union occurs in males. Only those sites of major forensic relevance are shown and the ages for females would be in many cases be slightly different. (Reproduced from Saukko, P. and Knight, B. (2004) *Knight's Forensic Pathology*, 3rd edn, copyright 2004, Hodder Arnold.)

themselves are fragile and along with the cartilage they decay rapidly if the corpse is left in an exposed position or an acid soil. Other methods include assessing changes in the pubic symphysis and the auricular (i.e. 'ear-shaped') surface of the ileum (sacro-iliac joint). Unfortunately, both of these methods have their limitations. The pubic symphysis is delicate and therefore liable to damage and decay and it is less reliable as an indicator of age in women than men (Klepinger, 2006). The

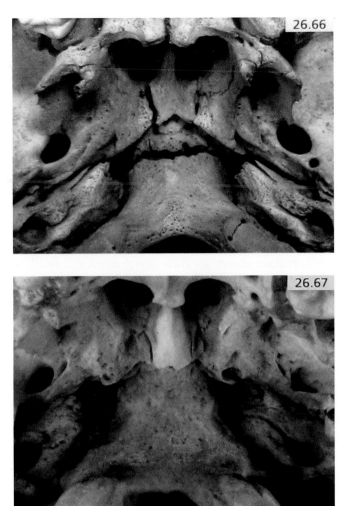

Figure 4.11 Fusion of the basilar suture (sphenoid-occipital synchondrosis) within the skull can provide an indication of age. (a) Before adulthood, the sphenoid and occipital bones are clearly separated by a cartilaginous plate. (b) In adults the sphenoid and occipital bones are fused. (Reproduced from Dolinak, D. *et al.*, (2005) *Forensic Pathology: Theory and Practice*. Copyright © 2005, Elsevier Academic Press.)

auricular surface of the ileum is a robust structure but it does not provide an especially accurate means of estimating age and above the age of about 50 progressive joint stiffening and fusion (ankylosis) further reduces its usefulness. Age-related degenerative conditions, such as arthritis or osteoporosis can provide a crude estimate of age but their onset and progression vary markedly between individuals and even children can suffer from arthritis (juvenile idiopathic arthritis is different from adult arthritis but still results in damage to the joints).

Facial reconstruction

Facial reconstruction from skull features is normally done manually by the application of modelling clay to a cast of the skull although computer-aided techniques are becoming more sophisticated and have the added advantage of being quicker, cheaper and the final images can be transmitted more easily between interested parties (De Greef & Willems, 2005; Wilkinson, 2004; Vanezis *et al.*, 2000). There are two basic approaches to the reconstruction process: the anatomical method and the tissue depth method (Kahler *et al.*, 2003). The former method requires a great deal of anatomical knowledge because the skin and all the underlying tissue layers must be built up gradually layer by layer starting from the skull surface. Consequently, these days, the anatomical method is usually restricted to the reconstruction of fossil humans and hominids for which there are no statistical data on facial features. The tissue depth method uses statistical data banks of average tissue depths between the bone and skin surface at various marker points on the skull. The tissue depths between the marker points are then interpolated and finally a 'skin surface' is applied. Obviously, the tissue depths at the marker points are affected, amongst other things, by build, gender, age and population characteristics but extensive databanks are now available for several population groups. Unfortunately, it is not possible to predict the shape of the mouth, the nose or the eyes and unless hair has been preserved there will be no indication of hair colour, or length. As mentioned previously, it can also be difficult to predict racial origin, and hence skin colour, from skeletal remains. Consequently, two reconstruction experts working independently from the same skeletal remains may produce rather different models. Indeed, owing to the limitations of existing techniques, it has been suggested by Stephan & Henneberg (2001) that facial reconstructions are limited to situations in which all other methods of identification have failed.

Surgical implants

Surgical implants are being used with increasing frequency and include a wide variety of objects ranging from the silicon bags of breast implants, to battery operated heart stimulators and titanium screws holding bones together. They can all potentially provide useful identification evidence (Fig. 4.12) although this very much depends upon the availability of past medical records and whether or not the implant carries an identification code (Simpson *et al.*, 2007). Bennett & Benedix (1999) provide a case study in which the presence of an osteostimulator helped in the identification process. The battery powered device is surgically implanted into bone to stimulate localized bone growth. In the case they describe, an osteostimulator was found in the lower back region of a dead woman. She had been shot in the head, placed in the boot of a car and the car set alight. The fire generated a high temperature over a prolonged time and the woman's body was so badly affected that skeletal and dental identification techniques provided only limited information. However, the police suspected who the dead person was and her medical records provided X-rays demonstrating the presence of an osteostimulator in the same region as that found in the dead body. Although the device lacked a serial number

Figure 4.12 The identification of the badly decomposed body of this woman was aided by evidence of past hip replacement surgery and the medical implant carried a unique identification code. (Reproduced from Dolinak, D. *et al.*, (2005) *Forensic Pathology: Theory and Practice*. Copyright © 2005, Elsevier Academic Press.)

it was felt that its make and position were so similar to those indicated by the medical records that a positive identification could be made.

Teeth

The study of teeth and the factors that affect their formation is known as odontology and a forensic odontologist is therefore a person who specializes in the study of teeth associated with dead (usually) bodies. The forensic importance of teeth arises from them containing a great deal of personal information coupled with being the most indestructible part of the body. Adults normally have 32 teeth: on each side of the mouth there are four incisors, two canines, two premolars and three molars. Each tooth has a characteristic morphology but they all consist of three regions: the portion above the gum line is called the crown, at the gum line is a constricted region called the neck, whilst the portion embedded beneath this within a socket in the jaw is called the root. The canines, incisors, and first lower premolars each have a single root whilst the first upper premolars tend to have two roots. The first two molar teeth tend to have two roots whilst the corresponding upper molars tend to have three roots. The number of roots attached to the third molar teeth can vary although most have a single fused root. Teeth owe their indestructibility through being composed largely of dentine – this is a highly calcified connective tissue that is responsible for giving the teeth their shape and rigidity. Calcium salts make up approximately 70% of the dry weight of dentine thereby making it harder than bone. In the crown region, the dentine is covered with a layer of enamel: calcium salts make up about 95% of the dry weight of this region so it is even harder than dentine. Within the root region, the dentine is covered with a thin bone-like layer called the cementum which attaches the tooth to the underlying bone via dense fibrous connective tissue called the periodontal ligament. At the centre of a tooth lies the pulp cavity that contains connective tissue, blood vessels and nerve endings.

Determination of age

We have two sets of teeth, or dentitions, during our life: the deciduous teeth (also known as the 'milk teeth', 'primary teeth' and 'baby teeth') and the permanent teeth. The deciduous teeth begin to emerge from the gums about 6 months after birth. Thereafter, further teeth emerge at set intervals until all 20 deciduous teeth are present by the time we are about 2 years old. The deciduous teeth start to be lost when we reach about 6 years of age and they have usually all gone by the age of 12. They are replaced by the permanent teeth – which include some additional molars. The wisdom teeth usually do not emerge, if they emerge at all, until after the age of 17. It is therefore possible to age a child with a reasonable degree of accuracy on the basis of the teeth that are present and their stage of development (Table 4.1) but with adults other features must be examined.

Our teeth start to form before we are born and mineralization usually commences before 16 weeks' development. Birth is a physiologically traumatic event for the baby and one of its consequences is to upset the production of dental enamel. The enamel is laid down as a series of lines called the striae of Retzius and at birth a 'neonatal line' is formed that is darker and bigger than the surrounding striae – this can be seen when a tooth is sectioned (Skinner & Dupras, 1993). The presence of a neonatal line within the first deciduous teeth or at the tips of first permanent

Table 4.1 Typical emergence dates of different types of teeth in normal healthy individuals

	Tooth	Maxillary (upper jaw)	Mandibular (lower jaw)
Deciduous teeth	Central incisor	7.5–12 months	6–8 months
	Lateral incisor	12–24 months	12–15 months
	Canine	16–24 months	16–24 months
	First molar	12–16 months	12–16 months
	Second molar	24–32 months	24–32 months
Permanent teeth	Central incisor	7–8 years	7–8 years
	Lateral incisor	8–9 years	7–8 years
	Canine	11–12 years	9–10 years
	First premolar	9–10 years	9–10 years
	Second premolar	10–12 years	11–12 years
	First molar	6–7 years	6–7 years
	Second molar	12–13 years	11–13 years
	Third molar (wisdom)	17–21 years	17–21 years

molars therefore indicates that a child survived birth and lived for at least a short time afterwards. This is therefore an important observation when the body of a baby is discovered.

Although it is possible to use dental characteristics to age children with a reasonable degree of accuracy, it is much more difficult for adults. Lamendin *et al.* (1992) derived a simple equation to estimate age based on measurements of periodontosis height and root dentine translucency ('transparency' in some texts). Periodontosis (also called periodontal recession and periodontitis in some texts though the latter is strictly an inflammatory condition) refers to the gum shrinkage at the base of our teeth that afflicts all of us as we get older and which is often accompanied by bacterial infection that causes inflammation. This results in the underlying tooth surface being exposed and staining and pitting of the tooth surface. The extent of this exposure is referred to as the periodontosis height. Root dentine translucency does not usually begin until we are in our 20s or 30s and is brought about by the deposition of mineral substances in the dentinal tubules. It starts in the root apex and then extends towards the crown and is best seen by placing the tooth on a glass-topped light box lit by an intense white light source – some workers use special ones called negatoscopes, that are normally used for viewing X ray film. Lamendin's equation was subsequently refined by Prince & Ubelaker (2002) who included a measure of root height and derived separate equations for men and women and for 'white' and 'black' Americans. Their equations are listed below:

White men: Age = 0.15(RH) + 0.29(P) + 0.39(T) + 23.17
White women: Age = 1.10(RH) + 0.31(P) + 0.39(T) + 11.82
Black men: Age = 1.04(RH) + 0.31(P) + 0.47(T) + 1.70
Black women: Age = 1.63(RH) + 0.48(P) + 0.48(T) − 8.41

Equations derived by Prince & Ubelaker (2002) for the determination of age from dental characteristics. RH = root height, P = (periodontosis height ÷ RH) × 100, T = (root dentine translucency height ÷ RH) × 100. Measurements are in millimetres.

Other workers have derived variations on these equations for other population groups. The technique works best for persons from about 30–40 up to about 70–80 years. Clearly, the effectiveness of this approach depends upon knowing the gender and population of the person from whom the teeth came and also that they were not affected by dental problems or medical conditions that affect the growth and health of teeth during life. In addition, different people examining the same tooth frequently arrive at markedly different age estimations indicating the degree of subjectivity associated with the measurement of the various tooth characteristics. However, the technique is cheap, simple, quick, and does not cause any alteration or destruction of the evidence.

Another approach to estimating age from dental characteristics is to count the number of bands found in the cementum: the cementum is initially produced just before a tooth emerges from the gum and is then laid down incrementally throughout life with growth occurring during the summer followed by a resting period over the winter. When a sectioned tooth is observed with polarized light a series of translucent and opaque bands are seen that represent the growth and resting phases respectively. A translucent and opaque band together are therefore equivalent to one year of life. Because teeth emerge sequentially at known ages, by counting the number of bands in individual teeth it is possible to estimate a person's age. Furthermore, by observing the degree of development of the last band to be laid down it is possible to estimate the approximate time of year at which death occurred (Wedel, 2007). Some authors claim that an individual's age can be estimated using this approach to within about 2–3 years whilst others state that there could be an error of 10 or more years. The technique is compromised by periodontal disease (i.e. disease affecting the gums and structures surrounding the teeth) and it is uncertain whether there are major differences in the ways that bands are laid down between populations.

A chemical approach to determining age from dental characteristics is to compare the ratio of left (L-) and right (D-) isomers of the amino acid aspartic acid within the root dentine (Ohtani *et al.*, 2003). We, like all living organisms, utilize only L-amino acids and when dentine is first formed it contains only L-aspartic acid. However, once the dentine is formed, the aspartic acid within it undergoes a process called racemization in which the molecules rotate until there is a 50:50 mix of L- and D-aspartic acid. Because the rate of racemization is known, by measuring the ratio of L- and D-aspartic acid it is possible to estimate a person's age to within about a year. The method is potentially compromised if the body is subject to high temperatures such as by being burnt or if it is not discovered until many years after death but is otherwise considered to be very reliable. Racemization of aspartic acid also occurs in the bones but it does not provide as accurate an estimate as in the dentine and there is a much lower correlation between the racemization ratio and age in women than there is in men (Ohtani *et al.*, 2007).

Spalding *et al.* (2005) have described a novel method of age determination based on the bomb curve radiocarbon-dating technique (Chapter 1). Their technique utilizes the absence of carbon turnover within the enamel once it is formed. This means that the ^{14}C level within the enamel reflects the level present in the atmosphere when the teeth were being formed. The atmospheric levels of ^{14}C are well documented: there was a rapid rise in the atmospheric levels ^{14}C when above ground nuclear testing began followed by an exponential decline once the tests were banned. The age at which the various teeth develop is also well known; therefore the ^{14}C levels indicate the year in which the enamel was produced whilst the tooth type indicates the age of the person when this occurred. For example, the wisdom teeth are the last ones to be formed, their enamel being laid down when a person is about 12 years old. If the ^{14}C levels of a person's wisdom teeth correlated with the atmospheric levels in 1974, then the person was probably born in 1962. Analysing ^{14}C levels in a variety of teeth improves the accuracy of the technique. This method still needs to be verified and the effect of factors such as diet assessed, but it offers a good prospects for determining the age of persons who were children, or born after, the start of nuclear testing.

Determination of gender and racial origin

It is difficult to determine a person's gender from the physical appearance of the teeth and in any case this is normally achieved from other morphological evidence.

However, if only the teeth are available, it is possible to extract DNA from the pulp and test for Y chromosome specific repeat sequences (Yamamoto *et al.*, 2000). Although there is some variation in tooth morphology between ethnic groups, it is very difficult to come to any firm conclusions based solely on these characteristics (Edgar, 2005). For example, persons of Asian and Native American ancestry tend to have shovel-shaped incisors – these are ones in which there is an indentation on the lingual side (i.e. facing the tongue) (Fig. 4.13a). However, this trait is not invariably present in persons of Asian descent and may also be found in other racial groups. Europeans often demonstrate specific features on their deciduous posterior premolars and permanent molars that are collectively referred to as Carabelli's trait (Fig. 4.13b). This feature is also highly variable between populations: Correia & Pina (2000) recorded Carabelli's trait in 85% of White North Americans but only

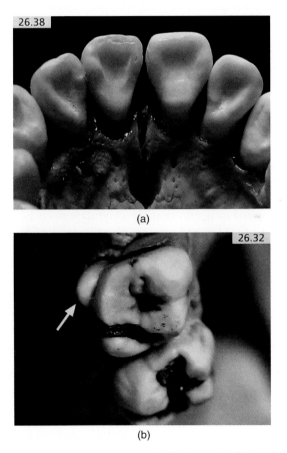

(a)

(b)

Figure 4.13 Dental characteristics can provide an indication of racial origins although they need to be treated with caution. (a) Persons of Asian and Native American ancestry tend to have indentations on the lingual side of their incisors (i.e. facing the tongue). (b) Persons of European descent often have a cusp on the lingual side of the permanent molars that is referred to as Carabelli's trait. (Reproduced from Dolinak, D. *et al.*, (2005) *Forensic Pathology: Theory and Practice.* Copyright © 2005, Elsevier Academic Press.)

13.5% of Portuguese. The nature of any dental work can, however, provide an indication of a person's origins. Typically, the materials used to prepare dentures and bridges and the quality of the workmanship can indicate when and where the work was done.

Identification of individuals based on dental characteristics

By noting the presence or absence of teeth, their appearance and the nature of any past dental work it is possible to build up a person's dental profile. This procedure is known as 'dental charting' and until the advent of DNA technology it provided one of the most accurate means for confirming a person's identity (Adams, 2003; Pretty & Sweet, 2001). It relies on comparing the chart of, for example, an unidentified dead body with those held on dental or medical records. Unfortunately, within the UK, there is no legal requirement for dental practitioners to retain dental records for a specific period. Further problems can arise if there were errors in dental charting pre mortem or post mortem, if the dead person had not visited a dentist for a long time, was a recent immigrant (and therefore had no records in this country), or the body was destroyed in some way such that only a partial collection of teeth was left for study (Whittaker, 1995). Consequently, even if records are available, there may not be an exact match between the dental chart of the dead body and that of the most likely candidate. The situation is alleviated if there are pre mortem radiographic records that can be compared with ones taken from the dead body. If the radiographs match then it is possible to confirm identity with a high degree of certainty. However, in the absence of radiographs, at this point, the examiner has to make a value judgement to decide if there is a rational explanation for any discrepancies. Clearly, the presence of a permanent tooth that the pre mortem chart stated had been extracted indicates that the candidate must be excluded. However, the absence of a tooth that the pre mortem chart states should be present could be explained by it being lost in an accident, a fight or extracted by another dentist subsequent to the records being made – it is now common practice for persons in the UK to go abroad for dental treatments. Although it is obviously preferable to have numerous points of concordance between pre-mortem and post mortem dental charts, there is no agreed minimum number required to confirm a positive identification. A computer programme called OdontoSearch (www.cilhi.army.mil) can facilitate chart comparisons – however, unlike IDENT1 or IAFIS, it is not a database from which a specific individual can be identified.

Despite the potential problems, dental charting remains a valuable method for identification. It is especially important when there are large-scale disasters such as plane crashes or explosions that produce numerous badly damaged unidentified dead bodies. For example, dental examinations were a central feature of the identification of the victims of the 'Bali bombing'. Bali is an island belonging to Indonesia that is a popular tourist destination. At about midnight on 12 October 2002, a huge bomb was exploded outside the Sari Club in the town of Kuta. The radical Islamic group Jemaah Islamiah with the intention of killing Americans planted the bombs. However, far more Australians and Indonesians were killed than Americans. The bomb was poorly manufactured but it still caused enormous devastation and a fierce fire followed. A total of 202 people are believed to have died and many of the bodies were torn apart by the explosion or cremated in the fire that followed. This made visual identification from appearance, clothes or possessions impossible in some cases. Many of the Australian victims were less than 30 years old and they had been brought up with fluorinated drinking water and a good dental service. Consequently, they often had little evidence of dental work that could be used to aid the identification process. However, the presence of dental braces and crooked, chipped or spaced teeth could still provide a characteristic pattern that could be compared to charts and photographs in which the victim was exposing their teeth whilst smiling.

Drug abuse and dental characteristics

Poor dental health, especially in young adults may be a consequence of underlying disease or neglect but can also arise through drug abuse. For example, methamphetamine abuse can give rise to a condition colloquially known as 'meth mouth' in which the person suffers from numerous dental caries, loss of tooth enamel and the deposition of calculus. Calculus, or tartar, is plaque (a sticky film of bacteria) that becomes hard through calcium salt deposition and is firmly fixed to the tooth – usually where the tooth meets the gum. Meth mouth results from a combination of the acidity of methamphetamine and the fact that the user feels their mouth to be dry and this is accompanied with a craving for sweet carbonated drinks thereby leading to the consumption of several litres of these every day. In addition, drug abusers usually neglect their dental hygiene and drug-induced teeth grinding and clenching results in excessive wear. However, tooth grinding is also a common nervous symptom and should not be taken as evidence of drug use. Solvent abusers often develop a characteristic 'glue sniffer's rash' around their mouth and nose and the tips of their nose may exhibit signs of frostbite as a consequence of inhaling aerosols. Drug users and drunks often lose and damage teeth as a consequence of falling or getting into fights whilst 'under the influence'. However, it is also a feature of persons indulging in physically dangerous contact sports such as rugby and boxing. Dental characteristics should therefore be evaluated in the context of other forensic evidence.

Dentures

Although dental health has improved considerably in most industrialized countries in recent years many people wear full or partial dentures. These can provide identification evidence, especially if the dentures carry a name or identification feature and/or there are good dental records of when and where they were manufactured. However, this is not the case for most dentures and even if there are such features it should not be assumed that the dentures belong to the person wearing them. For example, dementia sufferers living in care homes frequently misplace their dentures and these are acquired by others who use them as their own. The effectiveness of

Table 4.2 Summary of forensic evidence that can be obtained from human tissues

Tissue	Forensic evidence	Test
Skin	Identification	Fingerprints
		DNA profiling
		Tattoos
		Physical characteristics
		Scars
	Cause of death/sequence of events during an assault	Wound analysis
	Contact with chemicals (e.g. explosives)	Specific tests
	Drug abuse	Scars at injection sites
Hair	Identification	DNA profiling
		Physical characteristics
	Contact with chemicals (e.g. explosives, drug use)	Specific tests
	Geographical	Pollen
		Mineralogical
Nails	Identification of assailant	DNA profiling of material trapped under nails
	Geographical	Pollen
		Mineralogical
Bones	Identification	DNA profiling
		Physical characteristics
	Time since death	Isotope analysis
		Physical characteristics
	Geographical origin	Isotope analysis
	Cause of death/sequence of events during an assault	Wound analysis
		Diatoms
Teeth	Identification	DNA profiling
		Dental charting
		Physical characteristics
		Racemization
		Isotope analysis
	Geographical origin	Isotope analysis
Eyes	Identification	Retinal and iris scanning (live persons only)
		Iris colour (fades after death)
	Time since death	Specific chemical tests
Internal organs	Cause of death/sequence of events during an assault	Wound analysis
		Evidence of underlying disease

the fit is not an indication of ownership since dentures are often loosely fitting and may not have been changed for many years.

Future developments

Although fingerprints are generally considered to be highly reliable as a means of identification there have been several widely publicized incidents in which their use has been less than perfect. This has caused some concern and this is not allayed by the lack of statistical analysis that is a common feature of many other forensic indicators. It would therefore be helpful if some form of numerical analysis could be developed that would facilitate comparisons between prints. Admittedly, this has been attempted in the past and with little success. In addition, because it is now possible to lift someone's fingerprints and place them on an object they never handled there is a need to be able to distinguish between 'real' and 'planted' prints. This is because defendants will inevitably start to claim that their prints were 'planted' and in some cases this will quite probably be true.

Facial reconstruction is usually restricted to high profile investigations but as the costs of computing power and information storage decrease it is likely to be used in more routine investigations. Although it is likely to remain difficult to predict the shapes of the nose, mouth, or eyes computers do offer a way of quickly generating 'variants' on a theme (i.e. the basic facial features with different shapes of nose). This may be aided by developments in DNA analysis that enable one to predict morphological features (Chapter 3).

The analysis of amino acid racemization offers an intriguing means of improving age determination, especially for those over the age of about 20–25 when 'traditional' methods become increasingly unreliable. However, there remain relatively few publications on the subject and more data is needed on how post mortem environmental and taphonomic processes affect racemization.

Quick quiz

(1) Distinguish between plastic, visible and latent fingerprints.

(2) Explain why you would not use silver nitrate to detect fingerprints from windows of a house overlooking the sea.

(3) Briefly describe how tattoos can provide clues to a person's identity and the limitations of this form of evidence.

(4) Provide brief notes on three ways in which hair can provide forensic evidence.

(5) What are problems associated with measuring stature in both the living and the dead?

(6) How can one distinguish between the skeleton of a male and a female?

(7) Where would one find the 'line of Retzius' and what does its presence tell you?

(8) Explain why it is easier to distinguish the difference in age between children of 4 and 10 than between adults of 30 and 50 on the basis of their teeth.

(9) Give two reasons why the dental records held by dentists may not be an accurate reflection of an individual's dentition.

(10) What is 'meth-mouth'?

Project work

Title

The influence of age, gender and environment on the longevity of fingerprints.

Rationale

It is stated in the literature that the fingerprints of children do not preserve as well as those of adults. It would be interesting to determine whether particular features of fingerprints are more susceptible than others and how environmental conditions, such as temperature and humidity affected the results.

Method

Fingerprints of male and female volunteers of varying ages (preferably including babies to persons over 70) would be collected onto glass slides and strips of metal and paper. The volunteers would be required to have not washed their hands for an appropriate length of time beforehand – this would need to be determined in an initial series of experiments using different soaps. The slides and strips would then be stored in enclosed chambers at varying temperatures and humidities for different lengths of time before being be developed and compared.

Title

The influence of hair colour on uranium absorbance.

Rationale

There is concern over the illegal trafficking of unstable uranium isotopes for use in a so-called 'dirty bomb'. Measuring uranium levels in hair gathered during a surveillance operation (e.g. from combs and bathrooms) might provide evidence of involvement without alerting the suspect.

Method

Working with dangerous uranium isotopes is clearly not feasible but some forms are relatively safe to handle. Uranium is naturally present in the environment and we all contain low levels in our bodies. Volunteers of different ages, genders and hair colour could donate hairs that would then be analysed using atomic absorbance spectroscopy or EDAX–scanning electron microscopy. Total uranium content and its distribution along the length of different hair types could be assessed. A more detailed experiment could be done by supplementing the diet of rats of varying colours with low levels of uranium.

5 Wounds

Chapter outline

Objectives

Discuss how wound analysis can be used to differentiate between suicide, accidental death and homicide.

Compare and contrast the wounds produced by low- and high-velocity firearms and how one can determine the distance from which a victim was shot.

Describe how post mortem injuries may be distinguished from those that occurred before or at the time of death.

Definitions

Wounds can be crudely divided into those that are inflicted through the delivery of kinetic energy (i.e. a physical force), such as from a punch or knife thrust, and those

Essential Forensic Biology, Second Edition   Alan Gunn
© 2009 John Wiley & Sons, Ltd

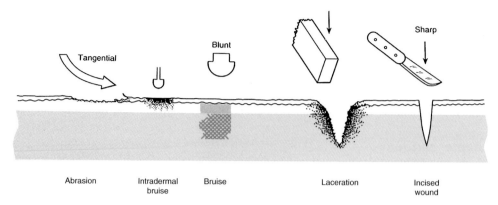

Figure 5.1 Diagrammatic representation of different types of wound to the skin. (Reproduced from Saukko, P. and Knight, B. (2004) *Knight's Forensic Pathology*, 3rd edn, copyright 2004, Hodder Arnold.)

caused by nonkinetic energy, such as burning or scalding. However, it is not unusual for a combination of both mechanisms to be involved, such as when a person is shot at close range. Wounds manifest themselves in a wide variety of forms and arise as a consequence of numerous natural and man-made factors so only a brief overview can be provided in this chapter and those seeking more extensive coverage are advised to consult a specialist medical textbook (e.g. Dolinak *et al.*, 2005; Saukko & Knight, 2004). Medically, traumas (injuries) resulting from physical forces can be categorized into those caused by blunt objects and those caused by sharp objects. Gunshot injuries can be considered as a special case. Blunt force traumas include bruises, abrasions and lacerations whilst sharp force traumas are divided into incisions, and penetration (stab) wounds (Fig. 5.1). Under English law, the legal definition of a wound is more restrictive: it is a situation in which the whole skin must be broken (this includes the skin within the cheek or the lining of the lip) (www.cps.gov.uk/legal/section5/chapter_c.html#10). Consequently, a slight scratch would not be a wound, but if the scratch bled, it would be. Similarly, a ruptured artery would not be a wound so long as the bleeding took place internally and not through a break in the skin. Injuries such as bruises, fractures and other internal injuries would therefore lead to prosecution under a different criminal offence. For the purposes of this section, wounds will be dealt with from a medical perspective. The trauma categories mentioned above are not mutually exclusive and a wound may exhibit more than one characteristic. For example, a bite may show both bruising and penetration of the skin surface. The nature and distribution of wounds on a body can provide a wealth of forensic information such as when and how they were inflicted and whether they were the result of foul play, an accident, self-harm, suicide or they were caused during the process of decay or even during the handling and storage of the body at the mortuary.

Blunt force injuries

Bruises

Bruises (also referred to as contusions and ecchymoses) result from the escape of blood from damaged blood vessels beneath the skin, within muscles, or the internal organs. The leaked blood causes the characteristic discoloration beneath the skin surface although this is not visible in deep bruises. Bruising is particularly obvious in persons with pale skin, such as those with red hair but may be masked by melan pigmentation in those who are darkly skinned. A bruise may not form at the injury site because the leaking blood flows via the path of the least resistance and then collects where it can travel no further. Consequently, a broken jaw may give rise to bruising on the neck. Bruises are commonly formed from forceful contact with a blunt object – such as would occur during a fall, being gripped tightly, or being hit with a fist or stick. Although the size of the bruise often relates to the degree of violence with which the blow was inflicted, caution is always needed in their interpretation. For example, elderly people and chronic alcoholics bruise easily. The amount of subcutaneous fat, the tautness and underlying support of the skin along with the nature and abundance of its blood vessels are all further factors influencing the extent to which bruises develop. Consequently, it requires considerable force to cause bruising to the palms of the hand or soles of the feet whilst the buttocks and the regions around the eyes and genitals bruise easily. Similarly, women bruise more easily than men because they tend to have a thicker layer of subcutaneous fat and obese persons bruise more readily than those who are slim with good muscular tone.

Bruising patterns

The pressure exerted via the fingers in manual strangulation/forcible restraint results in a series of bruises slightly larger than the fingerpads themselves (Fig. 5.2). The pattern of bruising may indicate a single event or repeated strangulation both before and at the time of death. The latter scenario can result from sex games that go wrong, wife/partner beating or because the murderer strangles his victim to unconsciousness allows them to revive and repeats the exercise. The bruises may be accompanied by scratch marks from the fingernails of either the assailant exerting the pressure or the victim attempting to remove the hand (Fig. 5.3). Women tend to have longer nails than men and consequently in their attempts to defend themselves they may cause self-inflicted scratches that are more serious than those of their assailant. They are also likely to scratch their assailant and will therefore probably have fragments of their assailant's tissues/blood (and hence DNA) under their nails. Damage to blood vessels that occurs after death does not normally result in significant bruising because the circulation has ceased and therefore blood is not forced out of the vessels. Consequently, a suicidal hanging would usually result in a prominent ligature furrow but this would be lacking or poorly formed if a murderer manually strangled or smothered someone and subsequently attempted to cover the crime by suspending his victim from a rope. A further distinguishing feature might be the presence of counter-

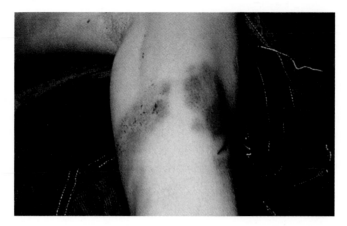

Figure 5.2 Bruising caused by forcible restraint. Note the fingerpad marks and the abrasions caused by the fingernails. (Reproduced from Shepherd, R. (2003) *Simpson's Forensic Medicine*, 12th edn. Copyright 2003, Hodder Arnold, London.)

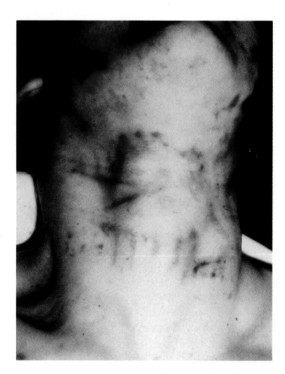

Figure 5.3 Bruising caused by manual strangulation. The abrasions underneath the chin and below the larynx were caused by the victim during her attempts to remove the assailant's hands. (Reproduced from Saukko, P. and Knight, B. (2004) *Knight's Forensic Pathology*, 3rd edn, copyright 2004, Hodder Arnold.)

pressure bruising in the strangled/smothered victim resulting from them being held down forcibly on a hard surface when the initial attack took place. This can result in bruising to the back, shoulder blades or pelvis (pinned face up) or chest, face, or knees (pinned face down). It should be noted that strangulation does not always result in tell-tale surface bruising; sometimes the bruising is only apparent in the underlying neck muscles. In cases of rape, bruising might be found on the inner thighs, especially if the attack took place whilst the victim was alive and conscious, along with counter-pressure bruising as outlined above and fingerpad bruising to the arms, neck or face as result of forceful restraint.

Deep bruises

Deep bruises to the muscles and organs normally result from extremely forceful blows or pressure and in some circumstances may prove fatal. These bruises are typically, although not always, caused by being hit with a wide blunt object such as a kick from a heavily shod foot. Perhaps paradoxically, deep bruises are not always accompanied by damage to the skin surface – a lot depends on the region of the body that was hit and the object used. For example, a kick to the abdomen may result in laceration of the viscera or rupture the spleen without causing bruising of the skin surface. Attacks involving kicking, however, often involve numerous blows being inflicted and these are typically also aimed at the genitals, face and kidneys so bruises would be found elsewhere on the body.

Ageing bruises

Recently formed bruises are dark red in colour owing to the presence of haemoglobin in the red blood cells. As the bruise ages, blood cells diffuse away from the wound site and are broken down as part of the wound healing process. The haemoglobin is converted into bilirubin by macrophages and hence the colour changes to yellow-green. Fresh bruises provide the most accurate impression of the object that caused the injury provided that they are examined shortly after infliction or if the recipient died soon after the assault.

Once they are formed, it is difficult to age bruises since the changes in coloration that take place during healing depend on many variables such as the size and depth of the bruise, the body part affected, gender, health, and age. Despite this, the presence of multiple bruises of varying stages of healing is strong evidence of repeated assaults such as would occur during child abuse, wife beating or prolonged torture. Even here, though, there are alternative explanations, such as underlying diseases of the blood/vascular system (e.g. leukaemia) and conditions affecting balance (and hence a propensity to fall over) that can result in the sufferer acquiring regular bruises.

Abrasions

Abrasions (scratches and grazes) are superficial injuries resulting from the body hitting or being dragged across a rough surface – or vice versa. They can also result

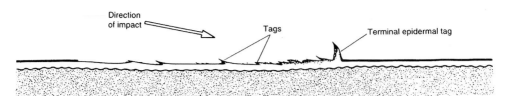

Figure 5.4 Diagrammatic representation of how one can determine the direction of a force (e.g. how a body was dragged) from an abrasion. Note how the skin tags are pulled towards the furthest edge. (Reproduced from Saukko, P. and Knight, B. (2004) *Knight's Forensic Pathology*, 3rd edn, copyright 2004, Hodder Arnold.)

from the skin being crushed and, depending on the depth, may or may not result in blood loss. Obviously, the amount of damage will reflect both the force and the nature of the surface striking the skin. If the surface is relatively smooth, it may result in such fine linear abrasions that the surface of the skin simply appears reddened – this is commonly known as a friction burn. Unlike bruises, abrasions only occur where the causative force was applied. Abrasions can indicate the direction in which a force was applied because torn fragments of skin are dragged towards the furthest edge of the wound (Fig. 5.4). They can therefore, for example, indicate how a body was dragged along a path and in this scenario, particles of soil, cement, or vegetation may become embedded in the wound thereby providing clues of where and when the event took place. Abrasions are a common feature of traffic accidents and can indicate how a body rolled or slid across the surface of a road.

Crush abrasions

'Crush abrasions' result from being hit with a blunt object with sufficient force to abrade the skin but not to overstretch and tear it. This can occur during a traffic accident from the impact of a car bumper or radiator grill or when a person is lashed with a plaited horsewhip (Fig. 5.5). The object causing the crush abrasion can leave an accurate impression in the skin that allows future identification. A record of the wound can be made using vinyl polysiloxane and this is then covered with isocyanate resin to produce a positive replica. This can then be observed using a scanning electron microscope (SEM) to reveal much finer detail than is possible with the naked eye or conventional light microscopy (Rawson *et al.*, 2000).

Ligature furrows

Hanging usually leaves a grooved wound in the neck referred to as a 'ligature furrow', although such marks may be faint or absent if a soft material, such as a ripped bed sheet, is used. In suicidal hangings, the ligature furrow seldom completely surrounds the neck but is angled with an obvious 'suspension peak' (also referred to as a 'suspension point') (Fig. 5.6). Exceptions to this rule can occur depending on the type of knot that was used and how the person was hung (Fig. 5.7). When

Figure 5.5 Crush abrasions caused by beating with a leather riding-crop. The victim was found bound and gagged in bed and had been murdered after being whipped. Note how the abrasions in the skin match the patterns on the riding-crop. (Reproduced from Ullyett, K. (1963) *Crime out of Hand*. Copyright © 1963, Michael Joseph, London, UK.)

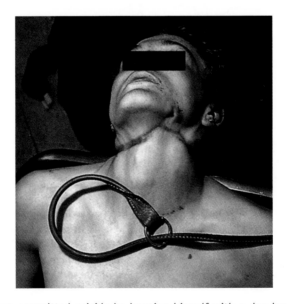

Figure 5.6 This man committed suicide by hanging himself with a dog lead. Note the ligature furrow around the neck, rising to form a suspension peak and the impression caused by the metal buckle. (Reproduced from Shepherd, R. (2003) *Simpson's Forensic Medicine*, 12th edn. Copyright 2003, Hodder Arnold, London.)

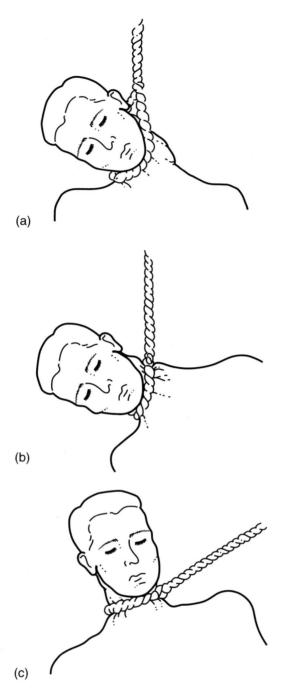

Figure 5.7 Diagrammatic representation of factors affecting the ligature furrow following hanging. (a) The normal method in which the noose is suspended above the victim's head and the furrow rises to form a suspension peak with a gap at the top where the ligature was not in contact with the skin. (b) The noose is tied with a slipknot. This causes the noose to become extremely tight and the furrow can encircle the neck. The furrow may be lower than that seen in (a) and may also be more horizontal. (c) The noose is suspended from a point that is only slightly above or to one side of the victim. In this situation the victim leans away from the point of suspension to tighten the noose thereby causing the furrow to be horizontal. (Reproduced from Saukko, P. and Knight, B. (2004) *Knight's Forensic Pathology*, 3rd edn, copyright 2004, Hodder Arnold.)

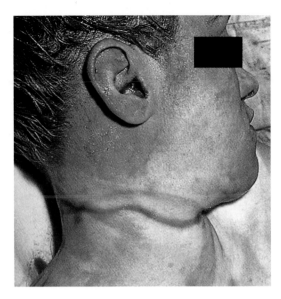

Figure 5.8 This man was strangled with a soft silk ligature. Compare the wounds to those in Fig. 5.6 and note how the ligature furrow encircles the neck without rising to a suspension point. (Reproduced from Shepherd, R. (2003) *Simpson's Forensic Medicine*, 12th edn. Copyright 2003, Hodder Arnold, London.)

a person is murdered by having a length of cord, wire or rope drawn tightly about their neck, the ligature furrow usually goes completely around the neck (Fig. 5.8). Complete encirclement may not occur if a collar or other clothing gets in the way of the ligature or if the ligature is applied by pressure across the front of the neck with the assailant standing behind. However, there is a strong chance of finding bruises and scratches from fingernails where the victim attempted to remove the ligature and in neither case would there be a suspension peak. This method of killing is sometimes referred to as 'garrotting' and a stick or similar device is often used to tighten the ligature – this together with any knots in the ligature can cause localized bruising.

Lacerations

Lacerations are caused when a blunt object hits the skin with sufficient force to overstretch it and thereby cause it to split and tear (Fig. 5.9). They are most commonly formed where there is only a thin layer of flesh above the bone – for example, on the scalp and shins – and the lacerations are accompanied by bruising or abrading of the surrounding tissue. The tissues underlying the skin surface vary in their strength and elasticity and consequently when tearing occurs, not all of them are broken. This means that if the cut surface is examined, tissue fibres will be found connecting the two sides of the wound – these are called bridging fibres. The amount of bleeding depends on the site of wound (scalp wounds bleed profusely) and the

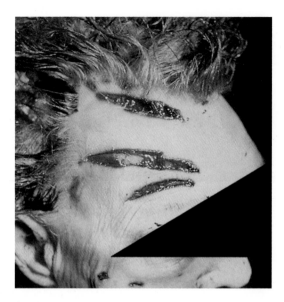

Figure 5.9 This man was beaten with a blunt steel bar that has caused several deep lacerations. At first, the wounds were thought to have been caused by an axe but the margins are crushed rather than cut. (Reproduced from Shepherd, R. (2003) *Simpson's Forensic Medicine*, 12th edn. Copyright 2003, Hodder Arnold, London.)

size of the laceration. Where the force is applied tangentially, such as a glancing blow from an iron bar or in a traffic accident, large areas of skin may be torn (flayed) away. Lacerations seldom provide a good impression of the object that caused them but owing to the force involved fragments may remain embedded in the wound. For example, a blow from a baseball bat or pickaxe handle might leave splinters of wood in the wound.

Lacerations to the genital and anal tract are often a feature of rapes and also of sex games that go wrong. The latter is often associated with men for whom the notion that bigger must be better is taken to dangerous extremes. The list of objects that have to be retrieved from 'back passages' is legion – including even concrete mix (Stephens & Taff, 1987) – which might be a cause for amusement were it not for the very real risk of fatal consequences. Sometimes men die of their own actions or that of their partners who then dispose of the object and deny all associations. The scene then has to be carefully examined to determine whether the victim's actions were voluntary or more sinister events took place.

Sharp force traumas

Incised wounds

Incised wounds – slashes or cuts – result from a sharp object cutting through the full thickness of the skin. They are distinguished from lacerations by the absence of

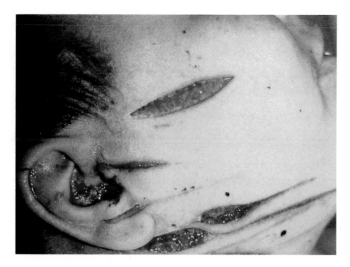

Figure 5.10 Typical incised wounds that are longer than they are deep and inflicted with a slashing weapon. In this case a knife was employed and the long 'tails' indicate that the slashes were directed downwards towards the chin. Note the absence of abraded edges to the wounds or of bridging fibres. (Reproduced from Saukko, P. and Knight, B. (2004) *Knight's Forensic Pathology*, 3rd edn, copyright 2004, Hodder Arnold.)

bridging fibres and from stab wounds by being longer than they are deep (Fig. 5.10). Typically, incised wounds are caused by knives or slashing weapons such as a machetes although they can be produced by any sharp object such as a broken beer bottle, razor wire or even the edge of a piece of paper. In fights, slashing weapons tend to be wielded at arms length and blows are aimed at the face, neck and hands. If the cutting surface is extremely sharp, it produces a wound with clean, well-defined edges. However, if the cutting surface is blunt then the wound has abraded edges and the surrounding tissue exhibits bruising. In neither case would it be possible to deduce much about the nature of the weapon used beyond its sharpness and it is highly unlikely that it would leave trace evidence behind. An exception might be where the weapon was fragile, such as a broken glass or bottle, in which case fragments can break off in the wound. The start of the wound is usually the deepest and can therefore indicate the direction of the blow – for example, whether the knife pushed forward, pulled backwards or used in a sawing manner. The wounds can be replicated, after a fashion, in the laboratory using a ripe melon as the body and a variety of knives as weapons and comparing the 'wounds' they inflict. Unless a major blood vessel is damaged, or they are deep and extensive, incised wounds might bleed profusely but they are seldom life threatening. However, large slashing weapons, such as swords, are much more likely to inflict fatal cuts or even decapitate or amputate the limbs of the victim. Both incised and stab wounds to the back are always highly suspicious.

Suicidal incised wounds

In the case of suicide, the victim often tests their personal resolve by cutting their fingers and/or making one or more parallel shallow 'hesitation cuts' to the wrist or neck before making the fatal incision. Superficial hesitation cuts or stab wounds are usually absent in cases of homicide although they may be inflicted when the assailant is attempting to scare/extract information from his victim or when the assailant is summoning up courage to inflict a fatal blow (e.g. Herbst & Haffner, 1999). Persons who intentionally cut their own throat usually begin the fatal cut high on the neck on the opposite side to the hand holding the knife. The weapon is then dragged downwards, passes through the common carotid artery and may sever the hyoid bone and thyroid cartilage. The weapon is usually found close to the body and may even remain in the victim's grip. Where the throat is cut during the course of a violent attack, the hesitation cuts are lacking although there may be more than one cut present. The cuts are usually forceful and delivered across the front of the throat and may begin from either side depending on whether the assailant was right- or left-handed and the attack took place from in front of or behind the victim. Typically, the larynx and all the major blood vessels are severed.

Case Study: The suicide of David Kelly

The diagnosis of suicide, especially by high profile individuals, often creates controversy. For example, during the Hutton enquiry into the death of the UK government biological weapons expert Dr David Kelly in July 2003, the official verdict was death by suicide as a consequence of blood loss following him slitting his left wrist. This view has been challenged by some doctors who consider that since only one artery – the ulnar artery – was cut this would have been unlikely to lead to sufficient blood loss to be fatal. Subsequent to the Hutton enquiry, further controversy arose when it was revealed that the knife that Dr Kelly purportedly used lacked his fingerprints. It is difficult to comprehend how Dr Kelly could have employed the knife without leaving his prints on it (he was not wearing gloves and none were found near the scene of his death) or why he might have attempted to wipe them off. When an artery is cut through completely, its thick muscular walls spontaneously contract and this reduces bleeding. This explains why people who lose a limb in an industrial accident or during a battle have been known to walk or crawl away, sometimes for long distances and also, possibly, why Dr Kelly's body was surrounded by relatively little blood. It should also be noted that a person slashing their neck or wrists might die as a result of an air embolism rather than blood loss. In this case, air enters the cut ends of the veins and travels back to the right atrium of the heart where it causes a cardiac arrest. Similarly, when the throat is cut, death may result from asphyxiation when blood enters the lungs.

Healed incised wounds

Healed incised wounds are a feature of past medical operations, such as removal of the appendix or a caesarean, and this may prove useful as an identifying feature. The effectiveness of the healing process is affected by numerous factors, such as the nature of the wound, the part of the body, general health and age. Interestingly, wounds inflicted during early gestation will heal rapidly and perfectly. A patchwork of healed and partially healed incised wounds is a common feature of self-harm. These wounds can be anywhere on the body although they are usually in places that can be hidden by clothing and are always restricted to sites that the victim can reach. Wounds to the mouth, nipples, eyes and genitals are common in sex-related crimes when an assailant wishes to mutilate his victim although persons suffering from extreme forms of mental illness will also harm themselves in this way.

Stab wounds

Stab wounds, also called puncture wounds, are those that are deeper than they are wide (Fig. 5.11). Unlike gunshot wounds, they are seldom perforating (i.e. they do not have both an entrance and an exit wound) unless the weapon is particularly long or it passes through a narrow part of the body (e.g. hand) or is wielded at a shallow angle. Virtually any thin, rigid object – even a pencil or a toothpick – can produce a stab wound if it is used with sufficient force. For example, Lunetta *et al.* (2002) relate a case in which a young schizophrenic man stabbed himself so hard through an eye with plastic ballpoint pen that it ended up lodged in his cerebellum. The man subsequently died of his wound but until he was autopsied it was assumed that he had been shot. Similarly, Grimaldi *et al.* (2005) describe a case in which a 72-year-old man committed suicide by stabbing himself in the chest with a pencil. Some criminals carry a screwdriver or chisel because unlike knives these are not

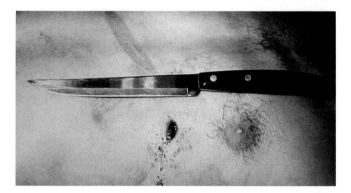

Figure 5.11 Typical stab wound and the knife that caused it. The edges of the wound have contracted and therefore make the wound appear narrower than the width of the knife. (Reproduced from Saukko, P. and Knight, B. (2004) *Knight's Forensic Pathology*, 3rd edn, copyright 2004, Hodder Arnold.)

obvious offensive weapons and can be explained away if stopped and searched by the police. However, they can be employed with potentially lethal effect. In addition, they are useful in breaking and entering premises. Cross-tipped (Phillips®) screwdrivers often leave a characteristic + shaped impression in the skin and especially in bone should it be hit.

Many weapons (e.g. knives, broken bottles) will produce both incised and stab wounds depending on how they are used. Obviously, the sharper the point, the easier it is for an object to cause a stab wound. Once the natural elasticity of the skin is overcome, the rest of the implement will follow easily. This can be demonstrated, albeit crudely, by pressing objects of varying sharpness and diameter against a ripe melon. Placing an electronic pressure transducer underneath the melon will enable force comparisons to be made.

The size and shape of a stab wound is not always an accurate reflection of the implement that caused it (Fig. 5.11). For example, the victim may attempt to twist away when the blade is inserted thereby widening the wound and giving it a triangular profile. Similarly, after inserting the knife the assailant may rock the blade or drag it in an upwards direction. In skilled hands this technique can have dramatic consequences: slaughterhouse workers use it to eviscerate an animal and can split the ribcage of a large pig from base to top in a single movement. Another complicating feature in the interpretation of knife wounds is the contraction of elastic fibres in the skin surrounding the wound. This can result in the wound being deformed and make it look as though more than one weapon was used in the assault (Fig. 5.12). Similarly, a single thrust can cause more than one wound if it passes out of flesh and then back in again. The depth of a stab wound may exceed the length of the blade that caused it. This is because some areas of the body, such as

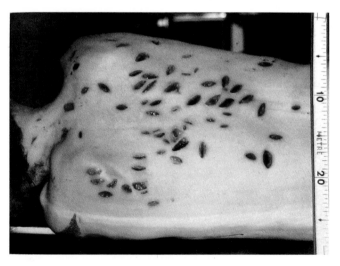

Figure 5.12 Multiple stab wounds inflicted by a single knife. Note how the wound shape and size varies as a consequence of the way the skin has contracted around the wounds. Multiple stab wounds like this are a common feature of sexually-motivated attacks. (Reproduced from Saukko, P. and Knight, B. (2004) *Knight's Forensic Pathology*, 3rd edn, copyright 2004, Hodder Arnold.)

the abdomen, can be compressed and therefore once the blade has been inserted up to its hilt, one can continue pressing inwards. Similarly, in children and young adults the ribs are relatively elastic and therefore the chest can be compressed. This is particularly the case if the victim is held tightly or forced against a solid object such as the floor or a wall. It can mean that an impression of the hilt is left at the wound site in the form of a bruise or a transfer bloodstain.

Blunt stabbing weapons, such as an unsharpened screwdriver, cause the skin to spit around the point of insertion and there may be associated bruising and laceration of the surrounding skin. Single-bladed knives, such as carving knives, often produce a wound in which one end has a sharp V-shaped profile and the other has a blunter end that has split and produced a triangle or spike: this is called 'fish-tailing'. Knives in which one edge is serrated, such as those used by fishermen, may produce wounds in which one edge is torn and lacerated although this depends upon the coarseness of the serrations. Stab wounds to the soft tissues may be difficult to identify once a body has entered the late stages of decay and after the body is skeletonized it becomes impossible. However, if during the course of the attack any of the bones were chipped – as is highly likely if the wound was to the chest – then this can raise suspicions about the cause of death (Bonte, 1975). During a knife attack, defence wounds to the hands and fingers are likely to occur and this can result in damage to the underlying bones.

Clothes offers some protection from stab wounds and buttons or objects carried in pockets can deflect blows and result in unusual wound tracks. Clothing can also provide an indication of the victim's position when they were stabbed. Consequently, the body of a stabbing victim should always be initially examined with the clothes in place. This enables the holes in the clothing to be compared to the underlying wounds. For example, if you raise your arms whilst wearing a coat or jacket, the material is also raised upwards. A blow to the upper chest at this time will pass through material that would cover the lower chest when the arms were held at rest. Consequently there will be an apparent discrepancy between the holes in the clothing and the stab wounds in the body. Similarly, the pattern of bloodstains on clothing is affected by the body's position. For example, blood seeping from a chest wound forms different trickle patterns depending on whether the victim is standing, sitting or lying on their back or sides. However, stab wounds do not always bleed profusely (at least not externally) especially if the stabbing implement is left in place.

Accidental knife wounds

Accidental incised and stab wounds are common in kitchens, slaughterhouses and anywhere that knives are used routinely. They also happen when friends, usually youths, are fooling around with knives or other sharp objects. Such accidents usually involve a single puncture or incised wound and if the victim was alone, the cause of the wound will be found nearby (although, the victim may stagger some distance before collapsing). Sometimes it is difficult to differentiate between an accidental wound and one in which an attack occurred – especially if there are no impartial witnesses. For example, in November 2000, a 10-year-old child, Damilola Taylor, was found dead in the stairwell of a south London housing estate. He had bled to death following a stab wound to his thigh that had severed an artery and cut through several veins. In a high profile trial, four youths were charged with Damilola's murder. A medical trauma expert witness called by the defence stated that the wounds were consistent with Damilola falling on a broken bottle rather than being intentionally stabbed. Severed arteries tend to constrict but by instinctively attempting to crawl home, Damilola may have opened the wound. The youths were acquitted at the end of a highly controversial court case. However, subsequently the forensic evidence was re-examined by a second forensic laboratory and their findings contributed to two of the youths being re-tried and found guilty of murder. The second laboratory discovered a spot of Damilola's blood on the trainer of one of the suspects and a further drop on the sweatshirt of another of the youths. In addition, fibres from Damilola's jacket were found on the suspects. This emphasizes the difficulty of interpreting evidence and how it is possible for even trained forensic scientists to miss valuable clues. For example, it subsequently transpired that the first forensic science laboratory would only have tested for the presence of fibre evidence if requested to do so by a senior police officer.

Case Study: Was it a trip or was it a crossbow?

In May 1991 in Leiden, Holland, a 21-year-old student returned home to find his mother dead. An autopsy demonstrated that a plastic Bic® biro had penetrated her right eye, entered her brain and lacerated the brainstem causing her instant death. An unusual feature of the wound was that the biro had perforated her eyeball whilst in most similar eye injuries the eyeball is pushed aside by the penetrating object. However, two medical experts stated that the most likely scenario was that the woman had stumbled whilst holding the pen and the fall had forced it through her eye and into her head. The police did not believe this but in the absence of further incriminating evidence the case was dropped.

After about 4 years, a witness came forward claiming that the student may have killed his mother by firing the pen from a crossbow whilst his psychologist (disregarding the supposed requirement to provide her clients with confidentiality) stated that he had confessed to the killing. The student was therefore arrested and at the trial he was found guilty and sentenced to 12 years in prison. An appeal was launched and as part of this experiments were undertaken to determine whether it really was possible to shoot a plastic ballpoint pen from a crossbow with sufficient force through the eye and into the skull. Initial experiments used the heads of dead pigs and a crossbow with a 10.9 kg tractive power. Unfortunately, these studies suffered from practical problems but it was demonstrated that even with the crossbow held in contact with the eye, the pen did not penetrate the eyelid or the orbita. The researchers therefore obtained ethical approval to use two human corpses that were destined for dissection training. The crossbows were held in contact with the eyes, which were open, so that pen would strike the eye with maximum force. However, no matter what angle the crossbow was fired from and even if a bow with a tractive power of 40 kg was used, the ballpoint pens never penetrated as far into the head as the one retrieved from the dead woman. Furthermore, the pens fired from the crossbows, with one exception, acquired an indentation mark from the drawstring and were damaged: a common feature being the central ink tube plus cone head telescoping from the shaft. By contrast, the pen retrieved from the woman's skull was undamaged and had no indentation mark. These results, together with other evidence, led to the student's release and subsequent acquittal (Bal, 2005; Rompen *et al.*, 2000). This case emphasizes the importance of an experimental approach to wound analysis and the importance of maintaining an open mind.

Homicidal stab wounds

Single stab wounds are often (though not always) a feature of homicides in which the victim is unable to resist owing to being asleep or incapacitated in some way, such as through drink, drugs, illness or old age. These single homicidal wounds are usually delivered with care and precision and the accused is therefore likely to be convicted of murder rather than manslaughter. This can be controversial in situations in which a woman who claims to have suffered years of physical abuse from her partner finally retaliates by stabbing him while he is in a drunken slumber. By contrast, if the victim is conscious and able bodied, the first blow is seldom incapacitating even if it is potentially fatal and the victim will attempt either to fight back or to flee. It is impossible to state how long someone would remain alive after receiving a fatal wound because there is a great deal of variation between individuals and sometimes a victim may continue to resist or run for several minutes after being subjected to a series of potentially fatal blows. Multiple stab wounds are a common feature of sexually motivated attacks and may be accompanied by mutilation. Sometimes there are large numbers of wounds as the assailant becomes frenzied and stabs his victim wildly. Altogether, sharp instruments are responsible for the majority of homicides in England and Wales. In recent years, the figures have varied from 27–33% of all homicides, and the proportion is higher in men (28–35%) than women (23–31%) (www.crimereduction.gov.uk/statistics38.htm).

Suicidal stab wounds

In suicidal stabbings, the victim often removes the clothing around the chosen wound site. This region is known as the 'elective' site or area and is commonly in the area of the precordium (i.e. that overlying the heart and lower chest) – and the blow is directed towards the heart. The knife may be partially withdrawn and reinserted several times and there might be several separate stab wound. However, there will not be any defence wounds and there may be evidence of tentative incised wounds either to the fingers or in the elective area. Sometimes the suicide victim may attempt to stab themself through the neck (rather than slitting the throat) and interpretation of such wounds is very difficult as this is also a very efficient way of murdering someone. The majority of suicides do not leave notes explaining their actions and even if a note is found, the possibility that it was written under duress

Table 5.1 Summary of wound types and their causes

Wound type	Results from	Typically caused by
Bruise	Internal blood loss from damaged blood vessel. May be superficial or deep.	Heavy blow. Restraint by gripping or pressing against a hard surface.
Abrasion	Superficial damage to skin surface. May or may not be accompanied by blood loss.	Contact with a rough surface. Dragging of body across an object or of an object across the body. Hanging, garrotting.
Laceration	Overstretching of skin resulting in tearing.	Glancing blow from heavy object.
Incised	Cutting of the skin and underlying tissues. Wound is longer than it is deep. Often superficial unless weapon is large, such as a sword.	Object with a sharp, rigid cutting surface, such as a knife or broken bottle.
Stab	Deep penetrating wound severs blood vessels and damages internal organs.	Rigid object with a sharp point, such as a knife.

or by the murderer attempting to cover up their actions should be considered. A stabbing victim may also be forced to stab themself during a struggle (e.g. the hand grasping the knife is turned against them) but in such circumstances there will probably be extensive bruising elsewhere on the body.

Bone damage

The skeleton remains long after the soft tissues have decayed and can provide valuable information on the cause of death. Similarly, evidence of bones that were broken some time in the past but had since healed can be a useful identifying feature. It takes considerable force to fracture or break the bones of an adult provided that they are not afflicted by a medical condition such as osteoporosis. Consequently, the presence of broken bones in a dead body always arouses suspicions. Living bone tissue is flexible and therefore blows delivered during life result in cracks and in fragments with irregular edges that tend to stay joined together. By contrast, after death, particularly once the bone has lost much of its organic matter and dried out, it becomes brittle and an applied force tends to cause it to shatter into numerous small regular fragments. In addition, the concentric and radiating fractures that are a common feature of certain wounds (e.g. gunshot) do not usually occur once the bones have dried out. The length of the post mortem period required to significantly affect the fracture characteristics of bone depends upon the nature of the bone and the taphonomic processes it experiences so can be expected to be highly case dependent.

It should be remembered that a broken bone might result from either direct force or forces transmitted from elsewhere in the body. The latter is often the case in the thorax where a single blow may result in multiple fractures within the ribcage as the forces radiate outward and create alternating points of tension in which one fracture can beget another (Love & Symes, 2004).

Hanging and strangulation

Fracturing of the cervical vertebrae is unusual in suicidal hangings unless there is a long drop and the noose is prepared with a hangman's knot. Consequently, an otherwise healthy adult male found with a broken neck may represent a case of homicide rather than suicide if he was found suspended from a length of electric cable attached to a low branch. By contrast, if the noose is composed of material that is both strong and relatively inelastic and the victim falls a suitable distance before it tightens, then they may be decapitated. The distance is dependent upon the weight of the victim: the heavier the victim the less distance they need to fall. It was part of the skill of a professional hangman to judge the drop required to avoid throttling the condemned person at one extreme and beheading him/her at the other. Tracqui *et al.* (1998) relate a case in which a young man jumped from the bridge above a canal with a nylon rope around his neck. He fell for 3.7–5.3 metres before the rope tightened and when it did he was instantly decapitated, his head sinking beneath the bridge and his body drifting over 200 metres downstream.

The hyoid bone forms part of the axial skeleton and two characteristics make it unusual (for a bone): it is a single U-shaped bone that does not have a partner, and it does not articulate with any other bone. It is found in the anterior region of the neck between the mandibles and the larynx and its function is to act as a sling to support the tongue and for some of the neck and pharynx muscles (Fig. 5.13). Damage to the hyoid bone, especially one or both of the horns of the 'U', is a characteristic sign of manual strangulation. It is far less likely to occur in hanging owing to the positioning of the rope and can therefore help to distinguish between suicide by hanging and manual strangulation. It is, however, notoriously difficult for even experienced pathologists to diagnose strangulation in some circumstances. For example, the joints of the greater horns of the hyoid bone are flexible in children and in young individuals this may lead the unwary to consider that they are damaged although the absence of associated bleeding would indicate that this is unlikely. However, it is also important to remember that absence of damage to the hyoid bone does not necessarily mean that strangulation did not take place. A great deal depends on how and where pressure was applied to the neck. In adults, manual strangulation is likely to result in fracturing of the thyroid cartilage (Adam's apple) but this may not occur in young children because their cartilage is more pliable. Care is needed here too because some people contain small cartilaginous nodules within the thyro-hyoid ligament and these can be mistaken for evidence of fracture of the superior horns of the thyroid cartilage. Again, the absence of bleeding would suggest that this is unlikely.

More women are killed by strangulation than men because there needs to be a considerable difference in the relative strength of the assailant and victim for it to

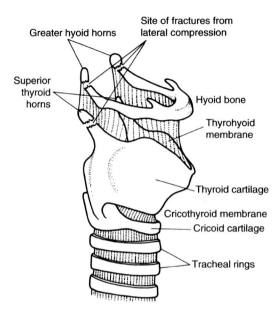

Greater hyoid horns

Site of fractures from
lateral compression

Superior
thyroid
horns

Hyoid bone

Thyrohyoid
membrane

Thyroid cartilage

Cricothyroid membrane

Cricoid cartilage

Tracheal rings

Figure 5.13 Diagrammatic representation of the larynx showing the commonest fracture points on the hyoid bone and thyroid cartilage as a consequence of lateral compression. (Reproduced from Saukko, P. and Knight, B. (2004) *Knight's Forensic Pathology*, 3rd edn, copyright 2004, Hodder Arnold.)

be successful. In England and Wales, 15–25% for all female homicides in recent years have resulted from strangulation (including asphyxiation) but for men the figures vary from only 3–7% (www.crimereduction.gov.uk).

Past fractures as identification aids

Evidence of ante mortem fractures, surgical interventions or bone deformities can be useful in the investigation process when an unidentified body is discovered. However, such traits should be used with care since even suites of them can be shared by many people. For example, Komar & Lathrop (2006) studied 482 skeletons in two forensic collections and found that 9.1% of the skeletons exhibited fractures to the right ribs, 3.7% had undergone cranial trephination and 17.4% showed signs of diffuse arthritis. They selected out of the collections the body of a 52-year-old white man with a broken nose and an orthopaedic repair to his left tibia and then looked for others who exhibited similar traits. They found that of the remaining 481 skeletons, 233 could be identified as belonging to white males, 133 of these were within the appropriate age range and of these 42 exhibited nasal fractures and of these three had undergone an orthopaedic repair to the left tibia. Consequently, within a population of 482, no fewer than four individuals exhibited the same combination of characteristics of gender, race, age range, fractures and surgical interventions. A forensic collection is probably not particularly representative of a population at large (though there is little evidence on this) but it indicates that distinguishing between individuals using suites of traits is not straightforward.

Toolmarks in cartilage and bone

Toolmarks are the result of the physical damage caused by the use of a hard object (i.e. a tool) on a softer surface, for example, when a chisel is used to carve out a piece of wood. In forensic pathology, toolmarks are typically seen in cartilage and bone as a result of indentations or cuts inflicted by a knife during an assault or by a saw employed to dismember the body after death. Observing small scratches, chips and other tool marks is not always easy with the naked eye. Consequently, where there is doubt, a scanning electron microscope (SEM) can be used to reveal fine detail (Alunni-Perret *et al.*, 2005). If the SEM is coupled to an analytical technique such as energy-dispersive analysis by X-rays (EDAX) it is theoretically possible to compare the ion spectra of debris found in the cut surface with that of the tool thought to have caused it although the applicability of this approach requires more research. Obtaining toolmark impressions from a suspected weapon is often done by pressing it against a soft metal such as lead or jewellery carving wax (Petraco *et al.*, 2005) and thereby rendering a series of grooves that act rather like a finger-print. These are known as negative impressions and can be compared with casts made of the grooves found in bone using a comparison microscope.

Additional aspects of wound interpretation

Handedness

In the absence of witness statements or other evidence it is difficult to determine whether or not an assailant was right or left handed on the basis of wound analysis because so much depends on the orientation of the victim and assailant at the time.

Defensive wounds

As their name suggests, defensive wounds are those acquired in the process of defending oneself from an assault and are therefore characterized solely by the behaviour of the victim that led to their infliction rather than the object that caused them. Usually, they are found on the hands and forearms although the feet and legs may be hurt if the victim is on the floor and attempting to kick away the attacker. For example, a single stab wound seldom incapacitates the victim immediately and in an attempt to ward off further blows they may grasp the blade resulting in incised wounds to the palms and fingers or receive stab or slash wounds to the fore- and/or upper arms. Similarly defensive wounds may take the form of bruises, abrasions lacerations and/or bone fractures to the limbs during an assault with a heavy blunt object such as a baseball bat or hammer or gunshot wounds in the case of a shooting (e.g. the victim raises his arms in front of his body immediately before being shot). Defence wounds therefore indicate how the assault progressed and that the victim was alive at the time.

Self-inflicted wounds on the assailant

Assaults are often chaotic events and the assailant may wield their weapon in a wild manner even if the victim is already dead. If the victim is vigorously defending themselves the scene is likely to be both noisy and violent. It is therefore not unusual for the assailant to wound himself, especially if he is using a sharp implement. In addition, the weapon may become covered in blood and therefore difficult to grip and if a knife without a guard is being used the hand holding it may slip down over the blade if it suddenly hits a hard object. Usually, the assailant acquires wounds in the form of cuts and stabs to the hands – the right and left hand are approximately equally likely to be injured – although other parts of the body may be hurt. Although some criminals carry guns not many of them are skilled in their use and consequently an assailant may accidentally shoot himself or a colleague. An assailant may also intentionally injure himself after perpetrating an assault in an attempt to justify the crime as self-defence or as part of a suicide attempt. In a study by Schmidt and Pollak (2006), 36% of assailants ($n = 58$) who used knives inflicted injuries upon themselves and 8.6% of the assailants did so deliberately.

These self-inflicted wounds are important because they result in the suspect's blood (and hence DNA) being left at the scene and the perpetrator may need to seek medical attention or otherwise have to explain to friends/colleagues/authorities how they came to acquire their injuries.

Asphyxia

The term 'asphyxia' is derived from the Greek for breathlessness and death from asphyxiation is a consequence of one or more of the following:

(1)   The cells of the body do not receive sufficient oxygen via the bloodstream;
(2)   The cells receive oxygen but are unable to utilize it;
(3)   The cells are unable to eliminate carbon dioxide.

Brain cells are metabolically very active and therefore have a high oxygen demand and they are also very sensitive to chemical changes in their environment. Anything that interferes with the blood supply to the brain therefore rapidly leads to a loss of nerve function thereby causing unconsciousness and, if prolonged, death.

Deaths attributed to asphyxia are sometimes divided into categories based upon the cause of the asphyxia (Table 5.2) although more than one cause may be involved. For example, an assailant might manually strangle someone whilst kneeling with their full weight on the victim's chest. It should also be borne in mind that

Table 5.2 Categories of asphyxia

Category of asphyxia	Example
Airway obstruction	Choking, smothering
Neck compression	Strangulation, hanging, garrotting
Chest compression	Pressing, being crushed by a heavy weight
Postural/positional	Position of body compromises ability to breathe
Environmental	Lack of oxygen in atmosphere, suffocation
Poisoning	Cyanide, carbon monoxide

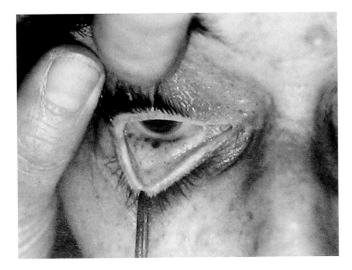

Figure 5.14 Petechial haemorrhages in the eye and surrounding skin following manual strangulation. (Reproduced from Saukko, P. and Knight, B. (2004) *Knight's Forensic Pathology*, 3rd edn, copyright 2004, Hodder Arnold.)

a single physical cause of death, such as strangulation, may have its lethal effect through two or more different physiological processes.

Pathological features of asphyxiation

Generalized or localized cyanosis is a common feature of asphyxiation as a consequence of darker deoxygenated blood imparting colour to the skin but is not always obvious. Small petechiae are often found on the surface of the conjunctivae although they may also occur elsewhere such as the eyelids or the skin of the face (Fig. 5.14). They are thought to arise if the venous return blood supply from the head is blocked whilst the arterial blood supply to the head remains operational. This could occur because the thick muscular walls of the carotid arteries coupled with their higher internal blood pressure makes them much harder to compress than the comparatively thinner walled jugular veins, and in addition the carotid arteries are located deeper within the neck and are therefore inherently less vulnerable. In this situation, the venous pressure in the head increases and can be sufficient to cause thin veins and capillaries to burst. Such petechiae are, however, not diagnostic of asphyxia and may be absent or result from other causes. For example, petechiae are commonly found in the eyes following electrocution or as a post mortem artefact when the body is left face downwards. In the latter case the blood pools in the facial region and as the swollen capillaries begin to leak, petechiae are formed.

Sometimes the reason a person died is obvious but at others there may be little apparent pathology or it is nonspecific and if the scene where the body was found does not suggest a cause, the death may go undiagnosed. Asphyxiation may be suspected but proving it is another matter. Some workers have therefore looked for molecular clues; for example, Ikematsu and his co-workers have measured changes in gene expression following simulated strangulation (Ikematsu et al., 2006; 2007). Their experimental design involved anaesthetizing mice and then either decapitating them or strangling them with a piece of string. Changes in gene expression were then monitored in the skin of the neck and brain at the time of death and for up to 30 minutes afterwards. Evidence of enhanced expression was found for several genes in both the skin and brain of the strangled mice indicating that the manner of death had specific effects. Clearly, this work is at the very early stages and begs the question of how long such changes in gene expression would remain measurable after death although it is becoming obvious that RNA is more stable than previously thought (Chapter 2). However, medical literature indicates that there are alterations in gene expression as a consequence of hypoxia resulting from natural disease processes (e.g. a heart attack) and it may remain difficult to distinguish between asphyxia resulting from a malicious act and death from natural causes using this approach.

Airway obstruction asphyxia

Airway obstruction may result from physical blockage of the mouth and nose or the airways themselves. Choking is a common accidental cause of death and the object can usually be found blocking the glottis and/or trachea. It is often seen in young children owing to their habit of putting things in their mouth and in elderly people who lack teeth and therefore the ability to chew food properly. Choking may also occur through an assailant forcing a gag into the victim's mouth – although, again, the cause of death will be obvious. Smothering occurs where the nose and mouth are blocked and thereby the victim is prevented from breathing. Typically, this occurs when the assailant covers the victim's head with a pillow or other material or covers the mouth and compresses the nose with his/her own hands. Unless already incapacitated, the victim will usually struggle and this can result in injuries to the nose and mouth. The scene should therefore be searched for bloodstained pillowcases etc. because in the absence of other clues such deaths can be hard to diagnose.

Neck compression asphyxia

Neck compression asphyxia results from the obstruction of the airways, the blood supply to/from the head or a combination of the two. Strangulation occurs when the asphyxia is induced by physically compressing the structures of the neck through the use of hands/arms/legs, a ligature (e.g. garrotting) or a hard rigid object such as a baton. Hanging occurs when the compression of the neck structures results from a ligature placed around the neck that is then tightened with the aid of all or part of the victim's own weight. Both strangulation and hanging can cause the formation of petechiae in the face and eyes but these are more commonly seen in cases of strangulation. This is probably because strangulation is usually a violent prolonged event during which the assailant battles to gain an effective grip and is not usually capable of occluding the carotid arteries. By contrast, hanging is most frequently a result of suicide and provided the noose is fitted appropriately, causes the immediate occlusion of both the jugular veins and the carotid arteries – and consequently there is less likely to be a build up in the venous return pressure in the head. Logic suggests that it is impossible for someone to strangle themselves because as soon as they lost consciousness they would also lose their grip and therefore cease to apply pressure to the neck. Despite this, there are a few isolated cases in the literature where the attempt was successful (Di Nunno et al., 2002).

Precisely how long a victim of strangulation or hanging remains conscious and then, ultimately, dies is highly case dependent because it depends upon the effectiveness with which the blood/air supply is disrupted, individual physiological differences, age, health etc. Some authors state that consciousness may be lost within as little as 5 seconds although it usually takes much longer than this. A video recording of a man committing suicide by hanging indicated that consciousness was probably lost after about 13 seconds and this was followed by convulsions and rigidity (Sauvageau & Racette, 2007). This indicates that compression of the neck induces more complex physiological events than simply reducing the supply of oxygen to the brain since most healthy individuals can hold their breath for at least 1 minute without any ill effects and many divers, admittedly usually with a degree of training, can swim actively underwater for much longer than this. The constriction of the trachea is unlikely to be a major factor in the cause of death since persons fitted with a tracheostoma (a breathing opening in the front of the neck) can commit suicide by hanging even if the noose is placed above opening and therefore, theoretically, the individual could continue to breathe.

Neck compression can cause the stimulation of blood pressure receptors (baroreceptors) in the carotid sinuses, carotid sheaths and carotid body and these then send nerve impulses to the tenth nucleus in the brainstem from which impulses are sent down the right and left vagus nerves (cranial nerve X) to the heart. Vagal stimulation of the heart brings about a slowing of the heart rate by decreasing the rate of spontaneous depolarizations in the autorhythmic heart muscle fibres. Heart muscle, like the smooth muscle of the small intestine, contains 'pacemaker' cells (in the heart they are called autorhythmic fibres) that exhibit unstable membrane potentials and are therefore self-excitable and generate repetitive action potentials. For this reason, both the heart and the gut can continue to relax and contract even if all the nerves serving them are cut. The reflex arc induced by neck compression (referred to as the vasovagal reflex) can therefore bring about the slowing of the heart and may stop it entirely in a condition known as 'vasovagal shock' or 'reflex cardiac arrest'. The extent to which this occurs is not known but it would explain why some individuals lose consciousness and die so quickly after the application of minimal force and/or without the development of petechiae or other obvious signs of asphyxia.

Autoerotic asphyxia is a very dangerous practice and involves the individual, usually male, in inducing temporary brain hypoxia and pain in the pursuit of sexual gratification. Some people report that hypoxia induces sexual hallucinations whilst a masochistic delight in pain is a well-known phenomenon. The practitioner usually has a means of regulating the degree of asphyxia but sometimes loses control of this and dies as a result. Hypoxia may be induced through hanging, the use a ligature, wearing of a mask or the use of a plastic bag – the only limit appears to be the imagination of the practitioner. The scene often contains images of a sexual nature and the victim is naked or wearing fetish clothing so there is little doubt about his original intentions (Janssen et al., 2005).

Chest compression asphyxia

Chest compression asphyxia (also referred to as crush or traumatic asphyxia) results from the inability of the victim to inhale owing to external pressure upon the chest. Such cases usually arise as a consequence of an accident – for example, in vehicle crashes or industrial accidents (Byard et al., 2006). It may also occur if a heavy assailant places their full weight upon the chest of their victim. Pressing was a barbaric means of execution in which increasingly heavy weights were placed on the chest of the victim over a period of days until they expired. Depending upon the nature of the weight and how it is applied there may be little evidence of damage to the chest (e.g. bruising, fractured ribs) and therefore the scene should be examined carefully. The victim usually exhibits a florid expression owing to extensive petechiae forming in the face and also in the eyes. Such findings are not diagnostic and could be caused a wide range of other causes including disease and poisoning.

Postural/positional asphyxia

Postural/positional asphyxia results from the victim finding themself stuck in a position that prevents them from breathing normally – for example, their neck becomes twisted. This can happen following a fall or similar accident and frequently involves intoxication through drink or drugs that prevents the victim from extricating themself.

Environmental asphyxia

Environmental asphyxia results from breathing in an air deficient in oxygen or from which oxygen becomes displaced. Sometimes the term 'suffocation' is used but this is a broad descriptor that is also commonly used to cover most other forms of asphyxia. Environmental asphyxia may happen accidentally such as becoming trapped in a sealed area (e.g. illegal immigrants being transported in a container), and from suicide/homicide as a consequence of a plastic bag being placed over the victim's head. In the latter situation a rubber band or similar constricting device is often used to improve the seal around the neck and if this is fitted tightly it causes a faint bruise. If the bag is removed before police/medical services arrive then this bruise might be the only clue suggesting how the person died. This might occur if the family of the dead person wish to cover up a suicide attempt for religious or insurance purposes or because a murderer wishes to cover up a crime. Such situations need to be treated with great care and sensitivity. When an individual commits suicide in this way (or is unable/unwilling to resist [e.g. an elderly sick relative]) there are no/few signs of struggle, the victim rapidly becomes unconscious and there are seldom any petechiae in the eyes or face because the blood supply to the head is not interfered with.

Poisoning

A wide variety of poisons can induce asphyxia either by preventing breathing (e.g. strychnine), interfering with the transport of oxygen (e.g. carbon monoxide) or the utilization of oxygen by the tissues (e.g. cyanide). The pathological consequences of cyanide and carbon monoxide poisoning are discussed in Chapter 1 but it is beyond the scope of this book to discuss poisoning in detail here and the interested reader should consult specialist literature such as Conn (2008).

Complex suicides

Sometimes the victim employs two or more sequential means of ending their life in so-called 'complex suicides'. For example, Nadjem *et al.* (2007) relate a case in which a woman ingested 107 pins, 37 nails, 9 sewing needles, 5 safety pins, 12 drawing pins, 7 matches split into two and 2€ cent coins before hanging herself from a doorknob with a headscarf. The swallowing of the sharp objects caused relatively little pathology and death was due to asphyxia. Altun (2006) reports three further cases including one in which a man had stabbed himself in the chest before hanging himself. The stab wound had not penetrated the thorax and was therefore unlikely to have been lethal. Persons who commit complex suicides such as these usually have a history of psychiatric problems and/or drug abuse but their complexity can raise suspicions that an attempt is being made to make a murder look like suicide. By contrast, cases of suicidal individuals attempting to make their deaths appear as murders are relatively rare although they do occur from time to time.

Case Study: Was it suicide or lynching?

The body of a 38-year-old man was found in his bedroom suspended from a doorway by a length of electrical extension cord, his arms were held in place behind his back by a pair of steel handcuffs, his ankles were tied together with coaxial cable and a sock was forced into his mouth as a gag and held in place by a bandana tied behind his neck (Adair & Dobersen, 1999). Death appeared to be due to asphyxiation and, as the victim had previously intimated to family members that he felt his life was at risk from a motorcycle gang, it also appeared suspicious. However, a more detailed analysis of the scene and witness statements soon established that the man took his own life. The absence of a suicide note had little bearing because they are lacking in many cases of genuine suicide. The front door to the apartment was dead bolted from the inside and there was no sign of forced entry, damage or theft of money or belongings. This made it difficult to comprehend how any assailants might have left the apartment and would mean that the sole purpose of their visit was to kill the victim. The handcuffs were of a type commonly used by magicians and escape artists and whilst outwardly similar to those used by police had a quick release catch on each cuff thereby permitting the wearer to rapidly release them without the need for a key. It was later established that the victim had bought the cuffs himself over a year previously and witnesses stated that he knew how to use them. The bindings to ankles and the gag were relatively loose and could have been tied by the victim. Therefore, although the victim was apparently 'tied up' it was physically possible for him to have done this himself and he could also have released himself had he wished. The area around where the man was hanging was undisturbed with no sign of any struggle and toxicological analysis failed to reveal any evidence of drugs that might have rendered the victim insensible. Furthermore, there were no signs that the man had been in a fight and his body was not swinging free of the ground – so he could have relieved the tension in the noose by standing upright. This indicates that if the man was lynched he was not only conscious at the time but unusually cooperative in the proceedings. Although the man did not have a history of clinical mental illness he had recently broken up from his girlfriend and had intimated that he should 'just kill himself' – such comments are not unusual in these circumstances but in this case they proved unfortunately prophetic. The unknown 'motorcycle gang' appeared to have no basis in fact and the reason why the victim chose to kill himself in this way will remain a mystery. However, the case does show how important it is to relate the pathological findings to those of the crime scene in order to establish the sequence of events that led up to a person's death.

Pathology associated with drug use

Use of illegal drugs, even hard drugs, such as cocaine, heroin etc. is common among all levels of society and, contrary to popular belief, many users continue to pursue their careers normally. Unless or until their addiction spirals out of control the user's colleagues, family and friends may be unaware of their activities. Consequently, finding drug residues in tissues at autopsy is not unusual and may have no forensic relevance beyond indicating that the person was in the habit of purchasing illegal substances.

Persons who regularly inject drugs – mainlining – often chose a vein in their arm or leg either repeatedly injecting at a single site or as series of tracks along the length of the vein. The presence of fresh bruising around the injection site indicates recent use. Excessive use causes thrombosis and scar formation and the vein can become unusable. If the addict wishes to disguise their drug use they choose injection sites that are easy to cover up, such as their feet or between the toes whilst those wishing to achieve particular effects or are simply running out of options may choose unusual injection sites such as the dorsal vein of the penis or under the eyelids. Those who die of a drug overdose are commonly found with a tourniquet attached (to raise the blood vessel and make injection easier) and the needle of the syringe still inserted. This indicates that death occurred very rapidly. Foam may be observed at the mouth and lips as a result of pulmonary oedema (fluid collection in the lungs) – this is most commonly seen in long term heroin and opioid addicts but is less common in novice users. However, pulmonary oedema can result from a variety of natural causes as well as poisoning so the presence of a so-called 'foam cone' is not diagnostic of opioid overdose. In addition, just because a known addict is found dead with a needle in their arm it should not automatically be assumed that they died of a drug overdose. For example, when Rachel Wheatear was found dead clutching a syringe in her bedsit in Exmouth, Devon in May 2000 it was assumed that she had died of an overdose. She was a known addict and her parents agreed that pictures of her dead body could be used in a national anti-drug campaign. It subsequently transpired that the syringe was clean and therefore unlikely to have contained the heroin that killed her. Furthermore, the bedsit was very tidy and as her mother stated that was completely out of character. At the time, no post mortem or full toxicology tests were performed and her body had to be exhumed in 2003 to confirm that she died of a heroin overdose. However, the circumstances in which she came to overdose remain a mystery and if she had been cremated the cause of death would also remain uncertain.

Body packing is the practise of swallowing packages of drugs, often in remarkably large numbers, in order to transport them through customs. These so-called 'drug mules' frequently bring attention to themselves on long haul flights by refusing free in-flight meals and drinks. Once they reach their destination they excrete into a bucket or similar container and thereby retrieve the packages: sometimes the packages are colour coded to distinguish their contents. Unfortunately, the packages often leak in transit resulting in a rapid fatal overdose and the carrier can also die of bowel obstruction and resultant peritonitis.

Body stuffing refers to the hasty swallowing of a drug in order to escape detection. The drug is usually not packaged to traverse the length of the gut and therefore the swallower risks the drug escaping from its wrapping and causing a potentially fatal effect. Many drug pushers keep small amounts of drug wrapped in cellophane in their mouths but if approached by a police officer spitting the drug out would mean that they could be linked to the package via the DNA in their saliva whilst swallowing the package would also carry its own risks.

Gunshot wounds

Gun crime increased enormously in England and Wales during the 1990s and into the early 2000s although the provisional figure of 9608 firearm offences for 2006–07 suggested that figures may be starting to decline again. It should be noted that a significant proportion of firearm offences involve the use of imitation weapons and relatively few are homicides. For the year 2005/06, the most recent for which figures were available at the time of writing, only about 8% of all homicides (61 out of 766) involved the use of firearms – and this includes the use of the firearm as a blunt instrument to bludgeon the victim to death (www.crimereduction.gov.uk).

Types of firearm

Firearms can be crudely divided into handguns, shotguns and rifles.

Handguns, as their name suggests, are usually relatively small and can be operated with a single hand. Handguns are usually designed to be used at ranges of up to 40 metres and are therefore not intended for long-range accuracy. Because they are easily concealed they are extremely popular among the criminal classes and most gunshot wounds in the UK result from handguns used at close range. Most handguns are either revolvers or semi-automatics. In revolvers, the ammunition is placed in a metal drum that revolves either clockwise or anti-clockwise and contains five or six projectiles. In semi-automatics, the ammunition is loaded into a clip that sits inside the handgrip of the gun and each time the gun is fired the spent cartridge case is ejected and a new cartridge is advanced into the firing chamber. Consequently, unlike in revolvers, it is possible for the gun to contain a bullet ready to be fired after the ammunition clip is removed. Accidental shootings can therefore arise when inexperienced or inattentive users forget to check their weapon before pressing the trigger. The inner surface of the gun barrel of both revolvers and semi-automatics contains grooves (rifling) that cause the bullets to spin along their long axis and this stabilizes their flight path. The rifling imparts a series of grooves onto the bullets and the pattern acts as a fingerprint for the weapon. It is therefore possible to determine whether spent bullets recovered from a crime scene originated from more than one gun and, if the suspect weapon is subsequently found to determine whether it was responsible for the shots.

Shotguns are most commonly used in sport and some farmers keep them for controlling vermin. They have long barrels and are smooth bored – that is, they lack internal rifling. They are therefore not particularly accurate but this is not important because they are primarily designed to fire numerous small projectiles that spread out in a fan-like fashion from the barrel. A sawn-off shotgun is one in which the barrel is artificially shortened thereby increasing the spread of the shot. Reducing the barrel length makes the weapon easier to conceal and although shotguns can only fire one or two projectiles (depending upon the number of barrels) before re-loading they are intended for use at close range and they can simultaneously incapacitate several people if employed in a confined space such as a room.

Because of the lack of rifling it is difficult to match spent projectiles to a particular shotgun.

Rifles are long barrelled weapons and, as their name indicates, their barrels are rifled. The combination of the long barrel and rifling increases their accuracy and some of them are effective at distances of over 1 km. They are designed to be fired from the shoulder and during the war in Chechnya Russian soldiers manning checkpoints would require Chechen men to take off their shirts to reveal whether they had the tell-tale bruising that can result from the recoil.

Types of ammunition

Although handguns and rifles tend to use different ammunition some handguns are designed to be able to use rifle ammunition and some rifles can use handgun ammunition. Rifle bullets are usually longer than those used in handguns and often have a sharp point and stepped waist. Full metal jacket bullets are those in which the lead core is completely surrounded by metal casing that is often made out of copper. These bullets do not expand on impact and cause deep penetrating wounds. However, if the bullet originates from a high velocity weapon and it strikes bone or a hard surface it may shatter into numerous small projectiles. Jacketed hollow point bullets have a metal casing but the tip of the lead core is exposed and has a hollow centre. These bullets expand on impact and this dissipates a lot of energy into the surrounding tissues. Consequently, they do not tend to penetrate as effectively as fully jacketed ammunition but they cause a lot of damage.

Soft point bullets are similar to hollow point ammunition but the exposed lead tip is designed to expand even more dramatically on impact: they therefore cause enormous tissue destruction. Hollow point and soft point bullets are used by some police forces in America and there are suggestions that they should be employed by UK police forces. Their purported advantage is that they cause such severe wounds that anybody shot by these bullets would be immediately incapacitated. At the same time, it is unlikely that they would travel through the victim's body or through doors etc. and thereby put at risk innocent bystanders or other police officers. Unfortunately, people who subsequently prove to be innocent of any crime are intentionally shot each year by the police and it would be doubly unfortunate if they were shot with bullets such as these that cannot do other than severely maim or kill. This ammunition is freely available in America and via the internet so it can be expected that criminals will avail themselves of its potential.

Dum Dum bullets are derived from standard ammunition that is doctored so as to make them prone to expanding and disintegrating on impact (Sykes *et al.*, 1988). They get their name because they were developed at a British army cantonment based in the region of Dum Dum on the outskirts of Calcutta (modern day Kolkata) in the 19th century (Spiers, 1975). Such bullets were banned for use in warfare by the first International Peace Conference at The Hague in 1899 and also at the 1933 Geneva Convention. Although the ban is still in force, the reasons they are not used more widely probably has more to do with their tendency to jam in self-loading firearms and the development of high velocity rifles and soft point bullets that can cause equally devastating injuries.

Shotgun cartridges usually contain lead or steel pellets (shot), the size of which is denoted by numbers (e.g. 5, 6, 7.5, 8) with the larger shot being used for bigger game although sometimes shotguns are used to fire single heavy projectiles called slugs. Shotgun pellets spread out as they leave the muzzle of the gun although if the gun is discharged close to the body the pellets will not have time to spread out and therefore cause just a single large entry wound. At a distance of 20 cm or more, the pellets cause separate entry wounds and the spread can indicate how far away the muzzle was from the victim.

Entrance and exit wounds

Gunshot wounds are penetrating injuries that involve an entrance wound and possibly also an exit wound (Fig. 5.15). The latter situation is most likely to occur if the bullet was fired from a high velocity weapon or from close range. It is important to be able to distinguish between entrance and exit wounds should a dispute arise whether the person shot was running towards or away from the shooter. If the person was shot from close range, the entrance wound will probably be surrounded by gunshot residues and may have an abrasion ring. Sometimes a bullet travels through the body in a straight line but if it hits a bone or it starts to tumble badly then its trajectory can be changed. For example, a bullet that enters the chest may leave via the neck or come to rest in one of the legs. If an exit wound is formed, it is typically larger than the entrance wound and bleeds more profusely .This is because when a bullet enters the body, the skin initially deforms and then, once the bullet has pierced and passed through, it retracts back. Consequently, the entry wound may appear to be smaller than the diameter of the bullet. However, one should not distinguish between an entrance and exit wound solely on the basis of relative size – for example, if the bullet fragments and only one part of it emerges the exit wound may be smaller than the entrance wound. By contrast, if several of the fragments emerge there may be a single entrance wound and several exit wounds.

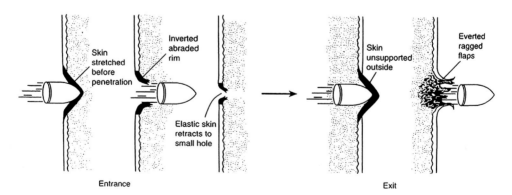

Figure 5.15 Diagrammatic representation of the formation of gunshot entrance and exit wounds. (Reproduced from Saukko, P. and Knight, B. (2004) *Knight's Forensic Pathology*, 3rd edn, copyright 2004, Hodder Arnold.)

If there is some impediment to the bullet leaving the body – for example, the body is pressed against a wall or the floor, or the victim wears a stiff trouser belt, then a 'shored' wound is formed in which the edges of the wound are abraded against the overlying object – such wounds can be difficult to distinguish from entrance wounds.

Range from which the victim was shot

An accurate estimation of range from which a victim was shot is important in reconstructing the scene of the event. For example, if a policewoman claims self-defence and she only shot the victim when he lunged aggressively towards her whilst the victim or onlookers claim that he was not being aggressive and was shot from a distance of at least 3 metres.

If the gun is pressed firmly against the flesh it forms a 'hard contact wound'. The edges of the wound are seared and blackened by the heat of the explosive charge coupled with soot and propellant being baked into the skin (Fig. 5.16). With no way of escaping, the hot gasses from the discharge are forced into the body causing the surrounding tissues to balloon and the skin to split. The latter is most likely to occur in shots to the skull and wherever else there is bone directly underlying the skin. The high temperature of the gasses burns the surrounding tissues, so blackening is found along the wound track. The wound also contains particles of soot and unburnt powder and, in the case of smooth bore shotguns, fragments of wadding.

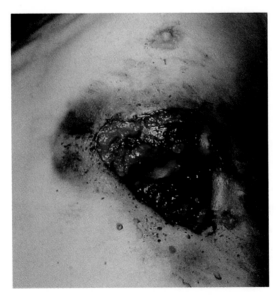

Figure 5.16 Suicidal shotgun wound to the chest. Note the massive disruption caused by the entry of gas and shot into the skin. There is a single gaping wound, the edges of which are blackened with soot. (Reproduced from Saukko, P. and Knight, B. (2004) *Knight's Forensic Pathology*, 3rd edn, copyright 2004, Hodder Arnold.)

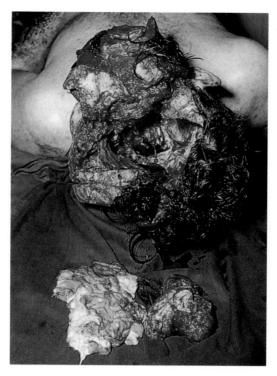

Figure 5.17 When a shotgun is fired into the head, the forceful entry of hot gas and shot into the cranium can cause the skull to literally explode as it is a rigid structure. In this case the suicide victim placed the barrel into his mouth and pulled the trigger. (Reproduced from Shepherd, R. (2003) *Simpson's Forensic Medicine*, 12th edn. Copyright 2003, Hodder Arnold, London.)

The presence of carbon monoxide (formed by incomplete combustion of the explosive charge) can cause the surrounding tissues to turn pink. The carbon monoxide binds to the haeme part of the blood pigment haemoglobin to form carboxyhaemoglobin and this is responsible for the cherry pink coloration. The consequences of gasses being forced into the body are particularly dramatic with shotguns and if these are discharged into the head it can result in the whole skull splintering and exploding (Fig. 5.17). This commonly occurs in suicides where the muzzle of the gun is placed in the mouth or underneath the chin. Where such appalling injuries are caused, it is essential to examine the whole body and the room in which it is found since it is possible that the person was murdered and then a fake suicide staged to mask the evidence. For example, was it possible for the victim to reach the trigger of the gun when holding the muzzle to their head? If their arms are not long enough, the suicide victim will sometimes use their toes or a prop to pull the trigger but if the body is fully shod and there is no prop then suspicions should be raised.

If the gun is held at a slight distance from the body, the gasses are able to escape, so the damage caused by ballooning is not seen. The circumference of the blacken-

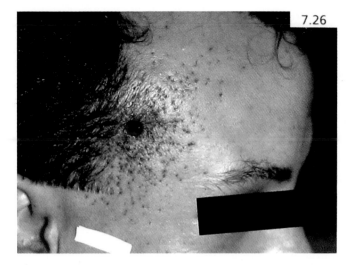

Figure 5.18 Gunshot wound to the temple. Note the circular entrance hole and powder tattooing surrounding the wound but limited amount of soot particles. This indicates that the woman was shot from an intermediate range. Precisely how far away would require experimentation with the same gun and type of ammunition that caused the wound. (Reproduced from Dolinak, D. *et al.*, (2005) *Forensic Pathology: Theory and Practice*. Copyright © 2005, Elsevier Academic Press.)

ing (smoke soiling) around the entry wound initially increases and then becomes more diffuse as the distance increases. Unlike hard contact wounds, it is possible to wipe away this blackening. The pattern formed by smoke soiling can indicate whether the gun was held at right angles or obliquely to the wound site. Smoke soiling usually ceases at a distance of about 1 metre. Among the gasses discharged from a gun are tiny fragments of extremely hot propellant and when they come into contact with the skin they cause a characteristic pattern of burns called 'powder tattooing'. The diameter of the powder tattooing indicates the distance between the victim and the muzzle of the gun but like smoke soiling it is not normally found beyond one metre (Fig. 5.18).

Shotgun wounds resulting from a gun fired from less than about 3 metres away commonly includes fragments of wadding. This can be useful in determining the type of shot and the gauge of the gun that was used but will not identify the individual gun. With increasing distance it is less likely that any pellets will pass through the body and above 20–50 metres the body might be peppered with shot but the wounds are unlikely to be fatal.

If the victim is shot through clothing it can affect the appearance of the wound. For example, soot and fragments of propellant may not reach the skin and consequently there may not be any smoke soiling or powder tattooing. However, marks may be found on the clothing and fragments of clothing material might be found along the wound track. As with stab wounds, aligning the entry holes in the clothing with those in the body can indicate the position of the victim (i.e. sitting, standing, arms raised etc.) at the time he was shot. Obviously, a great deal depends on the nature of the clothing, the distance and the weapon that was used. In addition, the use of silencers may produce unusual entry wound characteristics. However, there is a great deal of variation between guns and the type of shot used also affects the pattern. It is therefore essential to conduct tests on any gun that might have been used in a crime before firm conclusions are reached.

The influence of firearm velocity on wound characteristics

The seriousness of gunshot wounds depends to a large part on the amount of energy the bullets impart to the body and this in turn depends on the type of weapon, the characteristics of the ammunition, the distance and orientation of the victim from the muzzle of the gun, and the stability of the bullets' flight. Obviously, a bullet travelling at high velocity imparts more energy, and therefore causes more damage, than the same sized bullet travelling at low velocity. Similarly, a heavy calibre bullet will cause more damage than a small calibre bullet travelling at the same speed. Both low and high velocity firearms cause penetrating injuries resulting in direct damage to the tissues as the bullet traverses the body. Because of their greater kinetic energy, bullets from high velocity firearms are more likely to pass through the victim. In the case of low velocity weapons, the bullet forms a permanent cavity that is an accurate reflection of the amount of damage caused. By contrast, a bullet from a high velocity firearm has far more kinetic energy and when it hits a body it causes the formation of a temporary cavity that is much wider than the diameter of the bullet. This temporary cavity lasts only a few milliseconds and travels through the body with the bullet, collapsing behind it in a series of pulses: this causes a wide area of bleeding around a much smaller permanent cavity that may be little wider than the diameter of the bullet (Fig. 5.19). The consequences of the pressure waves vary between tissues: the denser the tissue the easier it is for the bullet to transfer its kinetic energy and therefore the greater is its potential to suffer damage. However, the risk of damage is reduced if the tissue is able to dissipate the absorbed energy

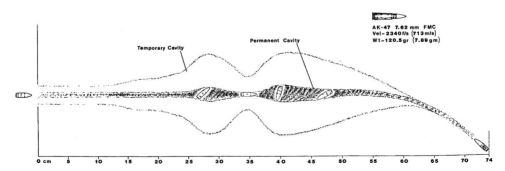

Figure 5.19 Diagrammatic representation of permanent and temporary cavity formation by a bullet fired from an AK-47 assault rifle. Note how the bullet tumbles during its flight. (Reproduced from Mahoney, P.F. *et al.*, (2005) *Ballistic Trauma*, 2nd edn, with kind permission of Springer Science & Business Media.)

through expansion and contraction. The tissues of the lungs, having relatively low density and also being highly elastic, are therefore the least likely to suffer harm in this way. Similarly, although muscle tissue is dense it is highly elastic and therefore protected to some extent. By contrast, dense, inelastic tissues, such as the brain, spleen and liver, may burst, as they are unable to expand and contract as the pressure waves pass through. Similarly, pressure waves can cause bones to shatter whilst the large intestine may rupture as the gasses within it expand. If they hit bone, bullets from both low and high velocity firearms are likely to cause it to fragment. This results in secondary projectiles within the body that cause their own wound tracks. It is also likely to alter the trajectory of the bullet itself. Some workers have attempted to use the direction of the bevelling produced in bones as the bullet exits from them as means of estimating the direction of fire but this is not particularly reliable (Quatrehomme & Iscan, 1997).

Differentiating suicidal, accidental and homicidal gunshot wounds

Differentiating between suicidal, accidental and homicidal gunshot wounds is not always easy. For example, in the case of suicide and accident, the weapon may be flung far from the victim's grip. Furthermore, if there is some time before the authorities discover the victim, the weapon may be missing because it is stolen or hidden by someone else. This might happen because a relative considered suicide to be a sin and therefore should be 'covered up', or a verdict of suicide might negate a life insurance claim or the firearm may have 'resale value'. These possibilities should be considered on the basis of the situation in which the body is found. The presence of powder tattooing can prove useful in determining the distance from which the person was shot – and hence whether suicide was physically possible. The number of gunshot wounds is not always a good indication since suicidal individuals may shoot themselves more than once. However, the location of the wounds is more helpful. Suicides tend to shoot themselves in the head, usually in the right temple if it is with a handgun – the choice of temple tends to be determined by whether or not a person is right- or left-handed but it is not an infallible predictor. Chest wounds may also result from suicide attempts. Gunshot wounds resulting from an accident can appear virtually anywhere – especially if the bullet ricochets before entering the body.

Detecting gunshot residues

When a person handles a gun, they will leave behind their fingerprints; if they fire the gun, then their hands and clothes will be contaminated with gunshot residues. These residues contain metals such as lead, antimony and barium that can be detected by flame atomic absorption spectrophotometry (FAAS) and scanning electron microscope energy dispersive X-ray spectrometry (SEM–EDX). Firing a gun will usually result in residues being found on the backs of the hands and the palms whilst merely handling a gun that was recently fired will transfer residues solely to the palms. The latter situation may happen if the weapon is passed to an accomplice or if the victim holds the barrel of the weapon in an attempt to deflect it (or in the case of a suicide, to steady it). The residues are rapidly lost, especially if the hands are washed. Similarly, some firearms (e.g. some rifles) deposit low amounts of residues, which may lead to false negatives, whilst others, such as revolvers produce large amounts that may even contaminate bystanders and lead to false positives. Further potential problems with gunshot residue analysis can also arise (Mejia, 2005). For example, it is claimed that a person entering a room shortly after a shot was fired in it may be contaminated with more residues than the person who pulled the trigger and promptly fled (Fojtasek & Kmjec, 2005) and that particles with a similar composition to gunshot residues can be formed by fireworks and within brake linings. In the latter situation, it is suggested that a person servicing the brakes of a car may acquire residues on his or her hands and clothing that could be mistaken for gunshot residues (Cardinetti *et al.*, 2004).

Taking DNA samples from the bloodstains found on knives, hammers etc. has been common practice for many years however, a new field of research is being developed called bullet cytology (Knudsen, 1993) in which tissue samples and DNA are recovered from bullets. Bullets from high velocity guns often pass straight through a victim, so this technique could prove useful in linking the two together. However, there have been few publications on the topic.

Bite marks

Bites are a common feature of physical and sexual assaults and both the victim and the assailant are likely to bite one another. In a survey of biting injuries reported in court cases in the USA, 17% of bite marks were found on men, of whom 52% were the accused, whilst 83% were found on females, none of who were the accused (Pretty & Sweet, 2000). Male victims of crime tended to be bitten on the arms and back whilst the assailant tended to be bitten on the hands and arms. Female victims tended to be bitten mainly on the breasts (40%), arms (13%), legs (13%), face (7%), neck (7%), and genitalia (7%). In violent attacks, parts of the ear, nose and other facial features may be bitten off entirely. Assuming that the bite site can be rendered in sufficient detail it can link victim and suspect together since we all have unique dental characteristics (Fig. 5.20). Unfortunately, this is seldom possible and Pretty & Turnbull (2001) relate a case in which an assault took place in the course of which the victim was bitten. Although the dental characteristics of the two assailants were different, they were also sufficiently similar for two forensic dentists to arrive at opposite conclusions about which of them had inflicted the bite. Contradictory views like this are not unusual and consequently bite mark analysis is not as reliable as a means of identification as it is sometimes alleged.

One of the problems of bite mark analysis is that bites inflicted in anger are painful and unwanted and the victim therefore attempts to pull away. This results in the teeth being dragged through and over the skin so the marks are not clear. Furthermore, the bites are inflicted on three-dimensional, curved surfaces of soft skin but the comparisons of bite site and dental characteristics are usually done from two-dimensional photographs. A further complication is that bites inflicted

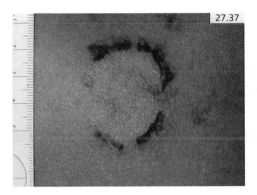
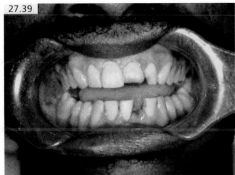

Figure 5.20 A man who was in the habit of biting the people he mugged left characteristic bite marks. (a) This bite was delivered to the scapula of one victim. It indicated that he had 'a missing or broken lower central incisor Figure and the upper arch would have broken or chipped maxillary central and/or lateral incisor teeth'. (b) This information led the authorities to focus on their search and one suspect was released. Compare the teeth of the man ultimately charged with the offences and the bite wound. (Reproduced from Dolinak, D. *et al.*, (2005) *Forensic Pathology: Theory and Practice*. Copyright © 2005, Elsevier Academic Press.)

whilst the victim is alive, cause bleeding, bruising and swelling and this can distort the bite site. However, these problems could be at least partially overcome through the use of computer-based image capture systems that enable three-dimensional records to be made (Thali *et al.*, 2003).

Not all bites are the consequence of assault. Love bites on the neck are very common, especially among teenagers. These, and other amorous bites are usually delivered slowly and the recipient does not draw away so aggressively. Consequently, there is not the evidence of dragging. However, during these bites, the skin is often sucked in and the tongue thrust forcefully against it thereby causing the formation of tiny red spots called petechiae in the centre of bite that result from blood leaking from damaged blood vessels. Furthermore, there may not be any evidence of tooth marks. Self-inflicted bites are commonly found on persons who become addicted to self-harm and may show similar characteristics. These bites will, however, not occur on areas of the body that are difficult to reach and there will probably be wounds of varying ages. Nevertheless, it is feasible for someone who bites themselves in this way also to be attacked and bitten by someone else and wounds of varying ages are often a feature of long standing abuse.

Bite marks are likely to contain the saliva and hence the DNA of the assailant and should therefore be swab sampled because this would provide the most reliable means of identification (Chapter 3). Unfortunately, in living subjects, the natural reaction of the victim is to wash the wound immediately. It is important to consider whether bites may have been inflicted through clothing in which case physical damage to the skin may be minimal but the clothing may yield the assailant's DNA.

Table 5.3 Distinguishing features of wounds associated with suicide and homicide

Wound type	Suicide	Homicide
Bruise	Not commonly seen in suicide attempts.	Common. Distribution provides evidence of assault and restraint.
Abrasion	Typically associated with hanging: a suspension point is usually obvious and there is no evidence of a preceding struggle.	Common. Caused during a struggle or when a body is dragged. Garrotting with a ligature: abrasion encircles neck and there is no suspension point.
Laceration	Not commonly seen in suicide attempts[a].	Common. Caused by being hit with heavy blunt object.
Incised	Slashing of wrist or throat. Evidence of tentative initial cuts at elective site. Incised wounds not found elsewhere unless past history of 'cutting'.	Common. May occur anywhere on body but often aimed at face and neck. Defence wounds may be found on hands and arms. Distribution indicates how assault progressed.
Stab	Typically aimed at heart. Clothing usually removed around wound site. Tentative initial cuts to test sharpness may be present. Multiple stab wounds possible.	Common. May occur anywhere on body. Stab wound delivered through clothing. Number may vary from one wound to many in a frenzied attack. Defence wounds may be found on hands and arms. Distribution indicates how assault progressed.
Broken bones	Not commonly seen in suicide attempts[a].	May occur during violent physical assault or shooting.
Gunshot	Typically contact or very close range and aimed at head or chest. May be more than one wound.	May be contact or from a far distance. Wound may occur anywhere on body. Wound analysis indicates position of assailant and victim.
Burns and scalds	Self-immolation is not a common form of suicide. However, localized burns and scalds may be found on a suicide victim from previous self-harm. Chemical burns to mouth and oesophagus are caused by suicidal ingestion of bleach or caustic soda.	Localized burns and scalds may result from torture before victim was killed. Burning and scalding not a common cause of homicide. Burning is a common means of disposal of a dead body. Presence of soot in lower respiratory tract and >10% carboxyhaemoglobin in blood indicates victim was alive when fire commenced.

[a]Suicide resulting from jumping from a tall building or in front of traffic or a train may result in severe abrasions, lacerations, broken bones and even amputations.

Burns and scalds

Medically, a burn is defined as a wound that results from dry heat, such as touching a hot plate, whilst a scald results from wet heat, such as exposure to steam or molten metal. However, in colloquial speech, a burn is often used to describe the wounds that results from all these situations.

There are several classification systems for categorizing the severity of burns, although the most commonly used is to divide them into three 'degree categories'. 'First degree burns' are those in which there is reddening of the skin and a fluid filled blister may form. These superficial burns usually heal without leaving any scarring within 5 to 10 days. In 'second degree burns' the epidermis of the skin is removed down to the dermis whilst in 'third degree burns' tissue lying beneath the dermis is also damaged, the underlying organs may be exposed and there may be blackening resulting from carbonization. Scalds can be categorized on a similar basis although carbonization would only result from exposure to molten metals or similar high temperature fluids. In terms of survival, the extent of the burns or scalds is of as much importance as their category. If over 50% of the body surface is affected, the chances of survival are poor, although it is difficult to generalize as there is a great deal of difference between individuals, and factors such as age, health and the speed with which medical assistance is administered all play a part in recovery.

The majority of people who die in house fires succumb to the inhalation of smoke and toxic fumes rather than being burnt to death. The presence of soot in the mouth and nasal passages is not a reliable indication that a person was alive at the time a fire was started because it could have entered passively. By contrast, the presence of soot in the lungs and air passages below the vocal cords is a very good indication, as is a saturation level of carboxyhaemoglobin in the blood in excess of 10%. Carboxyhaemoglobin results from the inhalation of carbon monoxide produced by incomplete combustion of flammable materials. The blood and tissues appearing cherry pink intimate the presence of large amounts of carboxyhaemoglobin in the body. However, the level of carboxyhaemoglobin needs to be confirmed by chemical analysis because low levels (less than 10% saturation) might be occur naturally from living near to a pollution source or from being a heavy smoker. It is also important to remember that being alive is not the same as being conscious and the victim may have died before the flames reached them. Similarly, the absence of either of these indicators does not mean that the victim must have already been dead: a great deal depends on the individual circumstances and the amount of smoke and carbon monoxide produced. It is difficult to distinguish between burns caused before, at the time of, or after death. In the case of flash fires or explosions, in which death may be virtually instantaneous, there might be extensive surface burning but little evidence of soot in the lungs or carboxyhaemoglobin in the blood.

A body exposed to high temperatures automatically adopts a characteristic 'pugilistic posture' in which the limbs become flexed. This occurs regardless of whether the person was alive or dead at the time of exposure (Fig. 5.21). As the skin and underlying tissues dry out, they contract and splits occur that can be mistaken for wounds. The drying out of the muscles induces them to contract and as the flexor muscles are larger and more powerful than the extensor muscles, they pull the body into the 'pugilistic posture'. The bones can fracture and break and a 'heat hae-

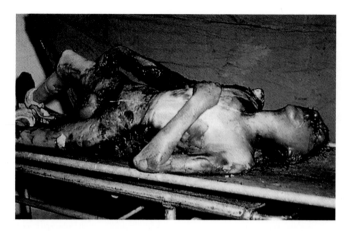

Figure 5.21 'Pugilistic posture' typical of a body recovered from a fire. (Reproduced from Shepherd, R. (2003) *Simpson's Forensic Medicine*, 12th edn. Copyright 2003, Hodder Arnold, London.)

matoma' may form in the skull – leading to suspicions that the victim may have been assaulted. A haematoma is a blood clot that forms outside of a blood vessel and this is often a feature of a skull fracture or a serious blow to the head.

If a body is burnt, it will inevitably destroy a great deal of forensic evidence. Consequently, it is common for a murderer either to remove the body and burn it elsewhere or to burn it *in situ*. An examination of the scene is therefore essential in determining the likelihood of a burnt body being the victim of a suicide, a homicide or an accident (Fanton *et al.*, 2006). It is rare for persons to commit suicide by burning in the UK although there is a suspicion that it may be under-reported because the coroner records an 'open verdict' rather than 'suicide'. Usually, the victim covers themselves in a flammable liquid, residues of which are detectable, a means of ignition is found nearby, there is no evidence of restraint or injuries caused before death and typically, the soles of the feet suffer little if any burning. In the case of accidental death, there is evidence of an innocent cause, such as a chip pan fire or smoking in bed and that the person had attempted to escape unless incapacitated through age, a medical condition, or intoxication. This would be indicated by the position of the body – usually the victim attempts to reach a door or window, or if cut off, to hide in a corner or cupboard. In a homicide, the fire is started deliberately and the victim may, if not already dead, be restrained in some way, such as being tied to a chair or confined to the boot of a car. The distribution of burns on the body and clothing can also provide an indication of the person's position. For example, extensive burns to the top of the head and face are suspicious in a reported kitchen accident involving clothing being set alight. In this scenario, the possibility that a flammable liquid may have been poured over the victim and then set alight should also be considered.

Most burns and scalds result from accidents but they are also a common feature of self-harm, the abuse of children and old people and deliberate torture (Greenbaum *et al.*, 2004). Both self-harm and the abuse of others often occurs over a long period of time, so there may be wounds of various types and varying stages of healing – an interesting case of this involving the use of a hot glue gun is discussed by González *et al.*, (2007). The distribution of the wounds will indicate whether the victim could have caused them themself; accidental burns are much less likely to be bilateral (e.g. to afflict both right and left hands) than those inflicted intentionally by a third party. Scalds often form trickle-shaped wounds indicating the flow of the fluid and hence the orientation of the victim at the time. Torture with hot implements usually takes place over a restricted period and is often accompanied with physical beating. If the implement is dry, such as an iron, its shape will be reproduced in the burn. The use of electric shock torture is sometimes difficult to demonstrate because although it is extremely painful, provided the electrodes are not applied for too long, there may be little associated pathology.

Chemical burns

Chemical burns are seldom encountered in a forensic context although occasionally people commit suicide through drinking acid, caustic soda, or some similar compound. In these circumstances the source of the poison is found nearby and there is extensive burning to the lips, mouth, and gastrointestinal tract. These chemicals cause extremely painful burns and so there are usually trickle burns down the body on both skin and clothing as a consequence of spillage and vomiting. Acid is often used as an offensive weapon, particularly in some countries/among certain communities (for example, Bangladesh) and whilst it causes horrific injuries it is seldom fatal. Dead bodies are sometimes found with ribbon-like burn patterns as a consequence of them being covered in bleach to remove forensic evidence (Adair *et al.*, 2007). The smell of bleach may still be apparent and the pattern reflects the position of the body when the bleach was poured on (e.g. it travels downwards and collects in natural hollows) (Fig. 5.22). The presence of bleach damage under and around the body would indicate that the bleach was employed at the site where the body was found.

Ageing of wounds

An accurate estimate of the age of a wound is important in many scenarios and some of these are listed in Table 5.4.

The formation of a wound results in the death of cells and if the victim remains alive it is followed by an inflammatory reaction in which cell debris and, if present, any fragments of the object causing the wound are removed. During this stage, immune cells (e.g. lymphocytes, macrophages and neutrophils) produce inflammatory chemicals called cytokines such as interleukin-1α, interleukin-1β and tumour necrosis factor α. The inflammatory substances cause the blood capillaries to dilate and become more permeable and thereby allow white blood cells (leucocytes), cytokines and the chemicals responsible for the blood clotting process to reach the injured region. This, so called 'vital reaction' is sometimes quoted as an indication that a person was alive when a wound was inflicted whilst its absence is taken as indicating that the wound was inflicted after death. Although the absence of leuco-

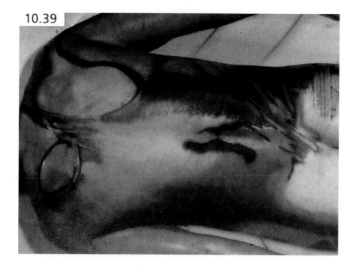

Figure 5.22 Bleach was poured over the body of this man after he was dead as part of a 'clean up' operation. Note the swirling patterns caused by the way the bleach has run and collected in the body's hollows. (Reproduced from Dolinak, D. *et al.*, (2005) *Forensic Pathology: Theory and Practice*. Copyright © 2005, Elsevier Academic Press.)

Table 5.4 Summary of some of the instances in which an accurate estimation of when a wound was inflicted or, in the case of multiple wounds, their sequence of delivery is required.

(1) If the injured person is alive to corroborate their account of when the wound was inflicted. For example, the suspect in a case of homicidal assault may have wounds to his knuckles but claims that these were received during a fight with another man either before or after the date on which the victim died.

(2) To determine whether all the wounds on the body were inflicted at the same time or if over a prolonged period how long this was. This can be important in cases of alleged child abuse and torture.

(3) If the victim is dead to determine whether the wounds were inflicted before, at the time of, or after death. For example, the body may exhibit evidence of many past injuries but these occurred months or years ago and may have no relevance to the cause of death.

cytes congregating on the surface of capillaries and infiltrating the wound region is a good indicator that the wound was inflicted after death it should be recognized that cellular processes do not cease immediately after the heart stops beating. Consequently, wounds inflicted immediately before, during and after the moment of death may not be distinguishable from one another by this means. Whilst the inflammatory process is ongoing, tissue repair commences with the stage known as 'organization' or 'proliferation' in which the blood supply is restored to the damaged region. In this stage, any blood clots are replaced with granulation tissue that contains delicate capillaries and proliferating fibroblasts (these cells secrete collagen, elastin and other proteins that provide structural stability). Finally, there is the stage of regeneration and scar formation although the capacity for this varies between tissue types and the nature of the wound. The sequence of these histological (cellular) and chemical changes can be used to age wounds but the accuracy is compromised by the complicating factors of age, health and nutrition.

Kondo *et al.* (2002) found that changes in the expression of ubiquitin (a low molecular weight regulatory protein) within the nuclei of neutrophils, macrophages and fibroblasts could be used to age wounds with a high degree of accuracy in living subjects but it is not known whether the technique would remain effective after death although Ohshima (2000) has described how the levels of a number of cytokines and their coding mRNA sequences might be used to age wounds found on corpses.

Post mortem injuries

Post mortem injuries can result from malicious and accidental events as well from as biological and environmental taphonomic processes and if there were no witnesses to the person dying, their interpretation can present difficulties. For example, an elderly person suffering a heart attack could fall down the stairs and suffer serious injuries. On discovery of the body it would be necessary to exclude the possibilities that the victim either first received a blow to the head or was pushed downstairs by an intruder and the shock induced the heart attack. This scenario might be resolved by a combination of bloodstain pattern analysis, fingerprinting and wound analysis. Incidentally, it is extremely rare for a person to suffer fatal injuries as a result of falling downstairs unless they were suffering from alcohol or drug-induced intoxication or from an underlying medical condition (Bux *et al.*, 2007).

Injuries caused days or months before death are usually recognizable from signs of healing but it is difficult to generalize about post mortem injuries (Byard *et al.*, 2002). Post mortem bruising may occur when a dead body is being recovered or handled in the morgue although these bruises tend to be small. Post mortem bruising is most likely to occur on the lower body surfaces because this is where blood pools. Decaying soft tissues are very delicate and easily damaged so it is normal to undertake a preliminary examination before the body is moved to the morgue. Clumsy handling can result in limbs becoming detached and bones broken. Vertebrates can cause extensive post mortem injuries up to and including consuming the whole corpse (Chapter 8). Invertebrates can also cause damage that may result in confusion at the time of autopsy. Feeding by dermestid beetles and their larvae produces irregular holes and tears in mummified skin that can be mistaken for wounds caused before death. These holes tend to have jagged edges, whilst those produced by maggots tend to have smooth edges. Both histerid beetles and burying beetles (Silphidae) will cause circular-shaped holes in a body. Bite marks inflicted by the cockroach *Dictyoptera blattaria* have been mistaken for evidence of

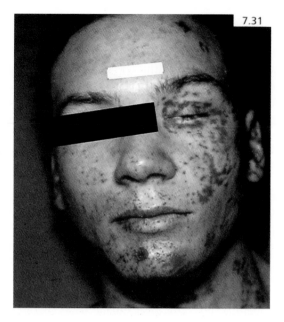

Figure 5.23 A bullet entrance wound is visible on the right hand side of this man's chin and there is associated powder tattooing indicating that he was shot at close range. On the left of his face, around his left eye and also on his neck are numerous irregular bright red marks that in places have merged together rather like a rash that were caused by ants. (Reproduced from Dolinak, D. *et al.*, (2005) *Forensic Pathology: Theory and Practice*. Copyright © 2005, Elsevier Academic Press.)

nonaccidental injury or even strangulation in cases of sudden infant death (Denic *et al.*, 1997). Ants can cause post mortem damage that resembles pre mortem strangulation or burns (Fig. 5.23) (Byrd & Castner, 2001; Campobasso *et al.*, 2004a).

Jayaprakash (2006) discusses a number of interesting case studies from India in which ants inflicted post-mortem injuries. For example, in one case the body of a young woman was found hanging by a rope with copious amounts of blood trickling vertically down her legs and dripping onto the floor where it formed a large pool. Initially, it was suspected that she had been raped and strangled after which the culprit(s) had attempted to cover up the crime as suicide. Four men were arrested and charged with her murder. However, subsequent analysis indicated that the blood was issuing from crater-like lesions caused by ants and not from wounds to the victim's genitals. These lesions were found on various parts of the body but those to the lower regions bled especially freely because blood collects in the dependent regions after death as a consequence of hypostasis. The pattern of the bloodstain tracks would have been quite different if the blood had issued from wounds caused before death/being placed in the hanging position. Although worker caste ants lack wings they have no problems locating a hanging body by climbing the surrounding walls and ceiling/branches and then descending the rope.

Future developments

Forensic pathology is a very difficult and complex subject that relies to a great extent on the personal experience and opinions of the individual practitioner. The reliability of many aspects of pathology would therefore be enhanced through the development of specific molecular markers. This is unlikely to be a simple process but subjects such as wound age are potentially amenable for this approach.

Quick quiz

(1) Distinguish between a bruise and an abrasion.

(2) A man is found with three stab wounds to the chest. What other evidence would you look for to determine whether the man was murdered or committed suicide?

(3) Under what circumstances can a knife form a wound track longer than the blade?

(4) A young man has been found suspended from a leather dog leash. His neck has been broken. Why is this death suspicious?

(5) Distinguish between the circumstances in which permanent and temporary cavities would be caused in firearm wounds.

(6) Write short notes on three causes of asphyxia.

(7) What is a Dum Dum bullet?

(8) What evidence would you look for to determine whether a person was alive or dead at the time a fire started?

(9) Distinguish between the characteristics of a self-inflicted bite, a love bite and an aggressive bite at evidence would you look for to determine whether a person was alive or dead at the time a fire began?

(10) State two means by which you could distinguish between a wound caused after death from one inflicted whilst the victim was alive?

Project work

Title

The influence of biological decay on bullet markings.

Rationale

When a bullet is fired from a gun, it acquires a unique set of grooves that can be used to identify the gun that fired it. Decaying tissues and microbial action surrounds a bullet that remains in a dead body and this environment may affect the preservation of these identifying characters.

Method

Used bullets fired from a single gun (the local gun club or Territorial Army brigade may be able to oblige) would be collected and placed in meat (e.g. minced beef or liver) that was then allowed to decay either above or below ground. This would therefore determine whether the presence of maggots influenced the results. The markings on the bullets would be compared at varying time intervals.

PART B Invertebrates and Vertebrates

6 Invertebrates 1: Biological aspects

Chapter outline

An Introduction to Invertebrate Biology
Invertebrates as Forensic Indicators in Cases of Murder or Suspicious Death
Invertebrates as a Cause of Death
Invertebrates as Forensic Indicators in Cases of Neglect and Animal Welfare
The Role of Invertebrates in Food Spoilage and Hygiene Litigation
The Illegal Trade in Invertebrates
Invertebrate Identification Techniques
Future Directions
Quick Quiz
Project Work

Objectives

Review the biology of those invertebrates that are most commonly associated with dead bodies.

Explain how invertebrates may be either a direct or indirect cause of human fatalities.

Explain how invertebrates may become involved in cases of neglect and animal welfare.

Discuss the role of invertebrates in food spoilage and hygiene litigation.

Discuss the illegal trade in invertebrate species.

Compare and contrast the benefits and limitations of different techniques for identifying invertebrates in a forensic context.

An introduction to invertebrate biology

Invertebrates are metazoan animals that lack a backbone. They include creatures as diverse as mites smaller than the nucleus of a large protozoan to giant squid several

Essential Forensic Biology, Second Edition   Alan Gunn
© 2009 John Wiley & Sons, Ltd

metres long. The arthropods are but one group of invertebrates albeit they are the most successful in terms of their abundance, biomass, and numbers of species. Arthropods are characterized by their possession of a hardened exoskeleton that is shed periodically to accommodate growth and they have specialized jointed appendages. Typical examples are scorpions, crabs and moths. It is the arthropods, and in particular the insects, that are the most common sources of forensic evidence. The study of insects is referred to 'entomology', a word derived from the Greek words *entomon* (an insect) and *logos* (science). Logic would therefore suggest that forensic entomology is the use of insects in legal investigations. However, many workers restrict the definition to the use of insects in murder and suspicious death investigations thereby excluding the numerous other legal cases in which insects are involved. Sometimes, the term 'medicocriminal entomology' is used to emphasize this restricted definition. This chapter will cover the basic biology of some of the most forensically important invertebrates but those seeking more detail are advised to consult Gennard (2007).

Insects account for 72% of all known species of animals and together with other terrestrial arthropods they can be found in virtually every terrestrial habitat, both natural and man made, from the polar regions to the equator. Within these regions, insects and other arthropods occupy virtually every conceivable ecological niche. Some are specialist herbivores, carnivores, detritivores, or parasites, whilst others are generalists that consume a variety of foodstuffs. Their lifestyles sometimes bring them into conflict with mankind because they consume the same food as ourselves, use us as their food, or occupy our dwellings. Knowledge of their biology is therefore useful in reducing the harm that certain species may cause either directly or indirectly (e.g. through spreading diseases) and in understanding how they can provide forensic evidence. However, the vast majority of arthropods and other invertebrates are harmless to mankind, many are extremely valuable as food, pollinators of our crops or biological control agents, and they all have an important role in the normal functioning of ecosystems.

Invertebrates as forensic indicators in cases of murder or suspicious death

The invertebrate species that provide forensic evidence in cases of murder or suspicious death investigations can be broadly grouped into those that are attracted to dead bodies, those that leave dead bodies, and those that become accidentally associated with the dead body and/or the crime scene.

Invertebrates attracted to dead bodies

Numerous invertebrate species are attracted to dead bodies including detritivores that feed on the decaying tissues, carnivores and parasitoids that come to prey on the detritivores, and coprophiles that tend to feed on faeces rather than the decaying tissues of the dead body.

Detritivores

Detritivores are creatures that consume dead organic material. Some detritivores are specialists but many of them also consume other foods to a greater or lesser extent depending upon circumstances. Similarly, even classic predators such as geophilomorph centipedes will consume dead organic matter.

A dead body represents a temporary source of easily metabolized organic matter that, unlike that of living organisms, is chemically and physically unprotected. In the initial stages, it therefore attracts highly mobile r-type species that are adapted for finding and exploiting such temporary food sources: r-type species are typified by short life cycles and the ability to undergo explosive increases in population under the right conditions. They rapidly consume the most easily degradable tissues until all that remains is material that does not have sufficient nutritive value to support their growth. Therefore, with time, the r-type detritivore species are replaced by species with longer lifecycles and lower reproductive rates that are able to exploit food that is more difficult to metabolize and has a lower nutritive value.

Blowflies

Flies (order Diptera), especially blowflies, are usually the first organisms to arrive at a corpse (Fig. 6.1). Sometimes they arrive within minutes of death and they are the species of greatest forensic importance (e.g. Arnaldos *et al.*, 2005; Byrd & Castner, 2001; Goff, 2000). Blowflies belong to the family Calliphoridae and are commonly called bluebottles or greenbottles. Numerous species of blowfly exist and it is not unusual to find several species on a single corpse. However, not all blowfly species utilize corpses and corpses may contain the larvae of many different species of Diptera other than those belonging to the family Calliphoridae. The common name 'blowfly' is derived from the noun 'blow' which means a mass of fly eggs.

Figure 6.1 Blowflies ovipositing within and around the nasal cavity of a sheep. The flies began laying eggs within 5 minutes of the animal being killed.

Food, a wound or a corpse covered in fly eggs is therefore said to be 'fly-blown' and the insect responsible is a 'blowfly'. The common names 'bluebottle' and 'green-bottle' may be derived from the Gaelic '*boiteag*' meaning a 'maggot'. If this is correct, a greenbottle is therefore a green fly that produces maggots (Erzinclioglu, 1996). Because many flies look superficially similar it is better to use proper taxonomic terms rather than words such as 'greenbottle' etc.

Blowfly life cycle

The life cycle of 'typical' blowfly species, such as those found on dead bodies, is straightforward. The gravid female fly (i.e. one ready to oviposit) lays her eggs on the corpse. She usually chooses one of the natural openings, such as the nose, ears and mouth, the eyes, or the site of a wound. Blowflies are attracted by the odour of blood and several species will lay their eggs on the wounds of an injured animal before it dies. The anus and genitalia may be used if these are accessible and eggs may also be laid on the vegetation underneath a corpse. Blowflies colonize corpses that are fresh or at the early stages of decay and although the adult flies may feed on a body that is dried out or skeletonized they would not lay their eggs upon it. Most adult blowflies are only active during daylight hours and when the temperature is high enough to permit flight. There is some uncertainty concerning the ability of blowflies to locate corpses at night or under low-light conditions but work by Wooldridge *et al.* (2007) indicates that *Calliphora vomitoria* and *Lucilia sericata* have only a limited ability to locate corpses during the hours of darkness. However, Wyss & Cherix (2002) found that *Calliphora vicina* were able to locate, colonize and grow upon remains found 10 metres inside a cave at a point where there was total darkness and the ambient temperature was a constant 5 °C. The unwillingness of blowflies to lay eggs upon remains at night may therefore be more associated with their natural cycles of activity rather than an inability to navigate when it is dark. There is a similar degree of uncertainty concerning the extent to which blowflies can exploit buried bodies. Most of the literature states that even shallow burial prevents blowfly colonization. This is because the adult flies do not burrow and they must usually have physical contact with the dead body before they will lay their eggs. Consequently, they are unlikely to lay their eggs on the surface and leave the larvae to burrow down to locate their food after hatching. However, *Lucilia sericata* will occassionaly lay eggs in the soil above buried remains and their larvae then burrow down up to 5cm to reach their food (personal observations). In some of these reports the maggots were not identified (e.g. Weitzel, 2005) so it is possible that they belonged to muscid and/or phorid flies some of which are often found on buried bodies or alternatively the maggots might have resulted from eggs deposited on the body before it was buried. In addition, blowflies might gain access to a body buried in a shallow grave if the overlying soil contains cracks, has a loose texture or there are tunnels to the body created by mice and rats.

Blowfly eggs are laid in batches that may number up to 180 and a female fly may lay a total of several thousand eggs over the course of her life. When an egg hatches a first instar larva emerges. At this stage it is small and delicate (Fig. 6.2) and it rapidly moves to where it can find optimum conditions – not too damp and not too dry. Blowfly larvae feed using chitinous mouth-hooks that drag material into their

Figure 6.2   Mature blowfly eggs laid upon vegetation underneath a body. The developing larvae can be seen within some of the egg cases.

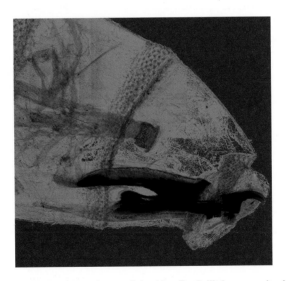

Figure 6.3   Anterior of the third instar larva of the blowfly *Calliphora vomitoria*. The cephaloskeleton has a large surface area for the attachment of muscles, thereby allowing the mouth hooks to be dragged back and forth with considerable force.

oral cavity (Fig. 6.3). They consume both the tissues of the corpse and the microbes that grow on it. In addition to the physical act of feeding, maggots also release enzymes and other substances that help to break down the underlying substrate. After approximately 24–48 hours, the larva moults to the second instar and after a further 24–48 hours, it moults to the third instar. The third instar larva feeds voraciously and rapidly increases in size and weight over 3–4 days (Fig. 6.4). Devel-

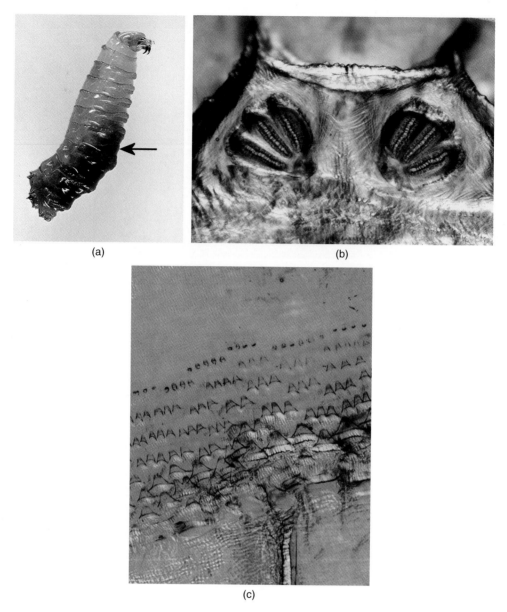

(a)

(b)

(c)

Figure 6.4   (a) Typical third instar blowfly larva (*Calliphora vicina*). The full crop indicates that this larva has not yet finished feeding. (b) Posterior spiracles of a third instar larva (*Lucilia sericata*). (c) Body spines on the third instar larva of *Calliphora vomitoria*. The body spines enable the maggot to grip the substrate. The shapes and distribution of body spines are often used as taxonomic indicators. For example, the body spines of *C. vomitoria* are larger and less pointed than those of *C. vicina*.

opment times are strongly influenced by temperature and some blowfly species develop quicker than others. Blowfly larvae tend to develop fastest when reared in groups and their development rate increases with larval density (Ireland & Turner, 2006). This is probably due to improved feeding efficiency from the combined effects of their digestive secretions and the heat generated when large numbers of third instar larvae are congregated together. In many species, competition does not appear to induce cannibalism although most experiments are done with groups of related individuals of the same age. Increased competition and starvation does, however, result in reduced size at pupation. The composition of the food can also affect development times and larval size (Green *et al.*, 2003) but a fresh, whole, human body will present maggots with more than enough food for at least their first generation to complete their development. Consequently, these factors are unlikely to be important in most PMI calculations (Chapter 7). Separated body parts, such as hands, may present a more limited food resource, especially if the maggot population is high. Although the larvae of many blowfly species are detritivores some of them will also act as cannibals/ facultative predators of their own and other species of maggot and their presence means that a degree of caution needs to be taken when interpreting PMI calculations. One of the best known examples of this is *Chrysomya albiceps* (Fig. 6.5) (Faria *et al.*, 2004). Although this is among the first species to colonize a dead body it is also possible that it could arrive

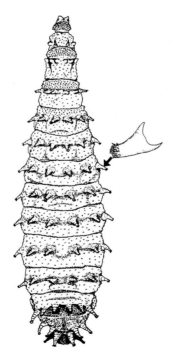

Figure 6.5 Diagrammatic representation of the third instar stage of *Chrysomya albiceps*. This species is predatory on other blowfly larvae. (Reproduced from Zumpt, F. (1965) *Myiasis in Man and Animals in the Old World*. Copyright © 1965, Butterworths.)

subsequent to other species but then kill off its competitors. *C. albiceps* is not present in the UK but it is found in central Europe and has been recorded as far north as Belgium.

Once a third instar larva has developed sufficiently, it empties its gut and (usually) leaves the corpse in search of somewhere to pupate. These maggots can be distinguished from those that are actively feeding because their gut is clear rather than dark – this can be useful in determining the age of a larva. Pupation usually takes place in the soil and its duration, like all the other stages, is highly temperature dependent and may last from just a few days to several months. As a maggot begins to pupate, its body contracts and when it moults the third instar cuticle is retained to form a protective puparium surrounding the pupa. The larvae of species such as *Lucilia sericata* and *Calliphora vicina,* may travel several metres from the corpse in search of a pupation site but others, such as *Protophormia terraenovae,* seldom move far, and may pupate among the clothing associated with the body (Gomes *et al.,* 2006).

After the adults emerge, they fly off in search of food. In addition to carrion, blowflies also feed on nectar (they can be important pollinators of flowers), honeydew, dung, rotting fruit and other decomposing matter. Blowflies are strong fliers and this flight is primarily fuelled from the glycolytic breakdown of carbohydrates – principally from the glycogen reserves in the fat body. They are therefore not capable of the long sustained flight of migratory insects such as locusts and some moths and butterflies which use lipid as their main flight fuel. Nevertheless, over time they may disperse considerable distances through a combination of directed and random flights. For example, in South Africa *Chrysomyia albiceps* were recovered up to 37.5 km from their release site after seven days (Braack & Retief, 1986) although the movements of *Lucilia sericata* in the UK appear to be much more modest in comparison (Smith & Wall, 1988).

In many species, the female flies need a protein meal before they can lay their eggs although this may not be necessary for the males to mature their sperm. It is said that female blowflies typically only mate once (this is unusual in all animals) although the male may mate many times (Erzinclioglu, 1996). Davies (2006) found that under field caged conditions only 20% of female *Calliphora vicina* had laid any eggs by 32 days after emergence. He found that there was a similar but not so pronounced delay in oviposition by adult *Lucilia sericata* and for both species their fecundity was highly variable and generally less than expected from the temperature the flies experienced. The adult lifespan varies between species and can be from a few weeks to several months. Those species that overwinter as adults may live for several months at this time of year although adults of the same species of fly emerging during the spring or summer probably have a much shorter lifespan. Most adult lifespan studies involve laboratory-reared insects and this can result in overestimates since in the wild the majority of flies would die from predation, starvation and disease before reaching their maximum potential age.

Fleshflies

Sarcophagid flies (Fig. 6.6), commonly known as fleshflies, often compete with blowflies during the early stages of decomposition. Some workers state that they

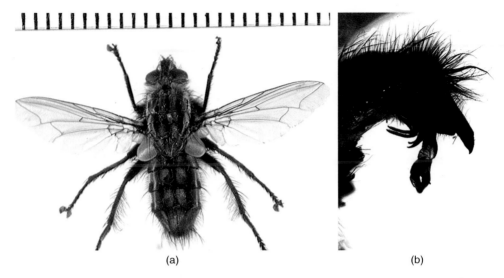

(a)   (b)

Figure 6.6 (a) Characteristic features of the adult fleshfly *Scarcophaga carnaria* (Sarcophagidae). Note the large pluvillae ('feet'), robust body, and grey longitudinal stripes on the thorax. In many species the eyes are bright red but this fades after death, as in this specimen. Unlike blowflies, the body of fleshflies is never 'metallic' and the abdomen often appears 'tessellated', i.e. has a pattern of black spots on a grey background. Scale = mm. (b) Male sarcophagids often have large complex genitalia and their characteristic shape is often used in taxonomy.

are among amongst the first insects to arrive at a corpse whilst others consider them to arrive after the blowflies have established themselves. The discrepancy probably results from differences in biology between the various species of *Sarcophaga* and the individual circumstance. Adult *Sarcophaga* species are reportedly more willing to fly during wet weather than blowflies and may, therefore, be among the first to colonize a corpse under these conditions (Byrd & Castner, 2001). Adult fleshflies are usually greyish in colour with longitudinal dark stripes on their thorax and bright red eyes, although the red colour fades after death. The tarsal claws and the pluvilli (tarsal pads) are usually large and give the fly an appearance of being 'big-footed'. The eggs hatch in the female fly's reproductive tract and she therefore lays first instar larvae. In addition to corpses, the larvae of various species of *Sarcophaga* are also found in dead organic matter and faeces etc. The larvae can be distinguished from those of blowflies etc. by their spiracles, which are partially hidden in a deep cavity, but they are otherwise very difficult to identify (Fig. 6.7).

Phorid flies (scuttle flies)

Phorid flies (Fig. 6.8) belong to the family Phoridae – a large group that contains over 2500 species and examples of virtually every lifestyle (Disney, 1994). They are all small flies, 1.5–6.0 mm in length, and have a characteristic 'humped' profile. The name 'scuttle fly' is derived from their rapid running behaviour. Owing to their small size, phorid flies can gain access to containers and rooms that would exclude

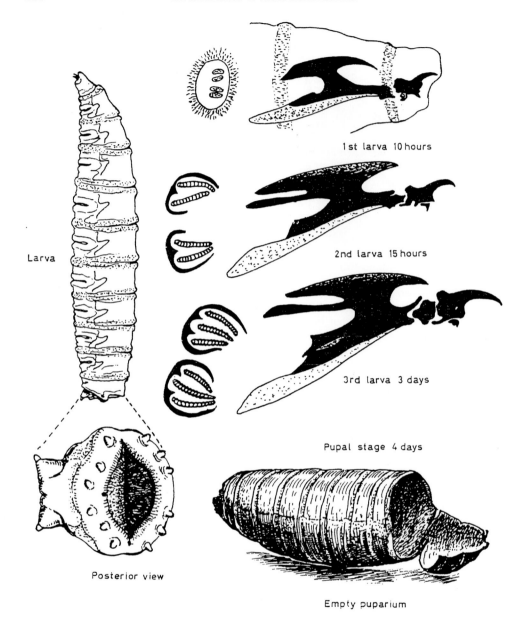

Larva

1st larva 10 hours

2nd larva 15 hours

3rd larva 3 days

Pupal stage 4 days

Posterior view

Empty puparium

Figure 6.7 Diagrammatic representation of the developmental stages of the fleshfly *Sarcophaga haemorrhoidalis*. Note the changes in the structure of the cephaloskeleton and posterior spiracles and how the posterior spiracles are not readily visible. *S. haemorrhoidalis* larvae are mainly found in faeces. (Reproduced from Zumpt, F. (1965) *Myiasis in Man and Animals in the Old World*. Copyright © 1965, Butterworths.)

Figure 6.8 Two adult phorid flies (*Megaselia scalaris*) and an empty puparium. Note the humped profile and relatively long legs of the adult flies and their small size. Each small square = 1 mm.

blowflies. The larvae of many species are detritivores and some of these feed on dead bodies. The coffin fly, *Conicera tibialis*, is the most well known of these and gets its name from the ability of the adult flies to detect the presence of a corpse buried over a metre below ground and then to crawl down through cracks in the soil to reach it. Successive generations of coffin flies can occur totally underground on a buried corpse although it may also be found on bodies left above ground. *Megaselia scalaris* is also often found on dead bodies both above and below ground but is also capable of causing myiasis and has been reported infecting wounds of living humans. It is an unusually adaptable insect and its larvae are willing to feed on everything from shoe polish to other insects. Indeed, not only can it go through successive generations underground on a buried corpse (Campobasso *et al.*, 2004b) but it is also allegedly able to do the same within the intestine of living humans (Disney, 2008). *M. scalaris* is found in many parts of the world although it is mostly associated with warm climates. It has been introduced into the UK but has to over-winter indoors. A number of other phorid flies are also found on corpses including *Diplonerva florescens*, *Megaselia abdita*, *Megaselia rufipes*, and *Triphleba hyalinata* and they all tend to be found during the later stages of decay as the body is starting to dry out.

Hover flies

Adult syrphid flies (family Syrphidae) are commonly known as hover flies and they are often associated with flowers, sap-runs and other sugary substances. Their name is derived from the ability of the adult flies to hover in a fixed position. Syrphid larvae exhibit a wide range of feeding strategies, including predation, herbivory and detritivory. Among the syrphids, the larvae of *Eristalis tenax* (drone fly) are the ones

1cm

Figure 6.9   Larva (upper) and pupa (lower) of the hoverfly *Eristalis tenax*. The telescopic breathing tube at the posterior of the body is a characteristic feature of this genus.

most likely to be encountered on a dead body, especially during the middle stages of decay when it has started to liquefy. Commonly known as 'rat-tailed maggots', the larvae have a long telescopic breathing tube that acts like a snorkel and enables them to breathe whilst the rest of their body is submerged (Fig. 6.9). However, when they have finished feeding, the larvae crawl away in search of a drier environment in which to pupate.

Piophild flies

Piophild flies are usually associated with corpses in the later stages of decay, when the body is starting to dry out although there are records of them appearing earlier. The adult flies are small and shiny-black, which in a certain light gives a bright green or blue sheen. *Piophila casei* is the best known of the piophilids, as it is also a common pest of stored products. The adults are 3.5–4.5 mm long and although their larvae may grow to 10 mm in length they are much thinner than the typical blowfly maggots. Its common name of the 'cheese skipper' results from the third instar larva's prodigious leaping abilities. It achieves this by grasping its posterior region with its mouth parts so that the body adopts a ring shape, it then pulls hard to create tension and on releasing its grip the insect is snapped into the air. Individual larvae have been recorded travelling up to 23 cm and achieving heights of 20 cm in a single leap. Both *Piophila casei* and *P. foveolata* have been recorded from human corpses. Life history data for *P. casei* is provided by Russo *et al.*, (2006).

Figure 6.10 A typical stratiomyid larva – this specimen was 9 mm in length. Unlike calliphorid larvae, stratiomyid larvae have an obvious conical-shaped 'head' and their body is leathery, flattened and divided into distinct segments. Stratiomyid larvae are common soil invertebrates but they can be found in large numbers within the skull and bones of dead animals after these are reduced to putrid dry remains.

Stratiomyid flies

Stratiomyid fly larvae are common soil invertebrates and they are often found on buried bodies or underneath bodies left on the soil surface. They have a characteristic flattened profile (Fig. 6.10) and a long conical head whilst their surface has a grained, leathery, appearance as a consequence of the deposition of calcium carbonate crystals in the cuticle. The adults are called 'soldier flies' because, in many species, they are brightly coloured and their scutellum (a shield-shaped plate on the dorsal surface of the thorax) is armed with backwardly pointing spines. Stratiomyids are not often mentioned in the forensic literature although in America *Hermetia illucens,* the 'black soldier fly', has been used to estimate time since death in the absence of data from blowflies (Lord *et al.,* 1994). Female *H. illucens* are usually considered to begin laying their eggs on a corpse once it is 20–30 days old and entering the advanced decay or putrid dry remains stage of decomposition. At this stage, blowfly maggot activity has declined or ceased entirely and in their place *H. illucens* larvae may become the dominant insect fauna on the corpse. However, in some circumstances it may start to colonize a corpse that is less than 7 days old and therefore still relatively fresh (Tomberlin *et al.,* 2005): this clearly has implications for any estimations of the PMI that are based primarily on the stage of development of this insect. *H. illucens* is not found in the UK although it is present in Continental Europe.

Trichocerid flies

Trichocerid flies are commonly called 'winter gnats' because adult mating swarms may be seen dancing above lawns, fields and hedgerows during the winter, even when snow is on the ground. The adult flies are also to be seen during the spring and autumn but are less common during the summer months. There are only ten species in the UK and the adults resemble small crane flies (daddy-long-legs). The larvae are long, thin and cylindrical with a well-developed head and a number of them are known to be detritivores. *Trichocera saltator* has been found on human corpses and they are most likely to be found during the winter period when blowfly activity is reduced.

Dermestid beetles

The beetles, order Coleoptera, comprise 25% of all animal species described to date and they are found in a wide diversity of habitats and exhibit a range of lifestyles – including the colonization of dead bodies. Dermestid beetles (family Dermestidae), commonly called larder beetles, hide beetles, carpet beetles, or leather beetles (Fig. 6.11), and their larvae are not found on fresh bodies but can quickly skeletonize it once it has started to dry out (Bourel *et al.*, 2000; Schroeder *et al.*, 2002). They are small insects usually less than 10 mm long and are common household pests that feed on organic matter in carpets etc. There are several species, although their life cycles and habits are fairly similar and their development times and the numbers of generations per year are strongly influenced by the environmental conditions. In the case of *Dermestes lardarius* the gravid female lays eggs directly on the food source in batches of 6–8 over a 3 month period, these hatch after 7–8 days and 5–7 larval instars follow. The final instar larva leaves the food source and pupates in a burrow nearby after which the adult insect emerges – this is 2–3 months after the eggs were laid. Adult dermestid beetles are often cannibalistic on their own eggs and larvae and should therefore be kept separate from these when collecting specimens (Archer & Elgar, 1998). Like most other insects, in the dermestids, the midgut epithelial cells are separated from the gut contents by a structure called the peritrophic membrane. This membrane is believed to facilitate the compartmentalization of the different phases of digestion and also protects the underlying epithelial cells from pathogens. In dermestids, the faeces are voided still enclosed within the peritrophic membrane so it emerges rather like a long, thin, pale Cumberland sausage (Fig. 6.12). The faeces, or 'frass' as insect faeces is often called, is dry and crumbly

(a)   (b)

Figure 6.11 Life cycle stages of dermestid beetles. (a) Adult of *Dermestes lardarius*. Length = 7–9 mm. (b) Unidentified larval dermestid. The larvae are sometimes called 'woolly bears' owing to their spiny body 'hairs'. Both adults and larvae are often found in houses and are common stored products pests. They are seldom found on remains until these have dried out.

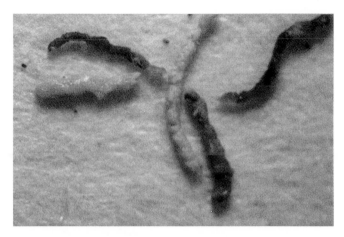

Figure 6.12   Faeces (frass) of dermestid beetles. (Reproduced from Gennard, D.E. (2007) *Forensic Entomology*, with permission from John Wiley & Sons Ltd, Chichester).

because dermestids have a highly efficient water retention system – this is partly why they are able to feed on dry, mummified remains. Some workers have suggested that the presence of these faecal strands might be used in the determination of the PMI although really they simply establish that dermestids are present or were present in the past even if the insects themselves cannot be located.

Pyralid and tineid moths

The caterpillars (larvae) of several species of pyralid and tineid moths, order Lepidoptera, are sometimes associated with corpses. For example, those of the common or webbing clothes moth *Tineola bisselliella* and the case-bearing clothes moth *Tinea pellionella* are common household pests that normally feed on stored clothing and carpets, stored grain, cereals and dry vegetable matter but they may also be found on corpses when the body has become dry and skeletonized. They are even able to break down the keratin found in hair and fingernails (Fig. 6.13). Similarly, the larvae of the pyralid moths *Aglossa caprealis* (Fig. 6.14) and *Aglossa pinguinalis* may be found on corpses that have dried out, although their normal environment would be on stored agricultural produce and decaying vegetation. A gravid female *Tinea pellionella* typically lays 40–70 eggs over approximately 24 days. The female moths do not fly actively whilst they are laying their eggs. Any moths seen flying around a room are usually therefore males or those females that have finished laying eggs. The eggs hatch in 7–37 days depending on temperature and the larvae often spin a protective silken tube around their bodies. The larval period may last anything from 2 months to 4 years depending on temperature, humidity and the nature of their diet after which the mature larva pupates within its protective tube and the adult moth emerges after 11–54 days.

Like many other insects, the larvae of tineid moths avoid light and are thigmotactic, i.e. they orientate themselves to vertical surfaces. Consequently, two bodies within a room, may be colonized, and therefore decomposed, at different

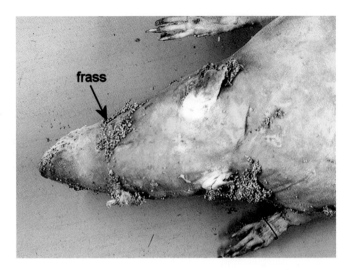

Figure 6.13 Damage caused to dry museum specimens by the larvae of clothes moths. Similar damage would be caused to mummified human remains. The hair has been removed and the damage to the ears could be misconstrued as pathology caused at the time or shortly after death. However, the frass (arrow) indicates past insect activity and close inspection of the cut surfaces would be needed to be certain of the cause of damage.

Figure 6.14 Adult of the pyralid moth *Aglossa pinguinalis*. Scale = mm. Adult pyralid and tineid moths do not cause any physical damage although adult tineid moths do have rudimentary mandibles.

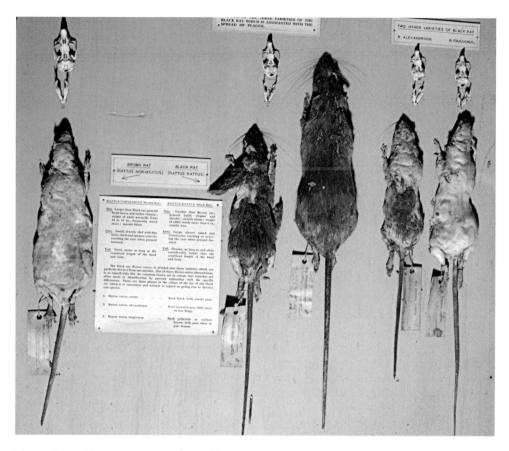

Figure 6.15 These museum specimens illustrate how bodies within the same room or environment may not decompose at the same rate. In this display case, the bodies of the rats closest to the display case were attacked first by the larvae of tineid moths (clothes moths).

rates (Fig. 6.15). A body at the side of a room will tend to get colonized before one in the centre and a body that is in full sunshine may be colonized later than one in a dark corner.

Mites

Mites (Acari) are frequently abundant on corpses, the species composition changing as the body goes through its varying stages of decay (Fig. 6.16). For example, gamasid mites (e.g. *Macrocheles*) tend to be abundant during the early stages of decay, and tyroglyphid and oribatid mites (e.g. *Rostrozetes*) are more numerous once the body has started to dry out. Some of the mite species feed on the corpse, others are predatory and feed on the detritivores etc., and some feed on the fungi and microbes that grow on the body. Investigating the potential of Acari as forensic indicators has been hampered by their small size and the scarcity of mite

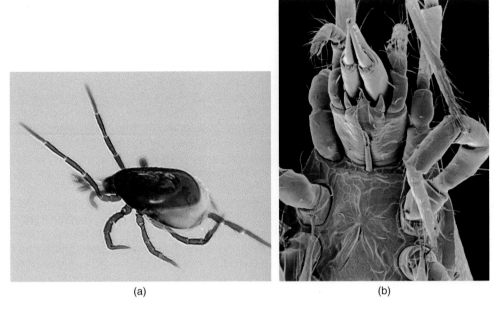

(a)   (b)

Figure 6.16 Unidentified predatory soil mites found associated with decaying remains. Both mites were about 1–1.5 mm long. (a) Light microscope photograph. Note how the body has swollen through absorption of water. Incorrect collection and storage (such as this) can radically affect the appearance of specimens. (b) False colour scanning electron micrograph showing the mite's complex mouthparts. The pincer-like chelicerae are used to grasp prey whilst the long finger-like pedipalps act as sensory structures.

taxonomists. Consequently, there is little published information on the mite fauna of corpses.

Earthworms

'Worms' are popularly associated with dead bodies. For example, one of the lines in the English folk song 'On Ilkley Moor Baht' at 'goes' then t'worms 'll cum and eat thee oop'. (These lines translate from the Yorkshire dialect as 'On Ilkley Moor without a hat' and 'then the worms will come and eat you up'.) However, the worms referred to in the song and in popular folklore are almost certainly the maggots of blowflies – anything small white and wriggly tends to get classed as 'a worm'. Earthworms are of course, a quite distinct group of organisms and there are about 26 species (plus some introduced species with a restricted distribution) in the UK (Sims & Gerard, 1985). Although earthworms will feed on decaying bodies, they are sensitive to changes in pH and would be repelled by the seepage of large amounts of acidic material into the soil. Earthworms, especially epigeic species (those living on the soil surface), may enter buildings so their presence on a corpse found indoors does not mean that it was previously buried or left lying on the surface of soil. Earthworms decay rapidly to a foul smelling mush once they are dead so, if a body containing earthworms is disposed of in the sea or a lake, they are unlikely to provide evidence of the body's previous location unless the body was found very quickly. Earthworms accumulate heavy metals, including arsenic, but it is not known whether they accumulate other toxins from a dead body.

Miscellaneous invertebrates

Many other soil invertebrates, such as nematodes, slugs, snails, Collembola (Fig. 6.17), Diplura, and millipedes may be found on corpses – especially those that are buried or left on damp soil but they are seldom reported as being helpful in forensic studies. This is partly a consequence of the difficulties associated with identification and partly because in many cases they develop much more slowly than blowfly larvae and are therefore less useful in determining the passage of time. However, monitoring the changes in the abundance and diversity of soil nematodes and Collembola has proved useful in studies on pollution ecology and the appearance of a large decaying body is, in some respects, a pollution incident, so it may be

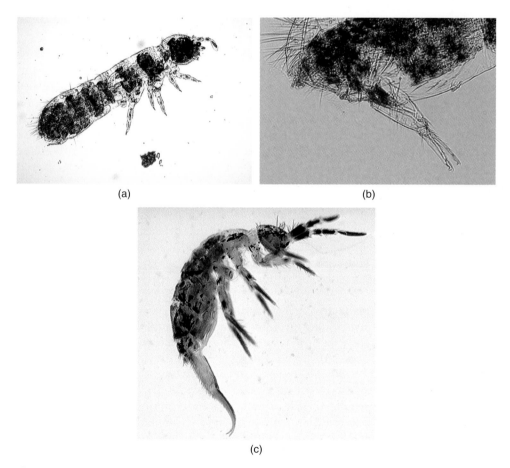

(a)   (b)

(c)

Figure 6.17   (a) Typical soil collembolan and (b) its springing organ (furca). Subterranean collembolans often have poorly developed springing organs or they may be absent altogether (c) Typical surface-dwelling collembolan found in leaf litter; note the much larger springing organ.

worth exploring how their populations change underneath or surrounding a dead body.

Coprophiles

Coprophiles are creatures that consume their own faeces or that of other animals. Although some invertebrates such as dung beetles are specialist coprophiles for which faeces is their only food, many also feed on other sources of decaying matter. For many invertebrate species, therefore, coprophagy and detritivory are simply options that are indulged in to varying degrees. Coprophilic species might be present when a body is coated with faeces (murderers sometimes desecrate their victim's body this way), when the gut contents are exposed through wounds or the decay process and when fouling of the clothing occurred before death, possibly through neglect or at the time of death through relaxation of the sphincters.

Flies belonging to the families Muscidae and Fanniidae

Flies belonging to these two families are usually small and brown or grey in coloration. Several species are commonly associated with faeces although as families they exhibit a wide variety of lifestyles at both the adult and larval stages (Skidmore, 1985). Some species, such as *Musca domestica* (housefly), *Fannia canicularis* (lesser housefly) (Fig. 6.18), and *Fannia scalaris* (Latrine fly), have adults that readily enter buildings and larvae that live in faeces or dead organic matter. Consequently, these species are common in unsanitary conditions and will infest soiled clothing, even if the wearer is still alive. They may be found on corpses, especially if it is soiled with faeces or the gut contents are exposed. Some species such as *F. canicularis*, *F. scalaris*, *Muscina prolapsa*, *Muscina stabulans*, and *Ophyra capensis* are found on buried bodies. In these species, the eggs are laid on the surface and the larvae burrow down after hatching. For example, female *Muscina stabulans* and *Muscina prolapsa* will lay their eggs in soil above baits that are buried up to at least 20 cm below ground and after hatching the larvae burrow down to reach their food (personal observations).

M. domestica requires relatively high temperatures to complete its life cycle and in the UK the adults usually do not appear outdoors until May–June. However, indoors they will breed continuously wherever the temperature remains high enough. The female flies lay batches of about 150 eggs within crevices of the larval food medium. The eggs can be difficult to see when laid on white material such as nappies or bandages. The larvae develop rapidly and the whole life cycle from egg laying to adult emergence may take as little as 6–8 days under optimal conditions. Consequently, populations may build up extremely rapidly. The larvae do not, as a rule, cause myiasis, although they may be found in blood-soaked clothing and hair. They are often found in clothing soiled by urine or faeces (Fig. 6.19) and this infestation may begin before the wearer dies. This is important to bear in mind when calculating the time since death.

(a) (b)

Figure 6.18 (a) Larvae of *Fannia cannicularis*. These are common soil invertebrates and often found on bodies at the later stages of decay. Note the lateral processes from the body that give the insect a 'hairy' appearance. These larvae grow to 5–6 mm in length and should not be confused with the much larger and more aggressive *Chrysomya albiceps* (Fig. 6.5) that has similar processes. (b) Adult *Fannia cannicularis*; adults are 6–7 mm in length.

Carnivores and parasitoids

Carabid and staphylinid beetles

Many beetles, especially certain carabid and staphylinid species are important predators of other soil invertebrates during both their adult and larval stages. Some species of carabid beetle are such effective predators that they are used as biological control agents but their impact on the corpse fauna is not known. The larvae of many carabid and staphylinid beetles are similar in appearance (Fig. 6.20) but can be distinguished from the shape of their legs. The carabid larvae have six-segmented legs whilst those of staphylinid larvae are five-segmented. In addition, the legs of many species of carabid end in two claws, whilst those of staphylinids have only one claw.

Burying beetles

Burying beetles (family Silphidae) are so called because the adults of several species bury dead animals, such as mice, upon which their larvae subsequently feed. They can detect a dead body at a considerable distance and are one of the first invertebrates to arrive at a fresh corpse. In those species that bury dead bodies, several individuals will co-operate in the burial and then fight to determine which one will lay her eggs on the corpse. They are incapable of burying an entire human corpse but may inter bits if it has been dismembered. Alternatively, they may chew out

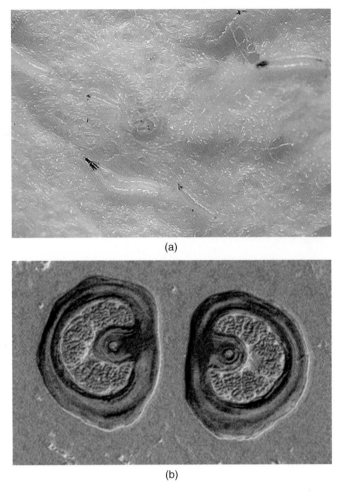

Figure 6.19 Larvae of the common housefly *Musca domestica*. (a) Larvae developing within a soiled nappy. (b) Posterior spiracles of third instar larva. Note how the structure of the spiracles is distinct from that of *Calliphora* and *Lucilia* (Fig. 6.4b)

hunks of tissue, which are then buried. After burying the corpse or body part, the female beetle lays her eggs above the carcass and then covers it with her hind gut secretions to inhibit the growth of bacteria and fungi. The eggs hatch after about 2.5–5 days and the female remains with the hatchlings. In some species, the male also remains and helps rear the young. Depending on the species, the adult(s) may feed the young directly from a depression they make at the top of the carcass. Larval development takes 5–8 days. After the final instar, the larvae crawl off and pupate for about 2 weeks. The adult beetles then emerge but they do not reproduce until the following year. Not all burying beetles actually bury corpses, for example, *Necrodes littoralis* usually breeds within the bodies of large vertebrates and has been recorded entering via stab wounds. Both adult and larval burying beetles will consume blowfly maggots and some burying beetle species act as transport hosts

(a)

(b)

Figure 6.20 Predatory beetle larvae. (a) Larva of a staphylinid beetle. (b) Larva of a carabid beetle. Although fearsome predators as both larvae and adults, the impact of these beetles on the establishment of maggots on corpses is probably small.

for mites belonging to the family *Poecilochirus* that feed on blowfly eggs. There is therefore a theoretical possibility that burying beetles may influence blowfly maggot numbers and age distribution.

Histerid beetles

Histerid beetles are commonly found on decaying animal and vegetable matter. The majority of species are predatory as both larvae and adults although they may also consume dead organic matter to a greater or lesser extent. Some species consume blowfly eggs and larvae although the consequences of this for the rate of decomposition are not known. Histerid beetles tend to arrive on corpses several days after blowflies and, during daylight hours, they are usually found underneath the body.

Ants and wasps

Ants (order Hymenoptera) are all social insects and they are found in large numbers in many parts of the world. In tropical countries ants can be the most abundant invertebrates found on a dead body and occur at all stages of decay. Many ant species are predatory and if a corpse is located close to their nest, they may slow the establishment of the blowfly population through removing eggs and young larvae. However, they may also feed on the tissues of the corpse and thereby speed the process of decay. The extent to which they accelerate or retard decomposition will depend upon the species, their abundance and the local conditions. Not all ant species are carnivorous and therefore likely to have an impact on decomposition so correct identification is important. There are few mentions of ants in the forensic literature although Goff *et al.* (1997) describe a case in Hawaii in which the PMI for a body discovered in a metal trunk was estimated from the time taken for an ant nest to develop within the trunk to the stage at which alate (winged) reproductive castes were being produced.

The larvae of all social wasps belonging to the family Vespidae are carnivorous and the adult wasps prey on a variety of other invertebrates that they bring back to the nest to be consumed. A large wasp nest may therefore have an impact on the surrounding invertebrate fauna (Archer & Elgar, 2003) whilst the adult wasps will also forage for scraps of meat from nearby corpses.

Centipedes and spiders

All centipedes (Chilopoda) and spiders (Aranea) are predatory and although they are often found on corpses their impact on the other fauna is not known. However, considering the enormous reproductive potential, rapid growth rates and powers of recruitment of many blowfly species it is unlikely that they exert a significant effect. Certain centipedes are found naturally in houses and outbuildings, whilst others – such as the long, thin geophilomorph species – are usually only found in the soil. Although spiders are commonly associated with spinning webs, many species actively hunt or stand guard at the entrance to their refuge and rush out to grab their prey as it comes past. A variety of species of spider may be found on corpses, sometimes because they are preying on the other corpse fauna but also because they are using it as a refuge or as an anchor for their web.

Parasitoid insects

Parasitoid insects are those that lay their eggs within the bodies of other invertebrates, usually other insects. The eggs hatch within their host, but this is not killed until the parasitoid has completed its larval development. Examples of parasitoid wasps attacking blowfly and housefly larvae or pupae include *Nasonia vitripennis*, *Alysia manducator*, *Muscidifurax raptor* and *Spalangia cameroni*. This can result in the slowing of the larval or pupal development rate and, ultimately, the death of the parasitized insect. *Nasonia vitripennis* lays its eggs in the pupae of a range of blowfly species. The adult female wasp (Fig. 6.21) uses her ovipositor to bore a hole through the puparium and lays her eggs on top of the developing pupa. This occurs 24–30 hours after the pupa has formed, i.e. the point at which the third instar larval cuticle separates from the pupal cuticle and forms the puparium. After hatching, the wasp larvae feed on the fly pupa, killing it in the process. The wasps pupate

Figure 6.21 Adult parasitoid wasp *Nasonia vitripennis*. This species lays its eggs within the pupal cases of blowflies but other species lay eggs within the larvae and can be seen searching for victims within a maggot feeding ball. Scale = mm.

within their host's puparium and the adults chew their way out after 10–50 days, depending on temperature. *Alysia manducator* has a similar life cycle but it lays its eggs directly inside the developing pupa. The forensic importance of parasitoids is that they can be locally abundant and therefore if blowfly larvae or pupae are being collected for rearing purposes a large sample size should be obtained to allow for parasitoid-induced mortality. However, because *Nasonia vitripennis* only lays its eggs on fly pupae 24–30 hours old, recording when the adult parasitoids emerge can be used to determine the post mortem interval (Grassberger & Frank, 2003). This is because if one knows how long the wasps take to develop it can be calculated when they laid their eggs and on that date the pupae would be 24–30 hours old from which point it would be possible to determine when the blowfly eggs were laid and therefore the approximate date on which the person died. The possibility of using parasitoids of phorid flies in a similar manner has also been suggested (Disney & Munk, 2004).

Invertebrates leaving dead bodies

Fleas and lice

Fleas, body lice, and head lice leave their host soon after it dies and the body temperature starts to decline. The sight of them crawling on top of a dead person's clothing is therefore an indication that they have not been dead for long. The so-called human flea, *Pulex irritans* is no longer a common pest in the UK and people are more likely to come into contact with the cat flea, *Ctenocephalides felis* and the dog flea, *Ctenocephalides canis*. Humans may harbour one or two fleas, usually acquired from their pets, and large numbers are only present in people suffering from neglect or living in unsanitary conditions. The same is true of the body louse *Pediculus humanus humanus*, but the head louse *P. humanus capitis* (Fig. 6.22) is extremely common, especially in children, even among affluent families. Crab lice are relatively common among the sexually promiscuous and those who have a relationship with someone who is. Although they are also commonly known as pubic lice, they may also be found on other coarse body hair, such as beards. There are unconfirmed reports that the numbers of pubic lice are declining as a consequence of the increasing popularity among both men and women of waxing to remove genital and body hair. Pubic lice move very slowly and are therefore less likely to be seen wandering away from a dead body. The potential of ectoparasites such as fleas and lice to act as a vector of disease needs to be borne in mind by persons working with dead bodies.

Invertebrates accidentally associated with a dead body

Miscellaneous insects

Invertebrates that are not associated with the decay process may become accidentally associated with a body through becoming trapped in its clothing or from using the corpse as a refuge. Alternatively, they may become trapped inside the vehicle or container in which the body is found. If these invertebrates have a restricted geographical distribution, are associated with very specific habitats, or are active at only specific times of year this can provide evidence of a person's association with a particular locality at a particular time.

Invertebrates as a cause of death

Within the UK, few native invertebrate species have venoms capable of causing a serious threat to human health. However, some may cause death by inducing an anaphylactic reaction in sensitive individuals or through stinging a person in the mouth or throat – this can result in rapid swelling that can cause asphyxiation (Chapter 5). In the UK and northern Europe most fatalities involving invertebrates are associated with honeybee or wasp stings and there is an average of 2–9 such cases per year in the UK. Honeybees leave their sting behind after it is used and this results in the bee dying. A bee sting can therefore be diagnosed from the physical presence of the sting or from finding the bee's dead body. Wasps and hornets do not leave their sting and unless the fatal event was witnessed their involvement may only be indicated by the presence of a nearby nest.

Invertebrates may also be responsible for human deaths indirectly by causing distraction or panic that results in a fatal accident. Spiders, wasps and bees can

(a)

(b)

(c)

Figure 6.22 Common human lice. (a) Human head louse *Pediculus humanus humanus*. (b) Egg ('nit') of *P. humanus humanus* attached to a scalp hair. The egg has hatched. (c) Human crab (pubic) louse, *Phthirus pubis*. These lice may also be found on other coarse hair such as the beard and eyelashes.

Table 6.1 Summary of insects commonly associated with dead bodies

Insect	Association
Diptera	
Blowfly adults	Lay eggs on corpse during early stages of decay. Feed on corpse at all stages of decay.
Blowfly larvae	Feed on corpse during early stages of decay.
Fleshfly adults	Lay eggs on corpse during early stages of decay. Feed on corpse at all stages of decay.
Fleshfly larvae	Feed on corpse during early stages of decay.
Muscid + Fanniidae adults	Lay eggs on soiled clothing and faeces. Feed on corpse at all stages of decay.
Muscid + Fanniidae larvae	Feed on faeces and soiled regions, also found at later stages of putrefaction when corpse is still moist but abundance of blowfly larvae is declining. Can colonize buried bodies.
Phorid larvae	Feed on corpse during early or later stages of decay. Can colonize buried bodies.
Stratiomyid larvae	Usually feed on corpse once body has dried out. Can colonize buried bodies.
Piophilid larvae	Feed on corpse during later stages of decay.
Eristalid larvae	Feed on soiled and liquefied regions.
Nematocera larvae	Feed on corpse during colder months when blowflies are inactive or their numbers are low.
Coleoptera	
Carabidae	Predators as adults and larvae. Feed on other insects and on corpse.
Staphylinidae	Predators as adults and larvae. Feed on other insects and on corpse.
Histeridae	Predators as adults and larvae. Feed on other insects and on corpse. Arrive a few days after the blowflies
Silphidae	Predators as adults and larvae. Feed on other insects and on corpse. Adults arrive during early stages of decay.
Dermestidae	Detritivores as adults and larvae. Feed on corpse once it has dried out.
Hymenoptera	
Vespid wasps	Adults predatory on other insects. Will remove flesh from corpse.
Parasitoid wasps	Adults lay eggs in larvae or pupae of Diptera.
Ants	Some species predatory on other insects. Will remove flesh from corpse.
Lepidoptera	
Pyralid moths	Larvae feed on corpse once it has dried out.
Tineid moths	Larvae feed on corpse once it has dried out.
Blattaria	
Cockroaches	Only found indoors in UK. Nymphs and adults will feed on corpse at all stages of decay.
Anoplura	
Head lice, body lice	Parasites that leave dead body as it starts to cool.
Collembola	
Springtails	General detritivores found underneath bodies left on soil, especially at later stages of decay.
Diplura	General detritivores found underneath bodies left on soil, especially at later stages of decay.

induce panic attacks in people with a phobia for them. They are therefore thought to be involved in a number of otherwise 'unexplained' vehicle accidents. A wasp or bee can enter the vehicle via a window, for example, and the driver subsequently loses control in an attempt to swat or avoid it. The driver need not be stung and the insect will almost certainly fly away unseen after the event. Should the accident result in the death of the driver, even if they were stung there may not have been time for an allergic reaction to develop and any sting-associated pathology could be missed at autopsy. Similarly, a person may run into fast-moving traffic in an effort to escape from a buzzing insect or a spider that has become entangled in their hair. Bumblebees are highly unlikely to sting but may induce panic reactions owing to their large size and loud buzzing behaviour.

Allergens present in the faeces and chitin of some invertebrates may induce strong reactions in sensitive individuals. For example, the house dust mite, *Dermatophagoides pteronyssinus*, is a common cause of allergy and can cause asthmatic attacks and other symptoms. Recent improvements in the ability to diagnose anaphylaxis at post mortem have indicated that some cases in which persons suddenly died of unknown causes might be linked to hypersensitivity to this mite (Edston & Hage-Hamsten, 2003). Similarly, locusts can induce potentially fatal reactions in sensitive individuals. Animal technicians who work with locusts on a regular basis must therefore take appropriate precautions. In countries where locusts occur naturally, human illness has been associated with their swarming behaviour. For example, during September 2003 in a town in central Sudan, eleven people were reported dead and over a thousand hospitalized with breathing difficulties as a consequence of asthmatic reactions to the sudden influx of large numbers of locusts.

Invertebrates as forensic indicators in cases of neglect and animal welfare

Myiasis and contamination

The invertebrates involved in neglect and animal welfare cases are typically those that infest wounds or which live on the body surface. The infection of wounds with the larvae of Diptera is known as 'wound myiasis' and usually involves the maggots of blowflies and fleshflies. If untreated, an infested wound is extended and more flies attracted to lay their eggs (Fig. 6.23). Wound myiasis caused by fleshflies can develop very quickly because the female flies deposit first instar larvae within and beside the wound. In addition to causing pain and distress, the condition may give rise to septicaemia (infection of the blood with bacteria) that is potentially fatal. Anybody with an open wound who is incapable of looking after themselves, such as the very young, very old, very ill, and those who are mentally incapacitated, is at risk of suffering from myiasis and the discovery of maggot infestation would usually be considered a sign of neglect. Wound myiasis is also a common problem amongst farm animals such as sheep and goats and pets such as rabbits but owners are seldom convicted under animal welfare legislation unless it has been allowed to develop extensively.

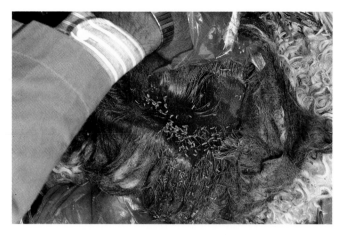

Figure 6.23 Severe blowfly strike in a sheep. The infection began as a breach strike and is extending to include the upper hind quarters.

Soiled clothing or dressings, unless changed regularly, will attract the attentions of flies including non-invasive species such as *Musca domestica* and *Fannia canicularis* and those such as *Calliphora vicina* that are capable of subsequently invading underlying tissues and causing wound myiasis. Fly eggs are usually white or pale yellow and are easily overlooked when seen against a similarly coloured background, such as a nappy or bandage. The eggs hatch quickly owing to the warmth of the body, and the larvae may be well developed before the infestation is noticed. Other invertebrate species also feed on soiled clothing or dressings but they are less useful for determining the length of time a person was neglected. For example, cockroaches retire to resting sites in crevices etc. after feeding.

Pharaoh's ants, *Monomorium pharaonis* and some other ant species will invade wounds and dressings and large numbers can arrive in a short period of time. Pharaoh's ants are extremely small and it is possible for a large colony to establish itself in a building (they do not live in the wild in the UK) before it is noticed. They are capable of transmitting a range of pathogenic bacteria including various species of *Staphylococcus*, *Clostridium*, *Pseudomonas*, and *Salmonella* and therefore their presence in hospitals is a serious matter.

Fleas, lice and mites

Fleas, lice and mites are common ectoparasites (parasites that dwell on the body surface) of humans living in unsanitary conditions and may sometimes be a sign of neglect. Most domestic and wild animals are infested with ectoparasites although provided the animal is healthy their numbers remain low and are not harmful. Sheep scab mite, *Psoroptes ovis*, is an exception and can rapidly develop into a potentially fatal infestation. It is a notifiable disease throughout the UK (i.e. it is a legal requirement to report the presence of the infection to the authorities) and sheep dipping to control it was compulsory at certain times of year whether or not the flock was infested. However, nowadays, sheep dipping remains compulsory only in certain regions. It is not an offence to own an infested animal but it would be an offence not to report its presence or to fail to treat the infestation having discovered it. Many factors influence the progression of sub-clinical sheep scab infestations, which are extremely difficult to detect, into the debilitating active phase – which may occur anything from a few days to over a year later. Consequently, it can be difficult to prove long-standing neglect since the sheep with the most extensive lesions may not be the ones with the oldest infestations.

The role of invertebrates in food spoilage and hygiene litigation

As societies develop, people become less and less willing to accept the presence of invertebrates in their food or the evidence that they have been present in the past. Consequently, even small numbers of invertebrates, their faeces, cast skins or bite marks can be sufficient to render a foodstuff unsuitable for human consumption. This can lead to the loss of large sums of money by the food retailer owing to an inability to sell the product or from legal proceedings resulting from being prosecuted for selling goods unfit for purpose. In addition to the problems caused by the physical presence of invertebrates, many of them are considered harmful owing to their ability to transmit diseases. For example, when insects walk over the food they can transfer pathogens such as bacteria, viruses, protozoa and nematodes. Numerous species of invertebrates are associated with food spoilage and only a few will be mentioned here.

Blowflies and other Diptera

Flies, and in particular blowflies, are notorious for spoiling meat and meat products and thereby rendering them unsuitable for human consumption. Spoilage may be caused by adult flies, their eggs and by the larvae. Adult blowflies commonly regurgitate their gut contents whilst feeding and defecate close to or on their food. Because adult blowflies also feed on faeces and rotting organic matter, they can passively transmit a wide range of human bacterial and protozoan pathogens (Fig. 6.24). Many pathogens pass unharmed through the flies' gut or are transmitted on the flies' feet. The stains produced by fly faeces are easy to spot on a pale background but easily overlooked on dark material such as fresh meat. Blowflies often leave large egg masses on meat, usually in damp crevices and under-surfaces rather than exposed regions. The meat is then said to be 'flyblown' and is unfit for human consumption. Meat that is left exposed for even a short period may become flyblown if measures are not taken to keep flies out of the shop or kitchen. Unless refrigerated, blowfly eggs usually hatch within 24–48 hours and the larvae develop rapidly. The larvae tend to feed close together and avoid the light so they may not be seen until after the meat is purchased and legal cases then arise if there is a dispute about when the meat became contaminated. These legal cases can usually be resolved by

Figure 6.24 A blowfly walked across this agar plate and bacteria have grown where the tarsi (white colonies) and proboscis (yellow colonies) came into contact with the surface.

correct identification and knowledge of the conditions under which the meat was kept from the time the animal was killed to the stage at which the dispute arose.

In addition to blowfly and fleshfly larvae, the maggots of various other fly species may be found in a wide variety of food products. For example, the larvae of *Piophila casei*, which are often associated with corpses at the late stages of decay, are also a pest of cheeses and hams. A number of Diptera species are pests of agricultural crops and their larvae may be found on a variety of crops (e.g. the onion fly *Delia antiqua*, and the cabbage root fly, *Erioischia brassicae*). Their presence is usually far less serious than blowflies on meat and the affected parts can be cut away so legal cases are less likely to arise.

Beetles

Many beetles, for example, the grain weevil *Sitophilus granarius* and the mealworm *Tenebrio molitor*, are common pests of stored products. Large populations can build up leading to damage to the grain etc. and the sacking it is stored in. The problems are becoming worse owing to increasing insecticide resistance among strains of various coleopteran stored product pests. Legal claims arise over disputes at what point a batch of stored product became infested with the beetles.

Ants and wasps

Ants, bees and wasps are often attracted to sweet substances, such as sugary drinks and foods whilst social wasps and some ant species are also attracted to meat products. The presence of ants, bees and wasps in food may therefore be a consequence of contamination before or after processing. Parasitoid wasps are often found in stored products as a consequence of their parasitizing pests such as beetles and lepi-dopteran larvae. Some parasitoid wasps have extremely long ovipositors that may lead to concern over the possibility of being stung – however, this is usually not a real risk.

Butterflies and moths

A number of lepidopteran species have larvae that are pests of agricultural crops (e.g. the cabbage white butterfly, *Pieris brassicae*) or stored products (e.g. the flour moth, *Ephestia kuhniella*). The presence of caterpillars on fresh produce is seldom a serious concern because they are easily seen and, along with their faeces, they can be washed off but the presence of stored product pests may give rise to disputes about when the infestation developed.

Cockroaches

Most cockroach species are large insects but because they are nocturnal evidence of their feeding activity is more obvious than the insects themselves. They are omnivorous and will eat everything from meat products to fruit, wax, paper and grain. Through contamination, they are capable of spreading a large number of human bacterial and protozoan pathogens, so evidence of their presence is a concern for public hygiene. Parts of cockroaches (e.g. legs, wings, egg cases) may occur in processed foods owing to the food being infested during storage or the processing routine. As with all invertebrates, food processing will affect the appearance of the insect body parts and therefore whether they were mixed before or afterwards although more work needs doing in this area.

Mites

There are numerous mite pest species but because of their small size they are often overlooked. Mites, such as the grain mite, *Acarus siro*, can be a serious problem in stored products, such as flour. The mite's faeces and cast skins cause flour, grain etc. to develop a musty taste and can cause digestive upset. People routinely handling contaminated foodstuffs can develop allergic reactions resulting in eczema and asthma. Infestations develop when the foodstuff is stored incorrectly, especially if it as allowed to become damp. Legal cases arise when there are disputes about when (e.g. processing, storage, transport) the food product became infested. This can be determined by estimating the mite population and then from knowledge of the biology of the mite and reference to the environmental conditions under which the product was stored it would be possible to estimate the time taken to reach this level. Legal cases may also arise when workers claim to suffer ill health through being exposed to infested grain. However, this would be a medical question and therefore beyond the remit of a forensic entomologist as many other substances to which the workers were exposed could also cause allergic reactions (e.g. fungal spores).

Table 6.2 Summary of major insect orders with forensic associations

Insect order	Forensic association
Diplura	Found on corpses
Collembola (springtails)	Found on corpses
Isoptera (termites)	Food spoilage, structural damage
Blattaria (cockroaches)	Found on corpses left indoors, food spoilage, nuisance
Psocoptera (plant lice)	Food spoilage
Anoplura (lice)	Neglect
Siphonaptera (fleas)	Neglect
Trichoptera (caddis flies)	Larvae found on corpses left in lakes, ponds or streams
Lepidoptera (butterflies and moths)	Found on corpses, food spoilage, structural damage, illegal trade in protected species
Diptera (flies)	Found on corpses, neglect, food spoilage, nuisance
Hymenoptera (bees, ants, wasps)	Found on corpses, neglect, food spoilage, cause of death, structural damage
Coleoptera (beetles)	Found on corpses, food spoilage, structural damage

Spiders

Spiders are all carnivorous and do not harm the food they are found in. However, large tropical spiders, such as tarantulas, are occasionally imported with exotic fruit and are capable of inflicting painful bites if disturbed and their setae (body hairs) are barbed and easily break off and can cause a severe rash if they penetrate the skin. Certain species of spider are prone to falling into foodstuffs that are left exposed to the air. For example, Wolf spiders, which do not use webs and actively search for their prey, and male house spiders (*Tegenaria* sp.) which emerge from crevices at night to search for females or water. Consequently, spiders are more likely to find their way into food after it has been purchased and placed in the kitchen.

The illegal trade in invertebrates

Some invertebrates are considered delicacies (e.g. lobsters, oysters) and consequently fetch high prices. This can lead to illegal fishing (e.g. taking above the allowed quota) and poaching that result in population declines and habitat destruction. In addition, fraud is widespread with consequences for lost revenue and the sale of incorrectly labelled goods/ food unfit for consumption. This is exemplified by the illegal trade in abalone in Australia, South Africa and a number of other countries (Tailby & Gant, 2002; Hauck & Sweijd, 1999). Abalone are large (shell size up to 230 mm) marine gastropod molluscs that occur on rocky shorelines up to a depth of 100 metres. There are 55 species of which 15 or so are cultured commercially. They are considered a delicacy in many parts of Asia and particularly by the Chinese who also ascribe various health-giving properties to it including, inevitably, aphrodisiac effects. A meal of the highest quality abalone such as 'Gon Bao Yu' can cost the diner up to $US 5000 in China and consequently abalone represents a high value, low volume product. Furthermore, abalone is easy to collect from the wild whilst farmed abalone cannot be produced in sufficient quantity to meet the demand. This makes the illicit trade in abalone extremely attractive to organized criminal gangs and has been linked to the Chinese triads in South Africa and to outlaw motorcycle gangs (among others) in Australia. The extent of the illegal trade can be judged from the fact that in the 18 months to the end of June 2003, the legal catch in South Africa amounted to 350 tons whilst 1200 tons were received in Hong Kong from South Africa during the same time period. There are no accurate figures for the current scale of the illegal abalone trade in Australia but one guestimate put it at about $US 300 million in 2004 (Morton, 2004). The abalone gathered by organized gangs is believed to be traded for drugs such as heroin and marijuana and has resulted in serious social problems and ecological damage. Control of the trade depends largely upon the monitoring of fishing activity and the inspection of exported abalone to ensure that it is correctly labelled and its weight is correctly declared. Forensic biology could play an important role in the process through the development of rapid means of identifying both the species of abalone and its provenance from tissue samples (e.g. Klinbunga *et al.*, 2004). This is because abalone that is being illegally traded may be mislabelled as another species of shellfish or coming from cultured stocks rather than the wild.

Large and colourful invertebrates are frequently acquired by collectors or sold for decoration. Some specimens change hands for large sums of money and there is a worldwide market which, for the most part, is perfectly legitimate. Unfortunately owing to habitat loss and excessive collecting from the wild, some species are declining rapidly. This increases their 'value' thereby stimulating further collecting and further declines in the population until the species enters an 'extinction vortex'. Some collectors are not content with obtaining one or two representative examples of a species and will amass as many specimens as they can – this can cause further serious damage to small local populations. Consequently, an increasing number of invertebrates are achieving protected status that makes it an offence to collect, trade, or import them. In the UK, the Wildlife and Countryside Act, 1981, names 14 species of insect and 13 species of other invertebrate. Under this act, it is illegal to capture, kill, or sell them, except under license. Possession of any of these species, alive or dead, is considered illegal unless they were acquired before the Act came into force or under license. Examples include the mole cricket *Gryllotalpa gryllotalpa*, the large blue butterfly, *Maculinea arion*, the swallowtail butterfly, *Papilio machaon*, the medicinal leech, *Hirudo medicinalis*, and apus, *Triops cancriformis* (Fig. 6.25).

Invertebrate identification techniques

Owing to their abundance and diversity of lifestyles, invertebrates can provide a wide variety of forensic evidence ranging from estimating how long a person has

(a)

(b)

Figure 6.25 Examples of protected invertebrate species in the UK. (a) The Large Blue butterfly, *Maculinea arion*. (b) Apus, *Triops cancriformis*.

been dead to determining whether the owner of a burger shop is guilty of failing to comply with local hygiene regulations. However, the common factor is that the quality of the evidence depends upon accurate identification and a thorough understanding of invertebrate biology. For example, if the species of blowfly involved in a murder case is incorrectly identified it could result in the estimated time since death being either too short or too long and thereby conflicting with other evidence. Similarly, if the insects found in a foodstuff are misidentified, it could result in the blame for the infestation being attached to the wrong party in the supply chain. Furthermore, if one does not understand the biology of the organisms that are to be used as evidence, then one may not collect them in an appropriate manner or come to false assumptions.

Taxonomy based on morphological features

Traditional taxonomists are a threatened species within UK academia and those who remain are an ageing population. The major museums continue to support their activities although funding continues to be a problem. However, their efforts have built upon the enormous amount of work done by our Victorian ancestors and the continuing activities of an army of gifted amateurs to ensure that the UK has one of the best described invertebrate faunas in the world. Identification keys are available for most groups but unfortunately, many of these are written for experienced professionals and can be extremely difficult to decipher. The Royal Entomological Society publishes a series of handbooks covering virtually all the insect orders in the UK whilst the publisher E.J. Brill produces the Fauna Entomologica Scandinavica series that also includes many UK insect species (e.g. Rognes, 1991). More 'user-friendly' taxonomic keys covering insects of forensic importance can be found in Erzinclioglu (1996) [simple keys to some blowfly adults and larvae] and Smith (1986) [excellent comprehensive keys to both adult and larval Diptera]. Taxonomic keys for invertebrates other than insects tend to be scattered throughout the literature and can be hard to track down.

In many species of maggot, the morphology of the anterior and posterior spiracles and the cephaloskeleton are important taxonomic features and are also useful in determining the instar (e.g. Szpila *et al.*, 2008; Smith, 1986; Zumpt, 1965). The anterior and posterior spiracles can often be seen without any prior treatment but to observe fine detail it is usually necessary to prepare slide mounts. Small larvae may be left whole but by the time they reach the third instar it is usually better to cut them into pieces. Cut off either the posterior or the anterior of the maggot as appropriate and boil it in 10% w/v sodium hydroxide to remove the internal soft tissues. This is best done on a hot plate. Wash the specimen in several changes of distilled water and dehydrate through alcohol steps (30% to absolute alcohol – 10 minutes in each), then clear in xylene (or similar clearing agent) and mount on a microscope slide in Histoclear (or similar mountant). Remember to place a cover slip on top of the specimen! If the specimen does not have to be preserved as evidence, it is not necessary to dehydrate it and it can be mounted directly in water after being boiled in sodium hydroxide.

The identification of invertebrate eggs is difficult and if possible it is best to let them hatch and rear the larvae to adulthood. According to Sukontason *et al.* (2004) a simple means of identifying blowfly eggs is to stain them in 1% w/v potassium hydroxide solution for 1 minute followed by dehydration in 15%, 70%, 95% and absolute alcohol (1 minute in each) and then permanent mounting on a glass microscope slide.

Molecular taxonomy

Owing to the lack of traditionally trained taxonomists, many scientists, and not just forensic biologists, are using DNA markers to identify species. Indeed, there is an international collaboration of scientists currently engaged on a scheme called the Consortium for the Barcode of Life that seeks to establish an identifying DNA barcode for all organisms. It is unlikely that a single locus will be suitable for all organisms but the mitochondrial gene for cytochrome c oxidase I (COI) has proved effective for distinguishing between many animal species. COI gene sequences and also those for cytochrome c oxidase II (COII) are a potentially useful means of identifying several blowfly species (e.g. Nelson *et al.*, 2007; Schroeder *et al.*, 2003) but there are risks associated with relying upon a single locus (e.g. Wells *et al.*, 2007) and some scientists remain sceptical of its potential (e.g. Will & Rubinoff, 2004). In addition, until more research is published there will remain concerns over whether all populations of a given species are identified by a particular gene sequence, whether there are problems caused by sub-species and species complexes, and if there is cross-reaction with gene sequences present in any of the other insects likely to be found in the same environment. Wells & Stevens (2008) provide a thorough coverage of the pitfalls of molecular taxonomy in forensic entomology and also its considerable potential.

Molecular taxonomy is an attractive prospect where maggots and other larval invertebrates are concerned. Taxonomic keys are not available for all insect larvae of forensic importance, rearing them to adulthood can be time consuming and success is not guaranteed. By contrast, because there are standard protocols for the extraction and analysis of DNA, these procedures could be done by a trained laboratory technician. Theoretically, one could measure the length of a maggot and then divide it into three portions. The anterior and posterior regions would be processed and preserved on microscope slides to determine the maggot's instar and enable later confirmation of identification by traditional means (if that is possible) whilst the central region would be processed to extract the DNA although a small proportion of this should also saved for confirmatory testing by an independent laboratory.

Future directions

There is a need for more information on the basic biology of most invertebrates of forensic importance. For example, the flight habits of blowflies, fleshflies and muscid flies in relation to different environmental conditions and in different parts of the country. There is still some uncertainty to the extent to which remains would be colonized by blowflies during twilight or night-time conditions. Similarly, it is uncertain how burial conditions affect colonization of remains by both blowflies and other Diptera and how burial influences their subsequent development. It would also be helpful to know the extent to which there are differences in developmental rates between populations from different geographical regions under 'normal' above-ground conditions. Most experiments on competition to date have involved varying the number of an individual species of the same age whilst most corpses contain mixtures of species and age ranges and this would also be an area of research worth exploring.

Molecular taxonomy offers great potential not only for identifying individual insects and other invertebrates but also the possibility of identifying their geographical origin should strain specific markers be identified. For example, strain/ population specific markers could be used to establish whether a body was moved and if so where from. Increasing use could also be made of genetic markers to identify illegal trade in organisms such as abalone, scallops, and lobsters. In addition, the potential of stable isotope analytical techniques could be investigated. These techniques are used to establish the provenance of all sorts of biological material but appear to be little used in forensic entomology.

Quick quiz

(1) State four sites where a female blowfly would typically lay its eggs on a corpse and explain why these sites would be chosen.

(2) How would you know when a final instar blowfly maggot had finished feeding?

(3) How do *Calliphora vicina* and *Protophormia terraenovae* differ in their choice of pupation site? Why is this of relevance in the collection of forensic evidence?

(4) Explain why although adult female blowflies are often seen resting and feeding on a corpse that is in the later stages of decomposition, they do not lay their eggs upon it.

(5) Explain why a buried corpse may contain thousands of maggots of *Conicera tibialis* but none of *Lucilia sericata*.

(6) At what stage of corpse decomposition do Dermestid beetles tend to appear?

(7) Explain how ants can affect the rate of corpse decomposition.

(8) What would the sight of numerous live body lice moving across the clothing of a dead person indicate? Explain your reasoning.

(9) Medical evidence indicates that a murder victim has been dead for only 24 hours but the blowfly larvae recovered from the body were estimated to be about 44 hours old. Is there a conflict of evidence? Explain your reasoning.

(10) Briefly distinguish between how a wasp may be either a direct or an indirect cause of human fatalities.

Project work

Title

The effect of burning on invertebrate colonization of a dead body

Rationale

Murderers often attempt to dispose of a dead body by burning. They are seldom successful but it does alter the chemical and physical nature of the body and this may impact upon the speed of colonization and the rate at which decomposition occurs. This in turn may require allowances when calculating of the minimum time since death on the basis of blowfly growth characteristics.

Method

The problem may be investigated in the laboratory and the field. In the laboratory, blowfly maggots may be reared on meat that has been subject to varying degrees of burning. The growth characteristics (e.g. length, instar) of the maggots should be monitored at least once a day along with the condition of the meat (e.g. change in weight). Blowfly adults may also be offered choices of burnt and unburnt meat on which to oviposit – this could also be tested using the classical Y-shaped choice chamber apparatus. The experiment could be replicated in the field, in which case the environmental conditions and also the invertebrate species composition should also be recorded. The experiment could be repeated at different times of year and also the meat might be covered in clothing material before it was burnt to determine whether this had any impact.

Title

The effect of competition on the growth characteristics of blowfly maggots.

Rationale

Blowfly maggots are usually in competition with one another and this may result in either an increase or a decrease in their growth characteristics. This in turn may require allowances when calculating of the minimum time since death on the basis of blowfly growth characteristics.

Method

Single and mixed species of maggots are allowed to develop on a fixed amount of food. The amount of food will be such as to ensure that the highest densities of maggots are at risk of consuming all the food before the end of their final instar. The numbers of maggots would be assessed daily to determine if and when cannibalism took place and the number of maggots that successfully pupate and emerge as adults. The length and instar of the maggots would be assessed at least once a day.

7 Invertebrates 2: Practical aspects

Chapter outline

Objectives

Describe how the PMI or time since infestation may be calculated on the basis of the species of invertebrates present and their stage(s) of development.

Discuss the limitations of PMI calculations based on invertebrate evidence, the factors that can influence them and how potential problems can be overcome.

Describe how invertebrates can be used to determine whether a body or object has been moved and to provide pharmacological and molecular evidence.

Describe how invertebrate evidence should be collected and stored for use in forensic analysis.

Essential Forensic Biology, Second Edition   Alan Gunn
© 2009 John Wiley & Sons, Ltd

Calculating the PMI/time since infestation from invertebrate development rates

If one has reliable information on the biology and development times of an invertebrate and how these are affected by temperature and other environmental conditions it is possible to back calculate to determine when the egg from which a specimen developed was laid. This is useful in determining when a person died, a wound myiasis developed or food became contaminated. The technique is dependent upon correct identification, the accuracy of the estimated environmental conditions the specimens experienced and the reliability of the experimental development data used to perform the calculations. Many forensic texts refer to the calculations as a determination of the 'minimum time since death'. It is called the 'minimum' time because the estimate reflects the least time it would take for the invertebrate to reach a particular stage of development – it is possible that the person might have been dead for longer but it is very unlikely that the initial infestation took place after the calculated date. In some texts the time is referred to as the 'earliest oviposition date' (EOD), which is more accurate and reflects the fact that eggs may be laid before a person dies (for example, in cases of myiasis) or some considerable time afterwards (for example, flies could not gain access to the body). The EOD and the PMI may therefore not be the same. The EOD should therefore be considered in relation to other PMI indicators but in the absence of other evidence it provides a guide on which an investigation can proceed. The term EOD is also more suitable than 'minimum time since death' because exactly the same calculations are used in other forensic circumstances such as cases of neglect and food contamination.

It is important to remember that even with the most rigorous collection and analysis of evidence, biological processes usually involve a great deal of 'noise' as a consequence of various complicating factors. Estimates of the EOD are precisely that – estimates – and the amount of variance around an estimate will increase with the length of time since a person died or since infestation occurred (Nowak, 2004; Archer, 2003).

Determining the age of blowfly life cycle stages

Blowflies are the most commonly used invertebrates to determine 'minimum time since death' calculations but the procedure would be the same for any other insect. Maggots are collected from all sites in, on, and around the body, ensuring that these include the oldest (largest) maggots and any eggs, pupae or adults that are also present. During the early stages of decomposition, the most mature maggots or pupae indicate the PMI because they will have developed from the eggs, or larvae in the case of fleshflies, deposited on the body immediately after the person died.

Eggs

If a body is discovered shortly after death blowflies may have begun laying eggs but these are not yet hatched. In this case, the eggs can be collected and stored in a controlled environment until hatching. Bourel *et al.* (2003) demonstrated that for

Lucilia sericata it is possible to estimate the time of egg-laying with an accuracy of about 2 hours by relating the time taken for the eggs to hatch to the environmental conditions they experience. It would be interesting to know whether this approach is suitable for other fly species because it would be particularly appropriate to disputes concerning neglect (e.g. time since a wound was dressed) or food hygiene cases in which blowfly eggs are likely to be the principle evidence. Some *Calliphora* species produce 'precocious eggs' – these are single eggs that are fertilized and therefore at a later stage of development than the unfertilized eggs immediately behind them in the fly's reproductive tract (Wells & King, 2001). Consequently, when egg-laying commences these eggs have a 'head start' and if the resultant larvae are collected they will suggest a longer PMI than was actually the case. How important this would be in practice is debateable since such 'precocious eggs' would be an extremely small proportion of the total laid, one should collect numerous eggs/ maggots and it is not a good idea to base conclusions on just a single specimen. Sarcophagid flies 'lay' larvae rather than eggs so it cannot be assumed that where both eggs and maggots are found together that they belong to the same species.

Larvae

The morphology of the posterior spiracles and the cephaloskeleton changes between instars and therefore provides an indication of the stage of larval development (Figs 6.7 and 7.1). Within a stage, an estimation of age can be made from measuring the length of the maggot because length increases with age until the larvae cease feeding and wander off in search of a pupation site – at which point the maggot starts to contract. Age cannot be judged on length alone since there is considerable overlap in the size ranges between instars (Anderson, 2000). Unfortunately, the method of collection and storage can result in the maggot shrinking. The amount of shrinkage will vary with the method used and the species and size of the maggot and this

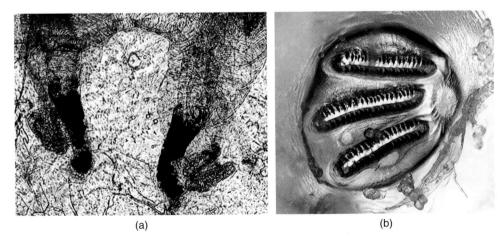

(a)   (b)

Figure 7.1 In blowflies, the morphology of the posterior spiracles changes with each larval moult. (a) In second instar *Lucilia sericata* two slits are present in each spiracle. (b) In third instar *Lucilia sericata* three slits are present in each spiracle.

could result in their age being underestimated (Adams & Hall, 2003). Ideally, therefore, maggots should be measured before they are placed in preservative. Otherwise, a reliable means of estimating the percentage shrinkage that has taken place needs to be made by experimentation and allowed for in any calculations. Another potential preservation artefact is head-curling and this too can make length measurement difficult. The risk of contraction and head-curling can be reduced by appropriate sampling techniques (see later) and, at least in *Calliphora augur*, there is a linear relationship between width measured between the 5th and 6th abdominal segments and body length (Day and Wallman, 2006) so width measurements could be used if head curling occurs. It is not known whether this approach would work in other species of forensic importance but it would appear likely.

Maggot dry weight increases with age until the larvae cease feeding and wander off in search of a pupation site although drying renders the maggot unsuitable for further morphological analysis. Maggot wet weight cannot be used because in live specimens it is affected by hydration and feeding status and in preserved specimens by the type of preservative used. Some workers measure the length of the crop in the final instar because this shrinks when feeding ceases. Age-related changes in the composition of cuticular hydrocarbons have been described in larvae of *Chrysomya rufifacies* (Zhu *et al.*, 2007). However, this method results in the destruction of the maggot and cuticular hydrocarbon compositions in other insects are affected by diet and environmental conditions and it is probable that the same will be true of blowfly larvae.

Pupae

Although the larvae of cyclorrhaphan flies (e.g. blowflies, fleshflies, and houseflies) provide the most accurate basis for estimations of the minimum times since death/time since infestation, it is possible to age their pupae, especially if pupation has recently commenced when the body is discovered. When the maggot moults to the pupal stage, the third larval instar cuticle is retained and serves as a protective puparium. During this process the insect contracts and the puparium darkens as it becomes tanned and sclerotinized over a period of several hours. The extent of the tanning is therefore an indication of age (Fig. 7.2). Within the puparium, the insect first enters the pre-pupal stage before metamorphosing to the pupal stage – this can be demonstrated by dissecting the pupa from its puparium. The possibility of determining the age of a pupa by assessing the activity of temporally expressed genes (i.e. those which are switched on or off at specific points in development) is being investigated but it is too soon to be able to say how applicable the technique will prove in a forensic context.

Adults

It is possible to distinguish newly emerged cyclorrhaphan flies from those that are older and, albeit crudely, how long they have been emerged. This may be useful where a body at an advanced stage of decay is discovered. Newly emerged flies are pale in colour, their wings may not be fully expanded, and the ptilinum (a sack-like

Figure 7.2   Pupae of the blowfly *Calliphora vicina*. Note how the cuticle of the final instar larva is retained to form the puparium. The degree of darkening of the puparium increases with age owing to tanning and sclerotinization. This reduces flexibility but makes the cuticle stronger and reduces water loss.

structure on the head) may either be expanded or not yet fully retracted behind the ptilinal suture (Fig. 7.3). This state lasts for only a few hours. It is difficult to determine the age of male flies but in females the ovaries are immature at emergence and the eggs develop over a period of days. Wings become abraded with time and have been used as an age estimate in ecological studies (Hayes *et al.*, 1998). The levels of pteridine eye pigments alter with age in a variety of fly species, including *Lucilia sericata*, but the technique is not used commonly in forensic studies. The different techniques that can be used to age adult flies are reviewed by Hayes & Wall (1999) whilst Smith (1986) provides an excellent description of blowfly larval stages and a series of forensic case studies.

Calculating the time since oviposition

(1) The accumulated degree hours/days method

Once the instar and length of a maggot is known, it is possible to determine how long it took to reach that stage under controlled conditions using published studies on rates of development and estimates of the conditions likely to have been experienced by the larvae. Correct identification is essential because even closely related species develop differently under the same conditions. There is surprisingly limited published information on the development rates of insects of forensic importance although data for several blowfly species can be located in Greenberg & Kunich (2002) and Anderson (2000). For most species it is not known whether individuals collected in different geographical regions would exhibit the same developmental characteristics although differences in other biological traits have been described.

(a)   (b)

Figure 7.3 Characteristic features of recently emerged blowflies. (a) The body is pale and although the wings are fully expanded the insect is not yet able to fly. (b) The body has started to darken but the ptilinum is still present – it will soon be retracted and stored behind the ptilinal suture.

For example, *Lucilia sericata* derived from Spanish populations tend to suffer higher mortalities but produce larger adults than English populations (Martinez-Sanchez *et al.*, 2007). Similarly, in *Calliphora vicina* larval diapause is influenced by the photoperiod experienced by the parent female and her genetic background. Female flies from Finland produce diapausing larvae in response to much longer days than flies from Scotland and that diapause lasts for longer and confers a greater resistance to low temperature (Saunders, 2000; Saunders & Hayward, 1998). This lack of development data is partly because compiling it is time consuming and the results are difficult to publish in high impact journals. It therefore tends to have a low priority in university research departments and the results, if published at all, are spread across a wide range of periodicals. Ideally, there should be an agreed protocol for gathering information on developmental rates and a central web-based databank in which workers could enter their findings.

The time period taken to reach the maggot developmental stages found on the body must be adjusted to fit the site where the body was found. One way of doing this is to convert the temperatures and times into accumulated degree hours (ADH) or accumulated degree days (ADD) by multiplying the time by the temperature (°C). Because the time required for development increases as temperature decreases, the total number of ADH or ADD required to reach any given stage should theoretically remain constant. For example, according to Greenberg & Kunich (2002), at 25 °C the average minimum duration of the developmental stages of *Calliphora vicina*, are as follows:

Egg stage =   14.4 hours
First instar =   9.6 hours
Second instar =   24 hours
Third instar =   158.4 hours

To calculate accumulated degree hours (ADH), multiply the sum time taken to reach a particular stage by (temperature − the base temperature). The base temperature is that below which the insect is unable to develop and is calculated for a laboratory culture by plotting temperature (X axis) against 1 divided by the time required for the insect to develop from being laid as an egg to becoming an adult (Y axis).

For example, the ADH taken for *C. vicina* to reach the 3rd instar is:

$$(14.4 + 9.6 + 2.4) \times (25 - 2) = 1104 \text{ ADH}$$

Hypothetical example

(1) Suppose that a body is discovered at 11.00 A.M. on June 23rd and that insect samples are collected at midday on the same day.

(2) The most mature *Calliphora vicina* larvae found on the body are just moulting from the second instar to the third instar.

(3) The total ADH required to complete the egg stage and all of the first and second instars is:

$$(14.4 + 9.6 + 2.4) \times (25 - 2) = 1104 \text{ ADH}$$

(4) To estimate the period of insect activity, work backwards from the time the maggots were collected (midday on June 23rd). There were 12 hours of development between midnight and midday and the mean temperature at the scene for that set period was 15 °C. This means that a total of 180 ADH were accumulated on June 23rd (12 hours × 15 °C).

(5) On June 22nd, the mean temperature was 14 °C, and this gives a total of 336 ADH (24 hours × 14 = 336 ADH).

(6) On June 21st, the mean temperature was 16 °C, and this gives a total of 384 ADH (24 hours × 16 = 384 ADH).

(7) Add the totals: 180 + 336 + 384 = 900 ADH.

(8) 1104 − 900 = 204 ADH still outstanding.

(9) On June 20th, the mean temperature was 16 °C, so each hour accounted for 16 ADH.

(10) Divide 204 by 16 = 12.8 hours of development on June 20th.

(11) Counting backwards from midnight, this means that the female flies must have laid their first eggs between midday and 11 A.M. on June 20th.

The above example is very basic and assumes that it is unnecessary to calculate separate ADHs for every individual hour of the day because the temperature of a whole corpse does not change as rapidly as air temperature. It also assumes that the relationship between developmental rate and temperature remains linear across all temperatures although this is not the case and especially so at the lower and upper threshold for development. Where there is a big difference between the temperature at which the insects were reared in the laboratory and that experienced in the field (such as the 9–12 °C in this example) a degree of caution is required in interpreting the results (Anderson, 2000).

To determine the temperature at the crime scene prior to the discovery of the body a recording device is left at the site for 3–5 days. The data is then compared to that from the nearest weather station using regression analysis. This enables a correction factor to be identified that can be used to estimate the temperature at the crime scene from the weather station records. More details on the procedure can be found in Gennard (2007). Some workers have attempted to improve the accuracy of their estimates by incorporating computer models of insect growth into their calculations (Byrd & Allen, 2001).

(2) The isomegalendiagram method

An isomegalendiagram is obtained by plotting temperature (*y*-axis) against time taken for a maggot to reach a given length (*x*-axis) (Grassberger & Reiter, 2001). It therefore consists of a series of curves (Fig. 7.4) from which one can read off, for example, that a maggot measuring 5 mm might have taken 60 hours to reach that size at 16 °C but only 24 hours at 25 °C. Although an extremely neat method, generating the data to prepare an isomegalendiagram is very time consuming.

Complicating factors affecting earliest oviposition date calculations

Regardless of the method used to calculate the oviposition date many factors can impact upon the accuracy of the estimate.

Factors that can result in an underestimate (e.g. person was dead for longer than the calculation suggests)

Fly access

In some circumstances flies may not gain access to a body until several hours or even days after death occurred. Alternatively, the person might die at a time of year when it is too cold for blowfly activity. If the body is stored in a freezer before disposal access may not occur until days or months after death occurred. In these

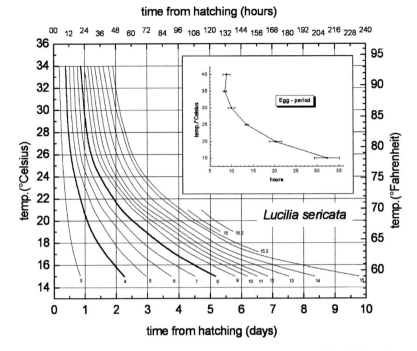

Figure 7.4 Isomegalendiagram for the blowfly *Lucilia sericata*. Time (*x*-axis) has been plotted against temperature (*y*-axis) and each line represents a particular larval length. One can use the graph to determine how long it would take a maggot to reach a certain length at a set temperature. The smaller graph shows the duration of the egg period at temperatures between 15 and 40 °C (Reproduced from Grassberger, M. and Reiter, C. (2001) Effect of temperature on *Lucilia sericata* (Diptera Calliphoridae) development with special reference to the isomegalen- and isomorphen-diagram. *Forensic Science International*, **120**, 32–36, Copyright 2001, with permission from Elsevier.)

Table 7.1 Summary of age indicators in cyclorrhaphan flies at different stages of development

Life cycle stage	Age indicator
Larva	Body length
	Crop length (post feeding stage only)
	Shape of cephaloskeleton
	Shape of posterior spiracles
	Shape of anterior spiracles
Pupa	Coloration of puparium (initial stages of pupation only)
	Stage of development of pupa within puparium
Adult	Ptilinum (expanded or retracted)
	Wings (not expanded in recently emerged, increasingly abraded with age)
	Hardening and darkening of cuticle immediately after emergence
	Development of ovaries
	Pteridine pigment accumulation in eyes

circumstances, discrepancies may be noticed between the time the person was reported missing, the state of decay and the stage of development of the blowfly maggots.

Insecticide sprayed on the body

The murderer may attempt to confuse the evidence by spraying the corpse with a substance that deters insects from colonizing the body. However, chemicals vary in their effectiveness at reducing blowfly egg-laying. For example, sodium hydroxide appears to have little effect whilst petrol or patchouli may delay egg-laying by one or more days (Bourel et al., 2004) but a great deal will depend upon the concentration and the amount of the chemical used. The presence of chemicals may be indicated immediately by their smell, their obvious physical presence (e.g. as a powder or crystalline deposit), or the damage they have caused to clothing and body tissues. They can be tested for by routine chemical analysis and it would then require experimentation to determine whether they might have affected insect colonization.

Larval/pupal diapause

Some fly species enter into a diapausing state (a form of dormancy) in response to conditions they experienced earlier in their (or their parents') development. This is usually either the larval stage or the pupal stage – the stage depends on the species. The development of diapausing larvae/ pupae is therefore prolonged and the adult flies may not emerge for weeks or months. This is most likely to occur in specimens collected during the autumn and winter periods as a consequence of a response to the reduced day length. Diapause is indicated by the lack of normal development even if the insects are placed in optimal conditions. It would require experimentation to determine the conditions under which a given population entered diapause and therefore when diapause might have commenced.

Predators

Vertebrate and invertebrate predators are often attracted to feed on the detritivores found on corpses and can slow the decomposition process by destroying the eggs and larvae of maggots, as well as killing adult flies arriving to lay their eggs (Archer & Elgar, 2003). In addition, some fly maggots feed on competitors within a dead body or a wound on a living animal. For example, the larvae of Chrysomya albiceps are well known for their predatory behaviour. It is therefore necessary to note any evidence of vertebrate predator activity and to collect all the invertebrates found in, on, and around the body and to estimate their abundance. One can then make judgements whether the predators were likely to have an impact. If maggots are being collected for rearing or will not be killed immediately it is important to segregate predatory species such as C. albiceps or at the least provide sufficient food to reduce the likelihood of predation/ cannibalism.

Factors that can result in an overestimate (e.g. person was not dead for as long as the calculation suggests)

Myiasis

If maggot activity commenced before the person died, the estimation of minimum time since death may be too long. For example, if a person is suffering from wound myiasis, they may ultimately die of their wounds or from the potentially fatal consequences of wound myiasis. Consequently, one should consider whether there is a mismatch between the medical or other evidence and that derived from entomological analysis. For example, the body is only just starting to exhibit rigor mortis but is infested with third instar blowfly larvae. It would, however, provide evidence that the person was wounded some time before they died.

Development of a maggot feeding mass

During putrefaction, large numbers of maggots collect together to form a feeding mass (sometimes called a 'maggot feeding ball') and this, together with microbial action, generates remarkable amounts of heat. This can raise the temperature in the centre of the feeding mass by more than 20 °C above ambient, the extent largely depending upon the volume of maggots and, for masses upon the ground, the underlying soil temperature. The internal temperatures of maggot masses above a volume of about 20 cm^3 is independent of ambient temperature and according to Slone & Gruner, (2007) the size of the body does not influence the mass temperature. There is some dispute in the literature about the impact of a large feeding mass on maggot development times and whether allowances should be made, and if so, how. Some workers consider no adjustments need be made to accumulated degree hours (ADH) calculations because it takes several days for a large feeding mass to develop. Indeed, work by Joy *et al.* (2006) indicates that the temperature of maggot masses does not differ significantly from ambient until the maggots reach their early 3rd instars. Furthermore, it is only at the centre of the maggot mass that the temperatures are very high and individual maggots are not thought to spend long there because it is potentially harmful – although proving this by tracking the movements of individual maggots within a feeding ball would present an interesting challenge! (One might start by rearing a sub-population of maggots on a diet containing a dye and then mixing them with a larger (un-dyed) maggot mass.) Anyway, if the largest maggots are those recovered from within a feeding mass and a record was made of the temperature within the mass, it would be sensible to calculate two minimum development times: one incorporating several hours development within the mass (determined on the basis of the mass's size and hence how long it has held an elevated temperature) and one without.

Minimum temperature for development

ADH calculations assume a constant, linear relationship between development rate and temperature and that below a certain temperature, e.g. 5 °C, development ceases. However, this assumption is questionable (Ames & Turner, 2003) and varieties of the same species of blowfly from different countries, or different parts of the same country, are known to exhibit different temperature sensitivities. If the temperature periodically falls below the supposed minimum for development, it would be sensible to calculate two minimum times since death firstly on the basis that development ceases and secondly that it continues. Ideally, larvae collected at the time should be placed at different low temperatures to record the effect on their rate of development and ADH calculations should be made with reference to laboratory-reared insects maintained at a temperature within about 5 °C of that experienced by the maggots collected from the body. However, owing to the cost and logistical difficulties, this may prove impossible.

Factors that can result in either an underestimate or an overestimate

Drugs and toxins

The effects of drugs and toxins on blowfly larvae are poorly understood with various studies indicating effects that have ranged from zero, to either speeding up, or retarding the rate of development (Bourel *et al.*, 1999). For example, maggots of the fleshfly *Boettcherisca peregrina* feeding on tissues containing morphine, develop faster, grow bigger and produce pupae that take longer to develop resulting in an error of up to 29 hours if the estimated post mortem interval is based on larval development and 18–38 hours if based on the duration of the pupal stage (Goff *et al.*, 1991). There are numerous drugs and toxins and numerous species of maggots associated with corpses, so it is impossible to generalize on the effects of their interactions. Furthermore, persons who die of drug overdoses may contain a cocktail of different chemicals and the drugs may not be evenly distributed between the tissues of the body. For example, persons whose death is associated with the drug ecstasy are often found to have consumed alcohol and other drugs, such as cocaine, amphetamines, and heroin. One can therefore only keep an open mind that chemicals within the body may have influenced maggot development to some extent.

Microclimate

Owing to local effects such as wind eddies and exposure the microclimate where blowfly larvae are feeding may be different from that a short distance away. Consequently, the maggots could experience a warmer or colder temperature than the one used for the development calculations and therefore develop faster or slower than expected. Local temperature measurements should therefore be taken from the exact site where the maggots are collected and one should be aware of possible complicating factors, such as snow acting as an insulator.

When the temperature drops below a certain level maggots relocate to beneath the body and aggregate together. By so doing they can facilitate the generation and retention of sufficient heat to continue feeding and growing even though the ambient temperature is well below their minimal developmental temperature. For example, a large mass of maggots can remain active upon a body placed in a typical morgue storage unit that operates at −1 °C to +4 °C for at least 40 hours and during that time cause considerable tissue loss (Huntington *et al.*, 2007). Consequently, not only would the time since death be underestimated if it was assumed that maggot development ceased once the body was placed in the storage unit but the loss of tissue could erase or obscure evidence of injuries relevant to the cause of death.

Case Study: How covering a body can affect maggot development

The body of 37-year-old man was discovered in a forest in Northern Germany at about midday on July 21st (Klotzbach *et al.*, 2004). He had received a fatal gunshot wound to his chest whilst another to his face had shattered his jaw and led to blood pooling within his oral cavity. Blowfly eggs and 2nd instar larvae were recovered from the body, stored in 70% alcohol and sent for analysis. It is difficult to identify accurately blowfly eggs and young larvae solely on the basis of morphological features and no larvae were collected for rearing to adulthood. The investigators therefore used the molecular technique of PCR–RFLP that compares the sequences of mitochondrial DNA that code for cytochrome oxidase I and cytochrome oxidase II (Chapter 6). The eggs were identified as belonging to *Lucilia sericata* and *Calliphora vicina* whilst the second instar larvae were identified as *Calliphora vomitoria*. This indicates how a variety of blowfly species may colonize a corpse at about the same time and therefore the need to collect multiple samples. The PMI calculated on the basis of the meteorological records from the nearest weather station and the instar number and size of the *C. vomitoria* larvae indicated that the man must have been dead for at least 5 days and had therefore been shot on July 15th. This, however, could not have been the case since the victim's girlfriend, who was considered a reliable witness, stated that the man was very much alive at 9.30 am on July 16th. Consequently, myiasis could not be an explanation for the discrepancy and an alternative reason had to be found. It was subsequently established that the larvae were collected from within the mouth of the victim and that his head had been covered by a leather jacket. The wound to the man's jaw would have attracted flies and the jacket clearly did not exclude them from gaining access. The jacket probably would, however, have provided a degree of insulation. Furthermore, the maggots growing within the oral cavity would also be protected from external conditions. Consequently, it was deemed likely that the maggots experienced temperatures at or above the maximum ambient temperature during their development. Recalculating the PMI on this basis indicated the man was probably shot on July 16th and this accorded with other forensic evidence. This case therefore indicates the problems that can arise if the entomological samples and data for their interpretation are not collected appropriately, the value of molecular taxonomic techniques and how insulation/ protection from the environment can influence maggot development.

Table 7.2 Summary of factors that can complicate the minimum time since death calculation based on blowfly larval development rates. Abbreviations: PMI = post mortem interval. EOD (earliest oviposition date) could be substituted for PMI in cases of neglect and food contamination

Factor	Reason
Restricted access of adult flies to body	Delay in egg laying so PMI estimate is too short.
Maggot feeding mass present	Elevated local temperature means maggots develop faster than expected from environmental temperature so PMI estimate is too long.
Many Diptera species present	Different species develop at different rates. Biggest may not be oldest, it may simply grow faster.
Food quality	Low food quality may delay larval development.
Larval /pupal diapause	Larval/ pupal development prolonged.
Pupal disease / parasitoids	Adults may not emerge.
Predators and competitors	Larval population has difficulty establishing itself. PMI estimate is too short.
Drugs and toxins	May speed up or slow down larval development.
Chemical fly deterrents	Delays egg laying so PMI estimate is too short.
Myiasis	PMI estimate is too long.
Body at advanced decay	No maggots or 2nd generation present.
Low environmental temperature	Uncertainty at what temperature larval development ceases.
Microclimate effects	PMI estimate is either too short or too long in relation to time of death owing to using incorrect development temperature.
Incorrect storage	Maggots shrink and therefore appear younger than they really are so PMI estimate is too short.
Laboratory development rate data unreliable	Larval development rate may be estimated incorrectly so PMI estimate is either too short or too long in relation to time of death.

Determination of the PMI using invertebrate species composition

The invertebrate fauna associated with a fresh body is different from that associated with one that is skeletonized. Similarly, the composition of the fauna is also influenced by the time of year, so the variety of invertebrates associated with a corpse in December will be different from that in August. It has therefore been suggested that one can estimate the PMI from the diversity of the invertebrate fauna. This could be particularly useful if the body is at an advanced stage of decay when there are few blowfly maggots present and these almost certainly developed from eggs deposited long after the person died – and therefore cannot be used to determine the PMI.

In practice, the invertebrate faunal succession varies markedly between individual corpses and environmental and local factors can influence when and whether certain insect species arrive (Archer, 2003). Consequently, the species composition is not a very robust method for estimating the minimum time since death or infestation.

Determination of the PMI using ectoparasites

Fleas, lice and ticks move away from a dead body when it starts to cool, so their presence would indicate that a person or animal had not been dead for long. Fleas can be revived if submerged for less than 24 hours and may therefore provide evidence of how long a body has been underwater. Lice tend to be more sensitive and die within 12 hours. The skin follicle mite, *Demodex* is a common parasite of humans and may sometimes cause pathology either directly or through acting as a vector of disease. It is unable to leave a dead body but it is capable of surviving for over 50 hours after death of the host. However, the lack of a clear relationship between survival rate and time since host death limits its usefulness in forensic investigations (Ozdemir *et al.*, 2003).

Determination of movement from invertebrate evidence

Identification of geographic origin

Because many invertebrates have a restricted geographical distribution or occupy specific habitats, their presence can be used to determine the past history of a person or object. For example, the discovery of a corpse in Kent containing an uncommon species of insect only previously recorded in the North of England would suggest that the body had been moved. Similarly, the presence of parasitic protozoa and helminths within the body can provide an indication of a person's origins or recent travels. For example, finding the eggs of the schistosome *Schistosoma haematobium* in the urine indicates that a person had been living in certain parts of Africa or the Middle East. More precise identification of geographic origin can sometimes be provided by genetic analysis since some parasite strains have localized distributions.

The invertebrates found on a body need not be a species associated with decay but could be predators or ones that used the body or its clothing as a refuge, or could have accidentally become trapped in the clothing of the person whilst they were alive. Similarly, the finding of an insect with a distribution restricted to Northern Europe in a foodstuff processed and packed in Indonesia would indicate that infestation occurred after the goods reached Northern Europe. The presence of insects can be used to determine the source of drugs such as cannabis. For example, the presence of foreign species can indicate the country of origin of the cannabis and therefore whether it was imported (Crosby & Watt, 1986). This is important because it affects whether or not the person is convicted of drug trafficking. During 2008, several packages of cocaine were washed up on beaches along the coast of Cornwall and Wales. The cocaine was protected within layers of plastic and then covered with sacking and was estimated to have a street value of over £1 million. Colonies of barnacles were attached to the surface of the packages and their size indicated that the packages had been in the sea for several weeks. Furthermore, the barnacles were of a species not native to cold European waters. It is believed that the packages were probably thrown into the sea somewhere in the Caribbean after the boat transporting them was about to be intercepted and that they then drifted across the Atlantic.

Murderers often move the body of their victim. Movement from outside to inside a building might be indicated by the presence of typical moorland or woodland species whilst the finding in summer of a dry mummified corpse containing dermestid beetles and tineid moth larvae in a deciduous woodland would indicate that the body was initially stored indoors. When a person dies indoors and the doors and windows are sealed to prevent the access of flies, the body may mummify. In this case, dermestid beetles and tineid moths, which are common household pests and may already be present in the room, will start to infest the corpse once it is dry enough. Mummification is unlikely to occur in an exposed corpse left outside in the UK because the climate is too damp. Furthermore, if there was no evidence of blowfly activity (e.g. puparia), it would indicate that the body was placed there a long time after death occurred.

Evidence of exposure

Shallow burial will usually prevent blowflies from colonizing a corpse. The exact depth will depend upon the soil type, moisture content etc. It therefore follows that if a buried body is discovered to contain the eggs, larvae or pupae of blowflies, the chances are high that the body was exposed to air for a period of time before it was buried. Similarly, if a body is found on the surface that contains little evidence of blowfly attack for the time of year and its stage of decomposition then it is likely that the body was maintained in a sealed protected environment until it was disposed of.

Movement by invertebrates

A corpse (or bits of it) might be moved by a murderer, by the actions of dogs, foxes, badgers etc., or by natural phenomena such as landslips and rivers overflowing. Invertebrates are not large enough to move a human body but they can rearrange their clothes. For example, in a somewhat bizarre experiment reported by Komar & Beattie, (1998) dead pigs of human size were dressed in women's clothing and their subsequent decay monitored. It was found that the maggot masses, acting in concert with bloat, caused the underwear and tights to be moved down the hind legs whilst the skirt was pushed up. The clothing was therefore moved into a position which is typical of that found in cases of sexual assault plus homicide. Similarly, the movement of large numbers of maggots from the body when they leave in search of a pupation

Table 7.3 Summary of insect evidence that suggests that a body was moved after death occurred

Movement of body	Insect	Evidence
Exposure followed by burial	Blowfly eggs and larvae	>24 hours exposure
	Histerid beetle larvae	~2–3 days exposure
	Piophilid fly larvae	~7 days exposure
	Dermestid beetle adults and larvae	>14 days exposure
Burial or placed in a sealed container followed by exposure	Blowfly eggs and larvae	Absent or stage of development younger than expected from degree of decomposition
	Phorid fly larvae	Abundant phorid fly larvae and adults but no evidence of blowfly species
Inside to outside	Blowfly larvae and pupae	Body mummified or skeletonized but no evidence of previous blowfly larval activity on the body or pupae in the vicinity
	Dermestid beetles and tineid and pyralid moths	Both present but no evidence of previous colonization by other insect species
Outside to inside	Blowfly larvae and other detritivore species	Presence of species that do not normally enter buildings.
	Any 'accidentally associated' species	Woodland, moorland etc. species accidentally trapped in clothing
Geographical (between different regions or countries)	Blowfly larvae / other detritivore species or any 'accidentally associated' species	Species found outside their normal geographical distribution

site can cause objects to be displaced. For example, there are unsettling reports of false teeth being transported on the backs of a seething mass of maggots.

Invertebrate evidence in cases of wound myiasis and neglect

Determining whether neglect has occurred through a wound, bandages or a nappy becoming infested with maggots requires a similar procedure to that used for estimating the minimum time since death (Benecke & Lessig, 2001). Egg and/or larval samples need to be taken, the time noted and a record made of the environmental conditions (temperature etc.) recorded. Neglect cannot be automatically assumed if evidence of insect activity is found because eggs may be laid immediately after routine cleaning and, in the case of fleshflies, because first instar larvae are laid, maggots are instantly present.

Case Study: Elderly neglect

Gheradi & Constantini (2004) describe a case in which the maggot-infested body of an elderly woman was discovered in the filthy flat she shared with her daughter. This resulted in her daughter being charged initially with 'concealment of a corpse' although she claimed that she had given her mother a meal the previous night. An autopsy later in the day of discovery demonstrated that the body's temperature was 34.7 °C, indicating that the woman had not been dead for long and this was corroborated by finding only the early stages of post mortem decomposition. However, the presence of third instar blowfly larvae and second instar fleshfly larvae on the skin and the diapers indicated that the woman had not been kept clean for several days and the daughter was subsequently charged with 'elderly neglect'.

Case Study: Was the patient infected before or after being admitted to hospital?

Hospitals are full of sick people and are therefore inherently risky places in terms of their potential for disease transmission. Of course hospitals do their best to minimize this risk but hospital-acquired infections, also called nosocomial infections, are not unusual among even the most hygienic of institutions. Where these nosocomial infections cause serious illness or death and it can be proved that the hospital was not providing good hygienic practices it is probable that litigation will follow. Consequently, it is important to be able to prove whether or not a patient was infected before or after they entered the hospital. Most of these infections involve microbes such as MRSA but occasionally they involve myiasis. Hira *et al.* (2004) provide an interesting case of this involving a young boy admitted to a teaching hospital in Kuwait following a traffic accident. On admission the child's face was covered in blood, he was badly injured and remained comatose in the intensive care unit over the following days. The nursing staff noticed maggots emerging from the child's nose on days 4 (96 hours) and 5 (120 hours) after admission and five of these were collected. They also described seeing a 'green fly' perched on the child's nose on day 5 but neither this nor some maggots seen on the child's ear were captured. The maggots were identified as *Lucilia sericata* and the eldest was a third instar, one was in the process of moulting to the third instar and the others were second instars. It was therefore concluded that the infesting larvae were coming to the end of their second instar. According to Greenberg (1984), *L. sericata* can be reared from egg laying to the end of the second instar in 50 hours at 29 °C and above this temperature their development rate is even faster. Consequently, even assuming a nasal temperature of 30 °C rather than a body temperature of 37 °C, any eggs laid on the child before he entered the hospital would have had ample time to complete their third instar and/or be vacating the wound in search of a pupation site. The infection was therefore most likely acquired in the hospital and although the 'green fly' was not identified, its presence in the intensive care unit indicated that it was possible for flies to enter the premises.

Detection of drugs, toxins and other chemicals in invertebrates

When a body is discovered and it is believed that drugs or poisons may have contributed to the person's death, there are standard procedures for testing for these substances. However, if the body is not found until the late stages of decay, there may be few or no tissues left for analysis. In these circumstances, it is possible to detect the presence of substances in the maggots and their puparial cases (Goff & Lord, 2001; Hedouin *et al.*, 1999; Pounder, 1991). For example, using LC–MS (combined liquid chromatography–mass spectrometry), one can detect nordiazepam residues in a single maggot allowed to feed on a diet containing drug residues equivalent to those that would be found in human skeletal muscle following a fatal overdose (Laloup *et al.*, 2003). However, Tracqui *et al.*, (2004) were unable to relate the concentrations of a wide variety of drugs found in dead bodies with those present in invertebrates feeding upon them. Furthermore, the invertebrates exhibited marked differences in drug residue concentrations between individuals recovered from the same body. Consequently, apart from noting the presence, they concluded that it would be unsafe to draw conclusions from drug residues found in invertebrates recovered from a dead body; the absence of drug residues in the invertebrates should not be considered reliable evidence that there were no residues in the person's body before they died.

Roeterdink *et al.*, (2004) demonstrated that blowflies reared on meat containing gunshot residues accumulate lead, barium and antimony although these metals bioaccumulate in the maggots' bodies to different extents. However, more work needs to be done to determine whether gunshot residues would affect the maggots' rate of development and whether forensically useful information could be obtained from the isotopic signature of the metals found within them.

Spiders' webs often trap particles being transported by wind currents. For example, in the Soham murder trial (UK, November, 2003) in which Ian Huntley was tried for the murder of two young girls, it was reported that dark discoloration of cobwebs was observed on light fittings close to the bin in which the victims' burnt clothes were found. The coloration of the cobwebs became lighter the further one moved away from the bin. Unfortunately, there does not appear to be a method for testing cobwebs for smoke exposure.

Obtaining human/vertebrate DNA evidence from invertebrates

It is possible to extract DNA from the crop of a maggot or beetle and thereby determine what it had been feeding on. This can be useful in cases where there are insects but no body, where there is a suggestion that the insects crawled onto a body from elsewhere, or that the insect samples have been switched (Campobasso *et al.*, 2005; Linville *et al.*, 2004).

Fleas and lice frequently transfer between humans and between humans and animals that come into bodily contact with one another. It is therefore possible to link persons together by analysing the blood meal present in the guts of lice or fleas (Mumcuoglu *et al.*, 2004). For example, crab lice, *Pthirus pubis*, may be transferred between assailant and victim during a sexual assault and DNA extracted from the blood meal present in their guts can be used to link the two people together (Lord *et al.*, 1998). Davey *et al.* (2007) demonstrated that it is possible to extract host mitochondrial DNA from individual body lice (*Pediculus humanus*) up to 72 hours after they ingested a meal. As they point out, one has to be careful when extrapolating studies conducted with laboratory-reared insects to the situation in the field. For example, in the case of blood-feeding insects, laboratory-reared strains often take longer and larger blood meals than their wild counterparts because there is no chance of them being swatted. Consequently, there is no selection for rapid feeding behaviour.

Determining the source and duration of invertebrate infestations of food products

After identifying the species of invertebrate found within a foodstuff and their stage(s) of development, it is often possible to determine when the infestation initially developed and therefore who was responsible for allowing it to happen. For example, the customer of a UK firm supplying frozen chicken wings to a Caribbean Island was refusing to pay for deliveries because after the wings were cooked they were found to contain 'living maggots'. The claim was considered fraudulent because although some flies will lay eggs on frozen meat (e.g. *Calliphora vicina*) even if the chicken wings had been contaminated by fly eggs during processing, the freezing process would have killed them. The transport of frozen foodstuffs is highly regulated and, until the customer removed the foodstuff's protective packaging, no flies should have been able to gain access to the chicken wings. The fact that living maggots were found in the wings after cooking says little for the standard of cooking and also indicates that there was sufficient time for flies to lay their eggs and for these to hatch – a process that would be expected to take several hours even in the Caribbean. The larvae were in their second instar and belonged to a species not present in the UK. Assuming that the chicken wings arrived in a frozen state, they must have been left unprotected in the kitchen. The only other alternative would be that the chicken wings were not transported properly and had become contaminated with eggs between arrival on the island and the kitchen.

When food is packaged, the presence of chewed holes in the goods can indicate whether the insects bored their way into or out of a bag or sack. Furthermore, the presence of only a few holes in the packaging and a population of very young insects plus one or two adults would suggest that the infestation was initiated from outside. The absence of holes would indicate that the infestation began before the goods were packaged and their stage of development could be used to calculate when this took place.

Collecting invertebrates for forensic analysis

Collecting maggots and other crawling insects

As with all other types of forensic evidence, it is essential that invertebrate samples are collected carefully with accurate records of where and when they were obtained and who collected them and that there is a fully documented 'chain of custody' otherwise the reliability of the evidence in court could be thrown into doubt (Amendt *et al.*, 2007). In a homicide investigation, representative samples of all invertebrates found on and around the body should be collected directly into separate containers (Fig. 7.5). Insects found on the surrounding vegetation or elsewhere in a room can be collected using a sweep net or knocked into pots as appropriate. Soft forceps and a fine brush should be used when samples need handling manually to avoid damaging the specimens. Records should be made of the exact position of all invertebrates found at a crime scene and their relative abundance. Specimens

Figure 7.5 Biological samples should be collected from around the body. Mature blowfly larvae and pupae may occur 3–6 metres away although in this case the body is fresh and therefore there would not have been time for them to have completed their development.

should be kept separate. For example, maggots collected from the mouth should not be mixed with those collected from the ears or anywhere else on the body. All developmental stages should be collected from the eggs and the smallest to the largest maggots, and any pupae that are present. Maggots should ideally be killed at the time of collection and a note made of the exact moment this was done. If they are killed later, a record should be made of the temperature they experience between the times of collection and the time they are killed. Sub-samples of all eggs, maggots and pupae should be reared to adulthood to confirm identification. Eggs and any larvae that are not killed immediately after they are collected should be provided with food to ensure that their developmental rate is not affected by a period of starvation or their size is affected by desiccation.

If samples are not taken at the crime scene, or precautions taken, the largest (and oldest) maggots might escape before the body reaches the morgue and it is possible for bodies to be contaminated with insects during transport, storage or within the mortuary (Archer & Ranson, 2005). For example, if several insect-infested bodies are being transported together or examined at the same time in the mortuary it is possible that insects could crawl, fly or be accidentally picked up and moved between corpses. Another reason for taking samples immediately is that dead bodies may be treated with preservative when they reach the morgue and this can cause the larvae to shrink.

Many species of maggot are usually present on a dead body so it is sensible to determine the PMI separately for several species and from larvae collected from different sites on the body to gain an accurate assessment. This is particularly important in countries with strong seasonal climates, such as the UK. For example, Erzinclioglu (2000) described a case in which the development stage of blowfly larvae found on a corpse discovered in spring suggested a minimum period since death of about 2 weeks. However, the presence of large numbers of the larvae of winter gnats (Diptera, Trichoceridae) indicated that the person had died much earlier during the winter.

If the body is at an advanced state of decay and mature blowfly larvae are present one should search for pupae in the surrounding soil. This may require soil core sampling up to 3–6 metres surrounding the body because some maggot species travel considerable distances when they are ready to pupate. Soil core samples should be hand-sorted to locate pupae and also extracted using a standard technique, such as Tullgren and Berlese funnels, to determine the presence of active invertebrates (e.g. mites, fly larvae, earthworms) (Jackson & Raw, 1966). Maggots move away from the light so they are most likely to be found in the darker regions. When large numbers of maggots are leaving a body they may form columns of sufficient density to leave visible tracks in the vegetation. They also orientate themselves to vertical surfaces, so they are also likely to be found close to walls and similar obstacles that they have met whilst crawling across the surface. It is important to check the surrounding region for other sources of post-feeding maggots and pupae (e.g. a dead rabbit) because they could give a false indication of the PMI (Archer et al., 2006). For example, if the oldest maggots associated with a body are in their final instar but post-feeding maggots or pupae are collected from the vicinity that originated from a nearby dead pigeon this could give an indication of a longer PMI than is actually the case.

Figure 7.6 Blowfly larvae may be reared to maturity inside plastic plant cloches such as this. The cloche should be stored inside an incubator or a temperature recording device included in the cloche if the development rate of the cloche needs to be recorded.

Crawling invertebrates associated with cases of neglect, food contamination and other forensic investigations should be collected in the same manner as that outlined above.

Rearing blowfly maggots

It is difficult and sometimes impossible to identify the eggs and larval stages of many Diptera. Consequently, where possible, a representative sample of the maggots should be reared to adulthood. Plastic plant propagators kept inside incubators at a known constant temperature make suitable rearing containers (Fig. 7.6). Care must be taken to ensure the maggots do not either desiccate or drown in excessive moisture. Maggots may be reared on liver or minced meat: enclosing the meat in loosely wrapped foil helps stop both the meat and the maggots becoming desiccated. Sherman & Tran (1995) describe a simple artificial diet composed of a sterile 1:1 mixture of 3% Bacto agar and pureed liver. This medium can be stored for long periods at room temperature, ensures replicable conditions, and reduces the smell. The rearing medium should be surrounded by a layer of dry, friable material (e.g. vermiculite) into which the maggots can move when they are ready to pupate. Whatever rearing medium or procedure is adopted, it is important to ensure that other fly species cannot contaminate the rearing medium at any stage otherwise false conclusions could be reached. For example, phorid flies such as *Megaselia scalaris* are notoriously difficult to exclude and can not only lead to a false assumption that they were present on the corpse but also attack the other insects present.

Collecting flying insects

Flying insects are collected using a butterfly net with a fine mesh. The net is swept back and forth, rotating the bag through 180° at each pass. The insects are then transferred to individual tubes after collection. The insects should be killed immediately after collection to prevent them from damaging themselves. The species composition of flies visiting corpses is not necessarily an indication of its state of decomposition. For example, although adult female blowflies are most attracted to fresh corpses to lay their eggs, they also visit corpses at a late stage of decomposition to obtain a protein meal, or, in the case of males, in search of mates.

Killing and preserving techniques for invertebrates

Invertebrates that need to be preserved for further study or as evidence should always be killed quickly and humanely. The ideal killing bottle should have a wide mouth to facilitate entry and egress and have a secure lid. It is best to have several killing bottles to reduce the risk of mixing up samples and to speed up the collection procedure. Plastic bottles are lighter and less prone to breakage than glass. There are a variety of killing methods (see below) but whichever one is chosen, a record should be made of how, where and when it was carried out in case there are disputes about whether the process might have affected the results. If measurements will be taken from the specimens for comparison purposes then the killing method and preservative should ideally be the same as those used to generate the reference data.

Killing methods for hard bodied invertebrates

Ethyl acetate is a highly effective killing agent against all invertebrates but can damage some plastics – so care needs to be taken when choosing a killing bottle. Liquid ethyl acetate should not be allowed to come into contact with the insect so first a layer of plaster of Paris is used to coat the base of the killing bottle. Once the plaster has dried, ethyl acetate is added to the plaster of Paris. The prepared killing bottle lasts for several hours/days before it needs to be replenished. Insects should be removed from an ethyl acetate killing bottle as soon as they are dead.

Chopped cherry laurel leaves are rapidly lethal for adult Diptera and Hymenoptera, but generally less effective against Coleoptera and cockroaches. Cherry laurel (*Prunus laurocerasus*) leaves give off cyanide when cut or bruised, and a layer of chopped leaves at the base of a small container makes a highly effective killing bottle. The insects should be separated from the leaf fragments by a layer of tissue to make their retrieval easier. The leaves create a high humidity inside the container, which keeps the specimens in a relaxed state but this creates problems for specimens such as Lepidoptera in which the wings may become 'waterlogged' and if specimens are left in for too long they can become mouldy. Laurel leaves may be kept whole for over a week before they are cut up to create a killing bottle.

Carbon dioxide gas makes a good killing agent but this is really only suitable for invertebrates that have been transported to the laboratory – gas cylinders are a bit heavy to take into field situations. Some insects are more susceptible than others. Blowfly adults are killed quickly but cockroaches (as always) are more resistant. Similarly, placing specimens in the freezer at $-20\,°C$ is effective for adult hard-bodied insects. The temperature of the freezing compartment of a typical home fridge is not low enough to kill some insects and even after 24 hours they will revive and crawl away. It is not a suitable way to kill maggots because it may influence subsequent length measurements.

Killing methods for soft bodied invertebrates

Near boiling water provides an extremely quick and effective end for all inverte-brates (Fig. 7.7). It is the best means of killing soft-bodied invertebrates such as maggots and other insect larvae. However, it is not suitable for winged insects because the wings become waterlogged and it becomes difficult to set them after-wards. Specimens killed this way should be dehydrated and preserved as soon as they are dead otherwise they tend to swell. Maggots that are left in water after they have died will also melanize very quickly, i.e. they will darken and this can obscure some morphological features. Boiling water should not be used because it can cause the specimen to rupture and the internal organs will spurt out. Heating up water is

Figure 7.7 These maggots were originally the same size. The lower maggot was killed in near-boiling water whilst the upper, contracted, maggot was placed directly into 70% v/v ethanol.

not always a feasible option in field conditions but a couple of large thermos flasks can often supply sufficient water.

Maggots and other larval insects may be placed directly into tubes of the preservative K.A.A.. This consists of the following ingredients: 95% ethanol (80 to 100 ml), glacial acetic acid (20 ml), kerosene (i.e. paraffin) (10 ml).

Preservation of hard-bodied invertebrates

Large Diptera, such as blowflies, and most other insects, are normally mounted on entomological pins placed through the thorax. Small insects are often staged by pinning them to a Styrofoam 'stage' using small headless entomological pins. A large pin is then inserted through the stage and the label is attached to this pin. Male genitalia are often an important aid to identification. These may be extended using a fine pin, although it may be necessary to mount them separately or preserve them as a microscope slide preparation. Very small Diptera, such as Phoridae, and other small invertebrates may be stored in 70% alcohol. Woodlice, centipedes and millipedes are also best stored in 70% alcohol. Coleoptera (beetles) are traditionally glued to pieces of card using a water-based adhesive to facilitate removal. However, this method makes the observation of underneath features impossible. Coleoptera should therefore be either pinned through one of the elytra and mounted like Diptera or stored in alcohol.

Preservation of soft-bodied invertebrates

Soft-bodied invertebrates, such as maggots, should be killed before placing them in preservative, otherwise they will shrink and their morphological features become obscured. A typical procedure would be to kill the specimens by placing them into near boiling water until they are dead and then dehydrate them through increasing alcohol concentrations (30–80%) before storing them in acetic alcohol (three parts 80% ethanol: one part glacial acetic acid). Specimens preserved in alcohol alone

Table 7.4 Summary of killing and preservation methods for soft- and hard-bodied invertebrates

	Soft-bodied invertebrates	Hard-bodied invertebrates
Killing Methods	Near boiling water K.A.A. fluid	Ethyl acetate vapour Hydrogen cyanide Carbon dioxide Near boiling water (not winged insects)
Preservation Methods	Acetic alcohol 70% v/v alcohol	Pinning Carding (beetles) Acetic alcohol 70% v/v alcohol

tend to become brittle but the addition of acetic acid helps keep them soft. Regardless of the storage medium, specimens may shrink with time and this possibility should be borne in mind if reviewing evidence from cases that are several years old.

Preservation of invertebrate eggs

Invertebrate eggs are difficult to identify to species and, where possible, a proportion of them should be allowed to hatch and the young reared to adulthood. Blowfly eggs may be dehydrated through increasing alcohol concentrations (15–80%) before storing them in acetic alcohol (three parts 80% ethanol: one part glacial acetic acid) or placed directly into K.A.A.

Future directions

Although the use of molecular techniques in forensic entomology is mostly focused upon species recognition they also offer potential for the estimation of age through monitoring the expression of genes related to age-specific developmental processes. For example, it is currently difficult to age pupae after the puparium has darkened but the cellular changes and reorganization associated with metamorphosis will be accompanied by a sequence of upregulation and downregulation of specific genes.

The forensic potential of invertebrates other than Diptera is poorly exploited. This is partly because if blowflies are present there is little impetus to analyse the other invertebrates and partly because it can be difficult to identify many of them. Nevertheless, those species that naturally occur in large numbers in the environment and exhibit rapid changes in population could have forensic potential. For example, the analysis of soil nematode populations could prove useful for establishing how long a body rested at a particular site. This might be accomplished using similar molecular techniques to those employed to assess soil microbe populations (Chapter 10).

Quick quiz

(1) Why is correct species identification so important in forensic entomology?

(2) When collecting maggots from a corpse, why should one be sure to include the largest maggots present?

(3) State four morphological features that can be used to estimate the age of a blowfly maggot.

(4) How would you distinguish between a newly formed blowfly pupa and one several days older?

(5) Given the data listed below, calculate the accumulated degree hours necessary for blowfly species A to reach the second instar.
Rearing temp = 22 °C
Egg stage = 16 hours
First instar = 10 hours
Second instar = 25 hours
Third instar = 34 hours

(6) Briefly explain how the formation of a maggot feeding mass could influence the development rate of blowfly larvae?

(7) Briefly describe how the presence of drugs could affect the development of blowfly larvae and why this is relevant to calculations of the minimum time since death.

(8) Why should maggot length be calculated before the specimens are placed in preservative?

(9) Briefly explain how the presence of invertebrates can be used to demonstrate that the body of a murder victim was exposed for several days before it was buried.

(10) State two reasons why it might be useful to extract human DNA from blowfly maggots.

Project work

Title

How long can a blowfly maggot live underground?

Rationale

There may be a delay between a victim of crime dying and their body being buried. During this time the body may become infested with maggots but there is little published evidence on the extent to which they are capable of feeding and developing on the corpse once it is buried. This may have relevance once the body is unearthed and an attempt is made to reconstruct the sequence of past events.

Method

Bodies or meat samples are infested with blowfly maggots for varying lengths of time and then weighed and buried at different depths. After set periods of time, the samples are dug up, reweighed and the number of live maggots, their stage of development and the extent of decomposition is recorded and compared to control unburied samples.

Title

Does food processing affect the morphology of the insect surface cuticle?

Rationale

People often complain of finding insects in tinned and packaged goods but it is not always certain whether they were added after the packet was opened or they fell in at some stage in the processing.

Method

Cockroaches, tenebrionid beetles or adult blowflies would make suitable experimental animals. The insects may be killed using carbon dioxide and the morphology of the cuticle observed after varying treatments such as boiling, cooking in a pressure cooker etc. A scanning electron microscope will reveal the best detail.

8 Vertebrates

Chapter outline

Introduction
Vertebrate Scavenging of Human Corpses
Vertebrates Causing Death and Injury
Neglect and Abuse of Vertebrates
Vertebrates and Drugs
Vertebrates and Food Hygiene
Illegal Trade and Killing of Protected Species of Vertebrates
Identification of Vertebrates
Future Directions
Quick Quiz
Project Work

Objectives

Identify the vertebrates most commonly responsible for dismembering and consuming corpses.

Distinguish between damage caused by scavengers from pre and post mortem injuries inflicted by humans.

Explain the circumstances under which vertebrates cause death and injury.

Compare and contrast the identification techniques that can be used to link a vertebrate with a crime scene.

Explain the difficulties of proving cases of neglect and abuse of domestic and wild animals and how childhood cruelty to animals may be linked to violent behaviour in later life.

Describe the illegal use of drugs in racing and how vertebrates are used to import contraband drugs.

Differentiate between valid and fraudulent claims of food contamination.

Discuss the extent of the illegal trade in protected species of wild animals and wild animal products and describe the techniques that can be used to identify the provenance of an animal or its body parts.

Essential Forensic Biology, Second Edition   Alan Gunn
© 2009 John Wiley & Sons, Ltd

Introduction

Vertebrates (fish, amphibians, reptiles, birds and mammals) are nowhere near as numerous or as diverse as invertebrates but their larger size (usually) and, in the case of mammals, shared characteristics with ourselves, tends to make them the first thing we think of when asked to 'describe an animal'. In forensic biology, vertebrates tend to be as much the victims of crimes as sources of evidence that can be used to solve them and police forces often have special units assigned to deal with such incidents. For example, in the UK, the Metropolitan Police Force has a Wildlife Crime Unit based within New Scotland Yard which works in partnership with police forces at both a national and international level.

Vertebrate scavenging of human corpses

Many vertebrates will exploit human corpses as a source of food. In Northern European terrestrial ecosystems, this usually means dogs and other caniids, rats, pigs and birds such as crows, ravens, buzzards and jackdaws whilst in aquatic ecosystems various fish and seagulls are responsible. However, even vertebrates that are normally considered to be herbivores such as squirrels (red and grey), sheep and cows will gnaw on bones, especially if they are living in a nutritive poor environment (Fig. 8.1).

Figure 8.1 Bones gnawed by a grey squirrel (*Sciurus carolinensis*). Note the characteristic paired grooves caused by their incisors. Many rodents gnaw on bones either to sharpen their teeth or to reach the nutritious marrow.

Dogs and other large carnivores

Dogs (*Canis familiaris*) are well known for their scavenging activities. They will spread body parts over several metres and bury bones and limbs thereby making it impossible to recover a whole skeleton. Although dogs have a partiality for bones they tend to scavenge corpses that are fresh or only just starting to decay (Haglund, 1997). The ability of dogs to locate human remains even after burial can result in a body being unearthed despite the best efforts of a murderer to conceal it and this is exploited in the training of 'cadaver dogs' by police agencies (Lasseter *et al.*, 2003). Scavenging invariably results in serious damage to the victim's body (Fig. 8.2) and may lead an investigator to assume initially that a person who died of natural causes or suicide was the victim of a brutal killing. Feeding often begins on the head and neck and the loss of tissue along with the consumption of the thyroid cartilage and hyoid bone can make the diagnosis of strangulation, among other causes of death impossible. The arms then tend to be pulled off followed by the legs – this order probably reflects the relative ease with which they can be grasped and disarticulated. Clothing is seldom an impediment to the body's consumption and disarticulation. Indeed the scattering of the clothes and removal of the genitals/ genital regions of both men and women can raise the suspicion of sexual assault (e.g. Romain *et al.*, 2002). The damage dogs etc. inflict therefore needs to be distinguished from that caused at the time of death or by a murderer cutting up the body of their victim.

Bite wounds inflicted after death – like all such wounds – do not usually bleed to any great extent. Consequently, the surrounding bloodstains would be distinct

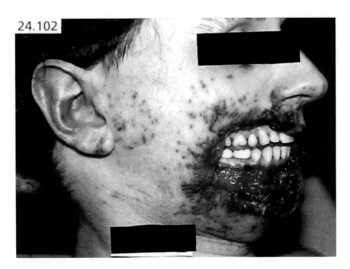

Figure 8.2 Post mortem damage caused by dogs. These wounds were caused overnight. Imprints from the incisors are visible around the margins of the wound. (Reproduced from Dolinak, D. *et al.*, (2005) *Forensic Pathology: Theory and Practice.* Copyright © 2005, Elsevier Academic Press.)

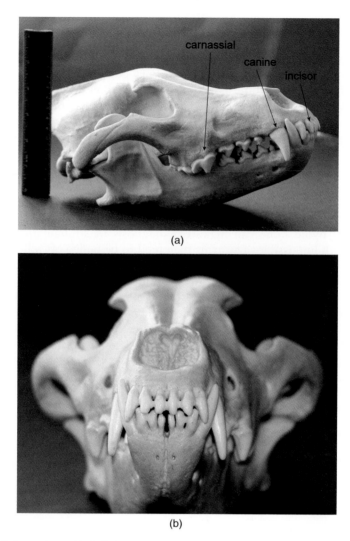

Figure 8.3 (a) Cast of a wolf skull (*Canis lupus*) showing typical caniid dentition. The incisors are poorly developed by comparison with the other teeth and used for nipping and tugging. The canines are large and robust and used for capturing and killing the prey. (b) The distance apart of the teeth can be used to judge whether the animal was the right size to have inflicted a wound. Bones are gnawed using the premolar and molar teeth.

from the spatter pattern emanating from a living person. Bite wounds sometimes indicate characteristic tooth marks (Fig. 8.3) whilst claw scratches can indicate the paw size. These may enable not only the species responsible to be identified but also the individual animal. The latter is especially the case where the person dies indoors or in an enclosed space in which the suspect animal(s) are also confined. In addition to the tooth marks, animals may also leave hairs from their muzzle whilst feeding, the individual characteristics of which, along with extracted DNA may be used for identification. The stomach contents of pets and domestic animals may also be examined to determine whether they had fed on a body but it is seldom possible to catch wild or feral animals for analysis.

Dogs and other carnivores produce four types of damage to bone: punctures, pits, scoring and furrows (Haglund, 1997). Puncture marks are usually found in thin bone such as the scapula and are caused by the canines and/ or carnassial teeth penetrating through the full thickness of the bone. The size, shape and distribution of the puncture marks can give an indication of the size of animal that inflicted them. Pits are indentations that are inflicted by any of the teeth when grasping onto bone whilst score marks result when the teeth are dragged along the surface of the bone. Furrows are deep channel-like grooves found along the length of the long bones such as the femur and are caused by the molars and premolars. When a dog spends a long time chewing on a bone it turns it over and over resulting in a mass of grooves and pits from which it is difficult to discern individual tooth marks (Fig. 8.4). Caniids and feliids always chew with one side of their mouth because their large canines make side-to-side chewing motions impossible. The different types of tooth marks can usually be distinguished from the cutting or sawing damage caused by human tools and they do not induce bevelling or concentric or radiating fractures such as those caused by gunshot wounds or trauma induced from a blunt or sharp implement. Archaeologists have done a lot of work on how to distinguish between the damage caused to bones by different animals, as opposed to humans, and also the effect of different types of cooking (e.g. Brain, 1981). On soft tissues, tearing and puncture marks can sometimes be matched with the tooth structure, dental formula or claws of the animal responsible whilst the presence of animal hairs in the wound and/or faeces in the vicinity provides further corroborating evidence. Domestic cats (*Felis catus*) do not cause as much physical destruction as dogs and

Figure 8.4 Bone gnawed by a dog. Note the mass of grooves and pits that make it difficult to discern individual tooth marks. Scale in millimetres.

their bite marks tend to be more dispersed and the tooth marks more defined (Moran & O'Connor, 1992).

Dogs and other caniids are not dainty eaters and they often swallow fragments of clothing from a corpse along with its flesh. Consequently, clothing fibres may be found in the animal's faeces. Experiments with hyenas and leopards (Pickering, 2001) indicate that fingers and toes may pass through their digestive system and be excreted in their faeces. Bone fragments and even whole teeth are often found in the faeces of dogs, foxes and other carnivores. Rings and other jewellery, especially gold, will pass through a digestive system largely unharmed but will usually show signs of acid etching. Therefore, an analysis of nearby faecal deposits and any dens, sets or burrows may yield missing bones and jewellery. The skeletons of young children do not survive scavenging attack as well as those of adults. Not only are their bones smaller and weaker, but also the epiphyseal plate, a band of proliferating and developing cartilaginous cells in the epiphysis (head) of the long bones is thicker in children and is easily chewed away by a carnivore. The bone shafts are then swallowed and are broken down more readily in the stomach because of their lower calcium content. Similarly, the sutures between the skull bones of infants and young children are movable thereby enabling the skull to be broken more easily than that of an adult.

When a dog is suspected of scavenging or attacking a human or other animal it is important that in addition to physical evidence witness statements should be obtained (where possible) of the dog's past behaviour, its behaviour at the time it was impounded and an assessment made by an experienced animal handler or behaviouralist over the following days to determine its mental state and relationship to humans. In previously well fed and otherwise normally behaved pet dogs and cats, post mortem scavenging behaviour is commonly thought not to take place until long after the death of their owner and the onset of starvation through being confined in a room or building with no alternative food supply. However, there is a great deal of variation between cases and it is possible that the body may be substantially consumed within a short period of time (Steadman & Worne, 2007). Rothschild & Schneider (1997) describe a case in which scavenging by an Alsatian dog took place within 45 minutes of the owner committing suicide by shooting himself in the mouth (hence a large wound was already present on the body). They discuss several possible explanations for the early onset of scavenging including aggressive behaviour caused by being confined and the dog, being a pack animal, attacking its owner at a time of weakness in an attempt to gain a social domination. However, they considered the most likely explanation in this case was that the dog attempted to help its unconscious or recently deceased owner, first by nuzzling and licking and when these failed to become panicked into attacking and mutilating him.

Rats and other rodents

When rodents gnaw on bones or other objects they leave paired parallel grooves with intermediate 'groins' – the width of the grooves indicates the size of the incisors and hence the size and probable identity of the rodent species (Fig. 8.5). Many

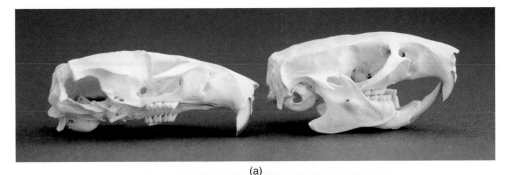

(a)

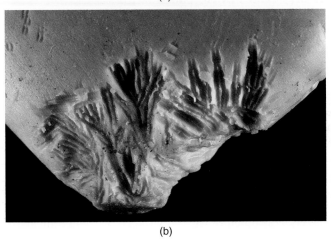

(b)

Figure 8.5   Rodent dentition. (a) Rat (*Rattus norvegicus*) skull showing the well-developed incisors. (b) Bar of soap chewed on by mice (mice are very partial to soap owing to its high fat
content). Note the indentations formed by the incisors. The indentations are too close together
to have been formed by rats.

rodents will feed on corpses, and porcupines are notorious for collecting all sorts
of objects in their burrows ranging from bones to tin cans. In the UK, field mice
exhibit a similar acquisitive nature. Grey squirrels (*Sciurus carolinensis*) prefer bones
that have dried out (Klippel & Synstelien, 2007) and there are no reports of them
feeding on fresh bodies (although when the opportunity presents they are predatory
and will kill small mammals and young birds). Rats will remove flesh from living
bodies and historical accounts of soldiers and prisoners living in unhygienic circumstances often mention rats nibbling at fingers and toes. Rats (mainly *Rattus norvegicus* in the UK) also favour soft, moist areas such as the eyelids, nose, and lips.
Consequently, rodent bite marks may be inflicted both before and after death upon
a person who dies of wounds, disease, or intoxication over a period of days and
they are commonly found on the bodies of homeless people or those living in squalid
conditions. Large numbers of rats are capable of overwhelming and killing a person
who is already comatose or too weak to defend himself but documented cases of
such instances are very rare. Unhygienic circumstances are not always a factor in

rodent scavenging because many people keep rats and other rodents as pets. For example, Ropohl *et al.* (1995) describe post mortem wounds caused by a free-range golden hamster (*Mesocritecus auratus*) that were so extensive that it was initially believed to be the work of a murderer attempting to scalp his victim. The hamster was identified as the culprit, because of its typical 'rodent signature' – rodents often leave characteristic faecal pellets (their shape and size varies between species) whilst feeding and their paired chisel-shaped incisors cause crater-like lesions with notched edges in soft tissue (Tsokos *et al.*, 1999). The hamster further incriminated itself by taking fragments of skin and tissue back to its nest – another typical rodent trait. Bite marks do not always cause tissue loss and by stretching the skin it is sometimes possible to see marks caused by the paired incisors.

Pigs

In mediaeval times, pigs (*Sus scrofa*) were allowed to roam freely and there are reports of them biting and even killing and eating babies. Nowadays this is no longer a problem in Northern Europe although the wild pigs and boar that can be found in some regions would probably be happy to exploit any dead bodies left in their woods. During experiments with domestic pigs that were fed fresh uncooked bones of sheep, cattle and pigs, Greenfield (1988) found that sows tend to chew briefly on the first bone they encounter before dropping it and moving on to another and repeating the process. After a short period of time the pigs concentrate on the smaller bones, especially the vertebrae, which can be picked up and carried around, and these are completely consumed. Large bones such as the femur are damaged, especially at the ends, but are not totally destroyed. Crime writers sometimes suggest that a good way to dispose of a dead body is to feed it to pigs but this requires the cooperation of the pig farmer because a whole body is unlikely to be entirely consumed overnight and it is doubtful that the pigs would be able to completely destroy all the bones.

Birds

A wide variety of birds will feed on a corpse or remove hair that is sloughing from its head to line their nests. Birds such as crows frequently begin feeding on the eyes of corpses and the tongue if it is extended. Over time they can cause sufficient tissue loss to make the cause of death difficult to establish (Asamura *et al.*, 2004). As sheep farmers will testify, birds do not always wait for an animal to die and any creature too weak to defend itself may be attacked. Similarly, sailors who have had to abandon ship and end up swimming in the sea have stated that gulls have attacked them and that some people drown in their attempts to avoid the birds. The lack of eyes and the presence of head wounds may therefore be a consequence of birds rather than human activity and may be caused either before or after death. Birds tend to produce stab-like wounds, the size and depth of which varies with the size of the bird's beak. Some birds, such as buzzards, employ a stab and tear technique when removing flesh. Birds will promote the spreading of the remains by removing scraps of tissue and small body parts such as fingers to be consumed elsewhere at leisure or to a nest site to be fed to chicks. Unless birds are seen feeding on a body and/or they leave their faeces, it may be difficult to implicate them with the wounds caused – although the lack of bleeding would indicate that it took place after death. However, bleeding may be extensive in the case of a body floating in water (Chapter 1).

Fish and other aquatic vertebrates

Many species of fish, both freshwater and marine, will feed upon dead animals and the consequences depend upon the species involved, their abundance and the duration of exposure. Like rats and mice, small fish tend to start feeding on the fingers and toes, earlobes, lips and nose and this causes numerous small crater-like wounds. Prolonged feeding can result in the loss of substantial areas of tissue. Large fish cause more serious wounds and sharks, such as the Tiger Shark (*Galeocerdo cuvier*) and the Great White Shark (*Carcharodon carcharias*) will bite off and swallow limbs whole. Interestingly, sharks can retain food undigested within their stomach for several weeks. Some sharks will feed on both dead and living organisms and therefore when part of a human body is discovered within a shark (which happens from time to time) it is not always possible to determine when the part was eaten or whether the shark was responsible for the person's death.

Vertebrates causing death and injury

Domestic and wild animals are seldom a cause of death in northern Europe but when fatal attacks do occur they generate enormous publicity and fear. By contrast, dog bites are common and may result in serious injury – and consequent litigation. In 2007 almost 3800 people in England attended a hospital accident and emergency department as a consequence of dog bites. This represents a 40% rise in the number of cases over the previous 5 year period. Wounds are usually inflicted to the extremities, especially when the attack comes from stray or feral dogs whilst pet dogs are more likely to attack the face or neck. The wounds may be life threatening if major blood vessels, such as the femoral artery are damaged. The wounds may also become infected with bacteria transmitted in the dog's saliva and in some countries there is the risk of rabies. In fatal cases, there is serious loss of blood and often damage to the hands and arms where the victim has attempted to ward off the dog(s). Young children and the elderly and infirm tend to be those most at risk. Domestic cats have finer and sharper teeth than dogs that enable them to penetrate bone despite their comparatively weaker bite. Cat owners commonly have fine scratch marks on the backs of their hands as a consequence of playing with their pets. Cats seldom attack unless they are cornered and feel threatened but when they do they can cause serious lacerations before escaping as quickly as possible. There are no reports of them causing death by attacking someone although tripping over a circling cat is a common cause of injury. For example, an elderly person may be found unconscious with a head injury and crime may be suspected until other evidence – such as the pattern of blood spatter – indicate a fall (although if the person does not recover or cannot remember what happened, this may be difficult to distinguish from a push by an intruder).

Large potentially aggressive dogs are becoming a popular fashion accessory with some sections of the community. In 2008, four breeds of dog were banned in the UK: the pit bull terrier, Japanese Tosa, Dogo Argentino and Filo Brasileiro. Sometimes the dogs are merely a prop to a fragile male ego but they are also used in illegal dog fighting and as a means of intimidation. The use of guard dogs to threaten prisoners was notorious in certain American prison camps in Iraq and Afghanistan. Proof of intimidation is difficult to establish if no physical injuries are caused, although the American prison guards were callous (and stupid) enough to photograph themselves committing the crimes. In addition to using dogs to bite and intimidate victims, some people have trained them to commit sexual acts and there are reports of them being incited to commit rape (Vintiner *et al.*, 1992; Schudel, 2001). Consequently, victims of sexual assault may have traces of nonhuman spermatozoa and it may require modifications to existing protocols for their detection.

In addition to the risks posed by normal pets and domestic animals, the opportunity and increasing popularity for keeping exotic animals has resulted in more people coming into contact with large and potentially dangerous creatures (Lazarus *et al.*, 2001). For example, there are currently more tigers in captivity than there are in the wild and between 1998 and 2001 27 persons were injured and a further seven killed by them in the USA (Nyhus *et al.*, 2003). Some of the deaths caused by captive wild animals are the result of people taking inadequate precautions when housing or handling them – these are often animals that are kept by private individuals. The keeping of exotic pets is usually a difficult, expensive and time consuming occupation and when the animals grow too large or aggressive it is not unusual for them to be dumped. For example, iguanas are popular pets in the UK where they are usually purchased when they are a few inches long. However, when mature they may measure up to 5 feet and can become aggressive – at which point they are looked on as a burden. The killing or dumping of unwanted exotic pets can lead to criminal charges although proof is difficult, especially if the animal was being kept illegally. Zoos have a duty to not only maintain potentially dangerous animals under safe conditions from which they cannot escape but also to enable them to be seen by members of the public whilst simultaneously preventing the naïve, deranged or suicidal from coming into contact with them. For example, climbing into a lion enclosure is sometimes used as a form of suicide (e.g. Bock *et al.*, 2000).

Neglect and abuse of vertebrates

Whilst pets and domestic animals sometimes attack and may even kill humans, they are far more frequently the subject of neglect and wilful abuse that result in the animal suffering and/or dying. Wild animals are also frequently killed for no other reason than personal gratification. To mistreat animals in this way has long been recognized to be a crime but identifying such activity has received a higher profile following the realization that childhood cruelty to animals is often linked to the development of violent behaviour towards humans in later life (Raupp, 1999; Lockwood, 2000; Dadds *et al.*, 2002). The diagnosis of neglect and abuse is a job for veterinary surgeons based on the clinical symptoms and where the animal is voluntarily brought to the surgery, they face the dilemma of reporting the owners to the police and thereby risking the animal suddenly 'disappearing' before it can be impounded or keeping quiet and attempting to treat the animal whilst encouraging the owners to behave more responsibly. Because the mistreatment and illegal killing of both wild and domestic animals can result in hefty fines and imprisonment, the person charged can be expected to mount an active defence and is likely to escape punishment on a legal technicality if the investigation is not conducted according to a recognized procedure and to the same standard as the forensic examination of a human crime victim (Cooper & Cooper, 2007; Merck, 2007).

Humans have always used a wide range of both vertebrate and invertebrate animals for their own sexual pleasure and Ancient Greek literature is full of strange couplings between humans and other animals – often claimed to be gods in disguise, which is a pretty feeble excuse. However, such activities were frowned on by most communities and if discovered could result in judicial proceedings in which both the man (it usually was a man) and the unwilling object of his attentions were condemned to death. Bestialism was a capital offence in the British Isles as late as the nineteenth century: John Leedham had the unfortunate distinction of being the last person in Derbyshire to be hanged for a crime other than murder when he was executed for bestialism outside the Derby New County Gaol on 12th April 1833. It remains a common and under-reported problem – probably as a consequence of its simultaneously ludicrous and unpleasant nature. In a survey of small animal veterinary practitioners in the UK, Munro & Thrushfield (2001) found that 6% of 448 reported cases of nonaccidental injury were of a sexual nature. As mentioned above, such activities are not only of concern for the distress caused to the animals but also for the possibility that they might lead to assaults on humans – although research is still needed in this area.

Wild animals do not always kill their prey quickly and cleanly, neither, contrary to popular belief, do they kill only sufficient to assuage their appetite. Consequently, one may find badly wounded wild or domestic animals that at first sight appear to have suffered at the hands of a sadistic individual. For example, in April 2004 numerous dead and dying frogs and toads were found in Aberdeenshire with their hind legs ripped off and this sparked a police investigation. It was subsequently discovered that the culprits were otters which bit off the hind legs of their victims and then skinned them. In the case of toads, this was done to remove their poison glands.

Vertebrates and drugs

Dogs have a keen sense of smell and they are used by police forces throughout the world to detect hidden drugs and explosives at airports, train stations and ferry terminals. Intriguingly, African Giant Pouched rats, *Cricetomys gambiensis*, have a similar ability to be trained to identify distinctive smells in return for a reward. At present, they are being assessed for their ability to detect tuberculosis in sputum and explosives in landmines (www.gichd.ch/pdf/publications/MDD/MDD_ch4_part2.pdf) but there are no obvious reasons why they could not also be trained to detect contraband drugs. The rats are said to be 'more mechanical than a dog and they are easier to transfer to different owners'.

In addition to detecting drugs, animals are also used to smuggle them – although the extent to which this is happening is not known. For example, in September 2003 at Schiphol airport, Holland, two Labrador dogs, in transit from Colombia, were discovered to have a total of 21 packets of cocaine sown to their stomachs. Suspicions were raised by the dogs' behaviours, one being aggressive whilst the other was weak. Both dogs had scars and X-rays revealed the presence of the containers. One dog survived the removal of the containers whilst the other had to be put down because they had fused to the stomach lining. Two persons were arrested when they arrived to pick up the dogs but the Colombian authorities have yet to apprehend the exporters and the veterinary surgeon responsible for implanting the drugs. The presence of scarring and unusual behaviour of any animal being moved between countries should therefore arouse suspicions.

Just as the use of drugs is a problem in human sporting activities, so is the administering of drugs to competition animals. The true extent of the problem is uncertain; The University of Ghent, Belgium, runs a doping control laboratory and according to their data between 1993 and 2003, the percentage of horse samples testing positive varied considerably between years (from 1.2 to 8.4%) whilst the percentage of human samples testing positive remained fairly constant (from 3.6 to 6.6%). Testing is now performed at many events and in the UK there is routine random testing of racehorses. Testers are also authorized to arrive at stables unannounced to collect their samples. The drugs involved are often those used in the treatment of disease but are being administered solely to improve performance: examples include erythropoietin (EPO) (which increases the red blood cell count), clenbuterol (a bronchodilator) and propantheline bromide (blue magic) (a muscle relaxant that also acts to increase a horse's lung capacity). Consequently, owners of competition horses have to be extremely careful over the medication their animals receive and to be aware of the risks of their animals receiving spiked feed or being inadvertently fed inappropriate food. For example, in America, owners of show horses are warned not to allow their children to reward them with drinks of coke because it might increase the caffeine levels above allowable levels. The use of 'downers' to reduce the activity of horses is also alleged to be common practice, although supporting evidence is not readily available. This might be done to make a troublesome horse more placid at the time of sale, to cause it lose a race, or to make it more manageable in the ring. The organization 'Ponies (UK)' has begun a programme of random dope testing following judges and stewards voicing concerns over the placid behaviour of some of smaller horses ridden by children. Bute (phenylbutazone), a nonsteroidal anti-inflammatory drug, used in the treatment of strains, sprains, and feverish symptoms is one of the suspects, although nonspecified herbal remedies are also thought to be involved. In 2002, a British trainer was fined £600 by the Jockey's club when one of his horses was found to contain traces of the 'stopping' drug acetylpromazine (ACP). Obviously the drug was not in a high enough concentration because the horse won by 11 lengths.

Vertebrates and food hygiene

Most food hygiene litigation involving vertebrates relates to rats and mice gaining access to stored food and damaging it through physically eating it and contaminating it with their faeces and urine. Even if the animals are not seen, their faeces have a characteristic shape and size that enables identification. For example, the faeces of brown rats (*Rattus norvegicus*) is usually deposited in groups (although it may be scattered) and tends to be spindle shaped, whilst the faeces of black rats (*Rattus rattus*) is scattered around and tends to be sausage or banana shaped. Food hygiene litigation tends to involve food hygiene inspectors rather than forensic scientists: there is seldom an issue over when the contamination occurred or who was responsible. For example, if rat faeces were found in the kitchen of a restaurant, rats must have gained access and it does not matter when they did or how many of them did, or where they came from, it is the proprietor's responsibility to ensure that they do not enter the building and he/she will be prosecuted.

Case Study: The mouse in the can of milk stout

Anecdotal reports of finding the tails of mice or rats in food products are commonly heard but many of these are probably apocryphal. They do, however, encourage people to attempt to falsely sue a food retailer by claiming to have found a rat or mouse, or a bit of one, in their food (e.g. Platek *et al.*, 1997). Williams (1996) describes such a case in which a man stated that having consumed a can of commercially produced milk stout, he found a dead mouse at the bottom of the can. Williams first determined that it was possible for the mouse to have gained access via the can's ring pull and then carried out a series of experiments in which dead mice were placed in cans of milk stout, which were then sealed on the factory production line. Some of the cans were then pasteurized (as normal) and others were not. Both sets of cans were then stored under ambient condition or refrigeration and opened after set periods of time. In all cases, the experimental mice were found to have undergone considerable decomposition within 1 month and after 3 months they were completely disintegrated. By contrast, the mouse obtained from the complainant's can, which was 3 months old, was well preserved and exhibited minimal post mortem changes. It was concluded that the complaint was fictitious. This case emphasizes the importance of carrying out carefully designed experiments to exclude all possibilities – for example, the effect of pasteurization.

Illegal trade and killing of protected species of vertebrates

As the wild populations of animals decline in many parts of the world, usually owing to mankind's activities, the need for effective conservation measures grows.

Table 8.1 Summary of vertebrate animals and their forensic relevance

Animal	Forensic relevance
Dogs	Cause of human injury or death
	Cause of traffic accident
	Victim of neglect or abuse
	Use in illegal baiting or dog fighting
	'Kidnapping'– pets, especially dogs, are sometimes stolen and held for ransom
	Doping (e.g. greyhound racing)
	Cause of post mortem damage
	Source of DNA linking a person to a locality
Cats	Cause of traffic accident
	Victim of neglect or abuse
	Use in illegal baiting
	Source of DNA linking a person to a locality
Rats and other rodents	Cause of human injury
	Cause of post mortem damage to flesh and bones
	Food spoilage
Domestic livestock	Victim of neglect or abuse
	Rustling
	Fraud (e.g. illegal movements across borders)
Birds	Cause of post mortem damage
	Victim of neglect or abuse
	Illegal trade in protected species
	Illegal killing of protected species
Fish	Cause of post mortem damage
	Fraud (e.g. mislabelling)
	Illegal trade in protected species
Wild mammals, reptiles and amphibians	Illegal trade in protected species
	Illegal killing of protected species
	Poaching

To be truly effective, these measures must be backed up with laws to ensure their enforcement and, consequently, litigation follows. On an international scale, wild animals are offered varying levels of protection under the Convention on International Trade in Endangered Species (CITES). Those species listed under Appendix I are fully protected and no trade is allowed, whilst those listed under Appendix II can be traded under a permit system. Individual countries also have their own laws that control which animals can be traded and hunted – for example, the shooting of songbirds is legal in some parts of Europe but illegal in the UK.

Three of the most common problems are the unlicensed trade in live animals, the unlicensed trade in body parts (skins, bones, meat etc.), and the intentional killing of animals for 'sport' or because they are perceived as a threat. There is a global market in live wild animals and their body parts that, for the most part, is perfectly legitimate and can provide employment to poor people in developing countries.

However, as the animals become scarce, their value increases and people are tempted into criminal activity – the illegal sale of plants and animals is estimated to be worth billions of pounds per year on a worldwide basis and often involves criminal organizations. In response, the laws have become tougher and in the UK the maximum prison sentence for illegal trading in wildlife is currently 5 years. Forensic science becomes involved at a variety of levels, for example, in determining the provenance of an animal, identifying bones or animal parts, and determining the cause of death.

Bushmeat

'Bushmeat' is the term given to the meat of wild animals that has been caught by villagers in central and western Africa and sold in the local market. For many years, the practice was 'low key' and was not thought to cause any harm to the local wildlife population – indeed, it was thought by some to be beneficial because it made the local people more aware of the need for conservation. However, in recent years there has been a large increase in the trade, partly owing to increased demand and partly through greater access to the forests through logging practices. It has become highly profitable: a single suitcase of bushmeat may have a street value of up to £1000 in the UK. Consequently, bushmeat is being seized with increasing frequency at airports on its way to African communities in the UK and elsewhere. The bushmeat is very often smoked and consists of species such as antelopes, bats, cane rats, pangolin, monkeys, and in Amsterdam a single seizure contained 2000 chimpanzee noses. It is estimated that in the region of 12 000 tonnes of meat and meat products are illegally trafficked into the UK each year. Most of this is derived from cows, pigs and sheep but a large proportion is bushmeat. Some of the bushmeat being imported includes that of protected species and much of it is not being transported correctly and therefore poses a risk of food poisoning. There is now a realization that it is a potential source of anthrax (Leendertz *et al.*, 2004) and zoonotic viral infections (Peeters *et al.*, 2002) and some people believe that the devastating outbreak of foot and mouth disease in the UK in 2001 originated from illegally imported meat (though not necessarily bushmeat). Despite this in 2006 only one seizure out of a total of over 35 000 resulted in a prosecution.

Identifying provenance

Identifying the provenance of an animal is important because in the UK, provided one is a registered breeder / seller with a license from the Department of Environment, Food and Rural Affairs, it is not illegal to sell an endangered species if it has been bred from existing UK stock. For example, tortoises and parrots were once common pets but since it became illegal to import them from wild populations abroad, their numbers here have declined and their value increased. Proof of provenance usually depends on the presence or absence of the appropriate documentation although the use of DNA technology and chemical analysis can prove useful in disputed cases. For example, there is a highly profitable and illegal trade in falcons and other birds of prey for use in falconry both within the UK and to supply the Middle Eastern market. Birds that are born and bred in captivity may be traded legally but these are insufficient to meet the demands.

Collecting and killing of wild animals

All species of wild birds, as well as their nests and eggs, are protected under British Law, although allowances are made to control pest species, such as feral pigeons, and for the shooting of game birds. Egg collecting was once a common hobby, especially amongst schoolboys. However, in the UK it became illegal to collect the eggs of wild birds in 1954 and since 1982 it has been an offence to even own the eggs of wild birds. Despite this, egg collecting remains a hobby of some people and they can have a serious impact on the populations of the rare species, such as many of the birds of prey. The eggs of such birds sell for high prices on the 'underground market'. Proving the provenance of wild bird eggs is less of a problem in ensuring a conviction because possession alone is a crime – although it always helps to know when and where the eggs were obtained. Wild finches and other songbirds are sometimes illegally trapped for sale as caged pet birds both within the UK and in the Mediterranean regions or to supply the gourmet food trade. As mentioned above, proving the provenance of these birds depends on documentation and, sometimes, DNA analysis. Although there is a legal trade in captive bred finches, those that are caught in the wild are thought to have brighter plumage and therefore command a higher price and will sell for up to £100 each.

Farmers and gamekeepers sometimes attempt to kill eagles and other predatory birds and mammals because they believe that they kill lambs and game birds (Stroud, 1998). This is usually done by shooting or providing poisoned baits. Therefore, the finding of a dead eagle, for example, especially one in outwardly good condition might be considered suspicious. It is the responsibility of a vet to carry out a careful autopsy and to submit tissues for toxicological analysis. If the evidence is to be used in a court of law, the autopsy must be performed with the same attention to detail as that of a human, with careful record keeping thereby maintaining a 'chain of evidence' (Cooper & Cooper, 2007). Sometimes it is important to determine the minimum time since death of a wild animal and this can be done using the same entomological techniques as those detailed in Chapter 7. Anderson (1999) has described a case in Canada in which insect evidence from two illegally killed black bears was used to establish the fact that the bears had died during the time the suspects were in the locality and this information helped in securing a conviction.

Identification of vertebrates

Sometimes it is only necessary to identify animals or animal products to species level – for example, when a protected species of parrot is allegedly included among a group of legally traded common species. Similarly, animal body parts from protected species are sometimes intentionally mislabelled or wrongly described in an attempt to avoid detection. Identification to species level is also required to detect cases of fraud in which low quality animal products are mis-sold and/ or adulteration occurs. At other times it is necessary to identify an individual animal with a high degree of certainty such as when a dog fatally mauls a child. Vertebrate identification techniques can be divided into three broad categories: physical techniques, molecular techniques and chemical techniques. Physical techniques are those that rely on the observation of the animal's characteristic features, its droppings, or of man-made artefacts, such as brands or tracking devices. Molecular techniques involve sequencing of the animal's genome and although protein and immunological techniques are also used on occasions they will not be covered here. Chemical techniques involve analysing an animal's chemical composition: some of these will not allow the identification of the species but can often provide good evidence of where it came from (i.e. geographical origin).

Physical techniques

Morphology

Living or recently dead vertebrates, especially the mammals, can usually be identified to species from their gross morphology. Once they are skeletonized many vertebrates remain easy to identify to species level provided that one has the skull but if there are only a few of the smaller bones available then identification may be limited to genus or family level. Once an animal has been processed for food it can be virtually impossible to identify it from morphological characteristics.

Identifying an individual vertebrate animal on morphological evidence presents much greater difficulties. For example, a dog or other animal responsible for an attack or for causing an accident that runs away afterwards would be very difficult to identify later solely on the basis of witness statements. Most people, when asked to describe a dog would be unable to provide more than a vague description. Statements such as 'it was a large brown Labrador and the accused is also a large brown Labrador' are not going to be much use as evidence in a court of law.

Scats

A number of mammals and other vertebrates produce characteristically formed faeces – called scats (Chame, 2003). These scats can be used to identify the presence of an animal in the locality and they are often used by ecologists to monitor the abundance and movements of elusive animals such as otters. In a forensic context, scats indicate that the presence of an animal (e.g. during a search for a pet held for ransom [increasingly common], evidence of badger baiting or illegal live animal trafficking) or the activities of scavengers. The scats also provide evidence of the animal's health and the food it consumed. The faeces can also yield DNA to confirm species identity and in some cases identify a particular individual animal. The UK has a relatively low diversity of mammal species and therefore many scats can be identified with reasonable confidence. However the size ranges of many animals overlaps and where animals share a similar diet it can make differentiation difficult – especially as scat form is affected by the animal's health and exposure to the weather. Where confirmation is required the scats should therefore be subjected to DNA analysis (Foran *et al.*, 1997).

Branding

Farm animals such as cattle and horses are often branded as a means of individual identification. Freeze branding is considered more humane than hot iron branding and results in white hairs growing at the brand site rather than those of the normal colour. Branding has the advantage of being cheap, permanent and impossible to conceal or shave away. Freeze branding can also be used on fish and results in the scales changing colour. Unfortunately, the effectiveness of branding is often compromised by poor record keeping and difficulties in transmitting information between interested parties. In addition, once the animal is slaughtered and skinned the means of identification is lost.

Tagging

Many farm animals in the UK are required by law to have identifying ear tags and there are a variety of tag types in use depending upon the species of animal. However, ear tags are easy to remove and replace. This facilitates rustling and the fraudulent movement of cattle within and between countries.

Microchips

Pets and high value domestic animals can have a microchip injected underneath the skin that provides a unique identifying number. The chip is read using a special scanning device. Although highly effective as a means of identification it does require the presence of the whole animal – alive or dead – and a scanner.

Hairs

Animal hairs are often characteristic of the species and can also provide information on coat coloration but their presence provides only weak evidence of an association with any individual animal unless supported by other evidence such as DNA. Hairs vary in length and colour between different parts of the body and this should be borne in mind when collecting specimens or interpreting evidence. Animal hairs may also originate from a fur garment or pelt – these hairs are often coloured, trimmed and lack a root.

   Dogs and cats are two of the most popular pets in many parts of the world and, as any owner will agree, their hairs tend to get everywhere despite strenuous efforts to keep rooms and clothing clean, so they are a potential source of forensic evidence

(D'Andrea *et al.*, 1998). For example, they can indicate that a suspect was present in a room or vehicle (dog and cat hairs are often found in the cars of pet owners), indulged in bestiality, or had contact with a particular animal. In cases of sheep worrying, badger baiting, illegal dog fighting etc., the hairs of the victim(s) may be found on the muzzle or coat of the accused dog or associated with the property/ possessions of the dog's owner. Similarly, in vehicle accidents animal hairs might be found on the radiator grill, bumper or elsewhere. In the latter cases there would be a high chance of finding bloodstains that might yield DNA to support the link. Just as pets and other domestic animals transfer their hairs onto humans, so humans transfer clothes fibres onto their pets and brushing their fur or coat can yield evidence of contact.

Hairs are characterized by a combination of the external and internal features. Scanning electron microscopy can provide exceptional detail of the scaling pattern and other external features of hairs along with their three-dimensional shapes although light microscopy is perfectly adequate for most analysis. In the latter case, a hair is placed onto a layer of wet nail varnish or lacquer on a glass microscope slide being careful to leave one end of the hair protruding beyond the varnish to facilitate its removal. Once the varnish is set the hair is gently removed thereby leaving behind its cast and this can be observed using a light microscope (Fig. 8.6). To observe the shape of the hair, sections are cut using standard histological techniques. The internal structure is revealed by mounting the hair on a microscope slide using either a temporary or a permanent medium and covering it with a cover slip. A thorough guide to the collection and analysis of hair and fibre evidence is provided by Jackson & Jackson (2008).

(a) (b) (c)

Figure 8.6 Different hair scaling patterns revealed by casts made in nail varnish. (a) Horse hair. (b) Cat hair. (c) Human hair. All images were taken at the same magnification.

Molecular techniques

Identification of species

Molecular techniques are extremely useful for species identification of animal tissue and fluid samples. For example, when bloodstains are found at a crime scene, food products are mislabelled or the body parts of protected species of wild animals are illegally sold/ trafficked. The Consortium for the Barcode of Life – an international collaboration of scientists that seeks to establish an identifying DNA barcode for all organisms – is currently focusing on generating sequence data for the mitochondrial gene for cytochrome c oxidase I (COI) as a means for distinguishing between all animal species. However, other mitochondrial gene sequences, such as those for cytochrome *b* and the hyper-variable displacement loop (D-Loop) are also commonly used for species identification. It is not necessary to sequence the whole of the mitochondrial COI gene and most workers limit their attentions to the 648 base pairs that make up the first half of the gene. The intention is that researchers could compare sequence data from their samples with that stored on computer search engines such as GenBank, BLASTn and BOLD and the degree of similarity would indicate the identity of the unknown sample. Mitochondrial gene sequences are particularly appropriate to this task since they tend to show greater differences between closely related species than do nuclear gene sequences whilst exhibiting relatively low levels of intraspecific variation. This is owing to mitochondrial DNA evolving faster than nuclear DNA and not undergoing recombination. In addition, because most animal cells contain many mitochondria a single cell will yield numerous mitochondrial genomes. This, together with the possibility of identifying species from relatively short sequence lengths makes mitochondrial gene sequencing particularly suitable for forensic analysis of samples in which the DNA may be degraded as a consequence of processing (e.g. in foodstuffs) or the animal decaying.

Many of the attempts at identifying animal species in food products have focused on sequencing mitochondrial cytochrome *b* gene (Teletchea *et al.*, 2005). This has proved particularly effective for the identification of fish and fish products for which conventional analysis is difficult. The need for this can arise when closely related fish need to be distinguished. For example, the fishing of bluefin tuna (*Thunnus thynnus*) is carefully regulated (particularly the Atlantic populations which are currently at critically low levels) but that of other tuna species is not and morphological distinguishing features can be removed intentionally or during processing (Lin *et al.*, 2005). Mitochondrial DNA has also proved useful for distinguishing between shark species. There is a big market for shark fins and jaws particularly in China and Asia and, although many species are protected, illegal shark fishing has lead to a catastrophic decline in many species. For example, populations of the scalloped hammerhead shark (*Sphyma lewini*), which was once considered at low risk owing to its wide distribution, have declined by as much as 98% in some regions. The processing of sharks usually takes place at sea – sometimes the fins are cut off and the animal is pushed back into the water to die. Consequently, there are few morphological features available to determine whether the fins came from protected species. Protection agencies are therefore now using DNA techniques, especially mitochondrial COI and mitochondrial cytochrome *b* sequences as means of identifying illegally caught shark products (e.g. Ward, D.W. *et al.*, 2005). Similarly, Marko *et al.* (2004) found that 77% of the fish sold as red snapper were actually other species – such widespread misrepresentation has serious consequences for the management and conservation of fish stocks. The need to distinguish between fish species can also arise to prevent fraud in which cheap farmed fish is presented as more expensive wild-caught fish of the same or different species (Kyle & Wilson, 2007).

Sequencing of the mitochondrial cytochrome *b* gene has proved useful in the detection of animal parts in traditional Chinese medicines. Rhinoceros horn is a commonly used ingredient in such medicines and whilst the horns themselves can be easily identified from their morphology, it is more difficult once they have been ground to a powder or made into sculptures. By amplifying and then sequencing a partial (402 base pair) fragment of the cytochrome *b* gene it is possible to distinguish between species of rhinoceros and to detect the presence of rhinoceros DNA even when powdered rhino horn is diluted with cattle horn (Hsieh *et al.*, 2003).

Pyrosequencing

A relatively new molecular technique called pyrosequencing has proved very effective for distinguishing between bacteria and it is being increasingly used in forensic science. Karlsson & Holmlund (2007) describe how it can be used to identify the presence of human DNA and also to distinguish it from that of other mammals. They amplified fragments of the mitochondrial genes for 12S rRNA and 16S rRNA by PCR, and then sequenced these using the pyrosequencing cascade system. Pyrosequencing works as follows. First a sequencing primer is hybridized onto the single stranded DNA template – in this case the amplified fragment of the mitochondrial DNA coding for either 12S rRNA or 16SrRNA. This is then incubated with the enzymes DNA polymerase, ATP sulphurylase, luciferase (which is chemiluminesent), and apyrase and the substrates adenosine 5′ phosphosulphate and luciferin. The reaction cascade is initiated by sequentially adding one of the four nucleotides – i.e. adenine, cytosine, guanine and thymine in their triphosphate form (i.e. adenine triphosphate etc.). Let us say that the first nucleotide to be added is adenine triphosphate. If the first unpaired base of the template is thymine then the adenine triphosphate will bind to it and in the process release its phosphate moiety as pyrophosphate (PPi). The enzyme ATP sulphyrase then combines the PPi with adenosine 5′ phosphosulphate to form ATP. The enzyme luciferase is then able to use this ATP to break down luciferin and in the process it releases energy in the form of light and this can be detected and measured. Finally, the enzyme apyrase breaks down any unbound nucleotides and ATP thereby allowing the next nucleotide triphosphate to be added. Consequently, by sequentially adding nucleotides one can identify the first unpaired base on the template since light will only be produced when they complement one another. In short, pyrosequencing works by identifying one base at a time through the release of light whenever a 'match' is found. Obviously, as a means of discriminating between species this technique is most reliable when there are many nucleotide differences between their DNA templates. Karlsson & Holmlund (2007) found a minimum of nine nucleotide differences between the sequences for humans and a variety of European mammals indicating that this would be a good way of distinguishing human from nonhuman blood or tissue samples. They also found that many domestic and wild animals could be distinguished from one another and Kitano *et al.* (2007) demonstrated that it could be used to identify many other vertebrate animals as well. Also in its favour, pyrosequencing is fast, accurate and can be easily automated. In addition it is also quantitative since the amount of light formed can be related to the amount of nucleotide base binding to the DNA template.

Limitations on the use of DNA for identification of vertebrates

Although DNA-based species identification holds great promise, its usefulness as a forensic tool cannot be fully exploited until standard protocols are agreed for the extraction and analysis of the gene sequences, quality control is assured and multiple sequences are available from voucher specimens of each species so as to indicate the degree of intraspecific variability (number of haplotypes) (Dawnay *et al.*, 2007). For example, if the data stored on the search engines contains errors through incorrect identification of the animal from which the DNA originated, contamination with human DNA or that of other animals, or the way the sample was processed it would lead to misidentification. Similarly, if sequence data for a given species was available for only a single haplotype it might lead to other haplotypes of the same species being considered to belong to a different species. There is also some concern over the extent to which fragments of mitochondrial DNA are translocated into nuclear DNA. These fragments, known as *Numts*, can become sequenced alongside or instead of the target mitochondrial DNA and thereby cause problems in interpreting the results. This is particularly the case in cats in which almost half the mitochondrial genome has been transposed into the nuclear genome. It is therefore essential to undertake appropriate precautions to ensure that the intended mitochondrial DNA gene sequence is actually the one that is amplified and sequenced. Problems of identification can also arise when attempting to distinguish between recently diverged species or hybrids of closely related species – although this is seldom a problem in a forensic context.

Identification of individual animals

The need to identify a specific individual animal usually arises when it is necessary to link an animal to a crime scene or to a suspect. Initial attempts at this tended to have limited discriminatory capacity although where obvious differences in sequence data were found they were sufficient to indicate the absence of a 'match'. For example, Schneider *et al.* (1999) describe a case in which a dog which was believed to have been responsible for causing a traffic accident was exonerated by comparing the sequence analysis of mitochondrial D-loop region of hair fragments found on the damaged car with those of samples obtained from the suspect dog. This technique was only suitable for excluding suspects rather than identifying culprits owing to limited polymorphism of the canine mitochondrial D-loop region.

Currently, STR arrays are available for the identification of individuals of several species of domestic animals and these are used in an identical way to those for identifying individual humans (Chapter 3). For example, horses can be identified using a combination of 12 STR loci and Tobe *et al.* (2007b) describe how these were used to confirm that a urine sample that had tested positive for an illegal drug came from a particular racehorse. Horseracing is notorious for alleged intentional doping and rivals making false allegations. Consequently, when a match was found between all the STR loci isolated from the urine sample and those of the hair samples taken from the horse this could be taken as good evidence that sample did indeed come from that horse.

According to Eichmann *et al.* (2004), canine STR typing is more effective when swabs are taken from severe bite wounds than from those that are relatively light. At first sight this appears odd because severe wounds bleed heavily and therefore the swabs would be badly contaminated with human DNA. However, severe wounds result from extremely forceful and often prolonged contact between the dog and its victim: consequently the dog's saliva is transmitted liberally into the wound and smeared onto the surrounding skin. By contrast, light wounds usually result from a quick snap or nip resulting in relatively little saliva being transmitted and, crucially, these wounds are more likely to be washed before medical attention is sought thereby further reducing the amount of canine DNA present. Analysis of DNA from dogs and other domestic animals is not only of value when the animal itself is the suspect or victim of a crime but may also be used as a means of providing a link between people or between people and a location. For example, dogs and cats are two of the most common pet animals and many of us unintentionally (and often unwillingly) carry with us evidence of that association. A review of canine and feline DNA analysis in a forensic context is provided by Coyle (2007).

Ivory

Ivory is another word for dentine – the main component of all mammalian teeth. However, most animals have relatively small teeth and consequently the majority of objects crafted from ivory are derived from the large tusks of elephants, hippos, walruses and narwhals. Some ivory is also derived from the tusks of mammoths (*Mammuthus primigenus*) that are retrieved from the permafrost regions of Siberia and Alaska. The trade in mammoth ivory is legal and large amounts are recovered each year in Siberia (36 tons were exported into the US alone in 2007). High quality elephant ivory is extremely valuable: in 2007 it traded for around $850 a kilogram and when it is considered that the two tusks (modified upper incisors) of an adult male African elephant (*Loxodonta africana*) can together weigh over 80 kg (an adult female's tusks weigh in the region of a more modest 18 kg) the existence of a large illegal market is hardly surprising. Indeed, although African elephants with combined tusk weights of over 200 kg existed in the recent past such remarkable individuals are now usually shot long before they reach their full potential. In Asian elephants (*Elephas maximus*), only the male has tusks and poaching has resulted in dramatic changes in the sex ratios of their populations. Some female Asian elephants have rudimentary tusks whilst many males in certain populations such as Sri Lanka lack tusks.

In 1990 a global ban on the trade in ivory was instigated by CITES and this was sufficiently effective to allow the elephant populations in some parts of Africa and elsewhere to recover from the brink of extinction. However, owing to the lack of aid for law enforcement, civil strife, corruption among officials and increasing logging (often illegal) facilitating access to forested regions coupled with the high price of ivory, elephant poaching is once again a serious problem in some countries. Perhaps surprisingly, the UK is one of the main sources of intercepted ivory entering the USA; the USA is one of the world's largest markets for ivory and ivory products. Although there are stringent UK regulations governing the sale of ivory and objects containing ivory, enforcement has proved difficult and whilst prosecutions do take place they are infrequent.

Because the trade in mammoth ivory is legal it is important to be able to distinguish it from elephant ivory. This can be done fairly easily from polished cross-sections. These reveal a series of lines called Schreger lines, in the dentine and where these lines overlap they form a cross-hatch pattern. In the outer region (i.e. adjacent to the cementum) the angles formed where the lines overlap are mostly acute ($<90°$) in mammoths but obtuse (90–$180°$) in African and Asian elephants. However, there is a degree of overlap in the 90–$115°$ range and therefore the average of several measurements should be used. Mammoth ivory also often exhibits brownish or blue-green discoloration owing to iron phosphate deposits and these are lacking in Asian and African elephant ivory. Alternative and more rapid means of discriminating between mammoth and elephant ivory are thermogravimetric analysis (Burragato et al., 1998) and Raman spectroscopy (Edwards, 2004) although these require specialist equipment. They do, however, have the advantage of requiring very small amounts of material and they are therefore ideal when the ivory has been worked into small objects such as personal seals or handles.

Case Study: Identifying the provenance of ivory

Up until recently, if a cache of illegal ivory was impounded determining its source depended largely upon following the paper trail of documents detailing its movements (e.g. when the container it was found in was loaded, who loaded it, where it was sent from etc.). However, the paperwork may be absent or unreliable and the trail may cease at the point of embarkation – and even if, for example, documents indicate that the ivory was initially loaded onto a ship in Kenya it does not mean that the ivory came from Kenyan elephants. Being able to pinpoint the source with accuracy is important because it means that resources can be focused where they are most needed, pressure can be put upon governments to take action and, hopefully, it will facilitate the identification and apprehending of the poachers and traffickers. Wasser et al. (2007) demonstrated that it is possible to use DNA isolated from ivory to predict its geographic origin. Recovering DNA from teeth and bone is difficult because it must be first released from within its mineral matrix and this must be done without causing excessive denaturation.

Wasser *et al.* (2007) felt that this was aided by using a freezer mill to grind the ivory down to a fine powder as this technique avoids heating the sample and thereby risking denaturation. A freezer mill works by placing the sample in a metal tube along with a steel impactor. The tube is surrounded by a magnetic coil and submerged in a reservoir of liquid nitrogen. The low temperature makes the sample brittle and prevents it heating when it is broken up by the forceful movements of the impactor that is driven by rapidly switching the surrounding magnetic field. The technique can be used for any hard samples and by producing a dust of fine particles it increases the surface area over which DNA extraction can occur. It is reportedly useful for preparing hair samples but, at least for bone, Loreille *et al.* (2007) found that a freezer mill failed to yield more DNA than grinding with a more conventional Waring MC2 blender cup. Once Wasser *et al.* (2007) had extracted the DNA from the ivory they analysed 16 nuclear STR loci that had previously been used in producing a geographic allele frequency map. The map was produced by sampling DNA obtained from faeces, tissue and ivory from 28 locations in 16 African countries. Incidentally, the elephant faeces used in this work was located with the aid of specially-trained sniffer dogs using the same training programme as that used to train dogs to locate narcotics. This enabled the workers to obtain samples without the need to track and locate the elephants themselves. With the aid of sophisticated statistical techniques Wasser *et al.* (2007) were able to predict the frequency of genes within areas of Africa that they had not sampled by taking into account the extent of interbreeding between neighbouring elephant herds. This meant that they could predict the likely source of ivory even if they did not have any DNA profiles from elephants living in the region from which it came. The value of this has been proved in several practical instances. For example, in 2005, shipping crates containing 6.5 tonnes of ivory were intercepted at the port of Singapore. To obtain this much ivory required the slaughter of around 6000 elephants and a complex logistical operation beyond the capabilities of small time criminals. The paper trail led back to Zambia but the Zambian officials claimed that only 135 elephants had been killed there in the previous 10 years and the ivory must therefore have originated from elsewhere. DNA analysis of 37 of the tusks indicated that the cache was derived from elephants living in the Zambian region rather than being formed from smaller numbers of elephants killed in a variety of different regions across Africa. Clearly, it is possible that the subsample of 37 tusks just happened to come from the Zambian region and the cache actually contained many more tusks from other countries but the authors point out that there were no obvious differences between the characteristics of the tusks that would make that a likely probability. Despite the bad publicity, nobody in Zambia has been charged with the crime although it did bring about changes to the law and the Director of Wildlife was replaced.

Chemical techniques

X-ray diffraction

X-ray diffraction can be used to distinguish rhino horn from that of other species and ivory from antler horn (Singh *et al.*, 2003). X-ray diffraction is based upon measuring the scattering (diffraction) of an X-ray beam after it hits a sample. It is a nondestructive technique and can reveal not only the crystalline structure of the sample but also details of its chemical composition and physical properties. The technique requires only small amounts of material and the sample is analysed as either a single crystal or a powder. In a forensic context the samples are almost always in the form of powders. The amount of sample required depends on the machine, the nature of the sample and the type of information required – it may be as much as a few milligrams or less than $1\,\mu g$. If the sample contains many different components and it is necessary to identify them individually, more material is required. It is commonly used for the analysis of paints, soils, drugs, bullets and gunshot residues, metals, powders that are alleged to contain anthrax spores, and a host of other situations (Rendle, 2003). The results are usually presented as a trace of peaks and troughs that is called a diffractogram. The diffractogram represents the photon counts of X-ray radiation as a function of the angle of diffraction. Diffractograms from, say, known rhino horn can be compared with those of an unknown to determine their similarity. Alternatively, individual components from within a mixture can be identified from the similarity of parts of the diffractogram to those of known standards (Davies, 2008).

Stable isotopes

Isotope ratio analysis of vertebrate tissues relies on the fact that the chemical characteristics of the soil on which an animal lives are reflected in the levels of chemicals and their isotopes present in their hair, bones and teeth. In this way it has proved possible to identify the geographical origin of ivory and bone (Stelling & van der Peijl, 2003). However, to be fully effective, this requires a comprehensive database of soil chemistry characteristics from those parts of the world the animal is thought to have originated.

Stable isotope ratios can also be used to establish feeding relationships, at least among aquatic organisms, because predators tend to contain higher levels of ^{13}C and ^{15}N than their prey. Farmed fish are often fed pellets containing high levels of protein and fat derived from other fish whilst their wild brethren have a more diverse and less concentrated diet. Consequently farmed fish tend to have higher levels of ^{13}C and ^{15}N and this enables them to be distinguished from 'wild' fish (see below).

Chemical composition

Sometimes an animal's chemical composition is affected by its diet and this can be used to identify the circumstances under which it was reared. For example, the fatty acid composition and mineral composition of fish are both affected by their diet and this can be used to distinguish between those that are farmed and those caught in the wild (Alasalvar *et al.*, 2002). This is important because wild caught fish are commonly believed to be more nutritious and have a better flavour and it is not unusual for such fish to sell for up to double the price of farmed fish. Between October 2005 and August 2006 the UK Food Standards Agency examined the fatty acid profile and isotopic ratios of ^{13}C: ^{12}C, ^{15}N: ^{14}N and ^{18}O: ^{16}O in a range of fish purchased from supermarkets and shops around the country. They found that 10% of sea bass, 11% of sea bream and 15% of salmon that were marketed as 'wild' had definitely originated from a fish farm. Although not necessarily representative of the whole retail industry the results indicate a potentially widespread problem with fish labelling. Similar worries concern the labelling of free-range and organically grown farm animals and crops.

Future directions

There is a need for rapid, cheap identification systems suitable for identifying individual livestock. The illegal movement of livestock and resultant fraudulent subsidy and compensation claims are big problems in the European Union and involve millions of pounds on an annual basis. In addition, the illegal movement of animals can present serious risks in spreading diseases, some of which can also infect humans. Furthermore, following natural disasters, such as flooding, animals can stray and in the absence of a means of permanent identification, there is the opportunity for theft and confusion.

The cost of molecular biological techniques is likely to decrease in future years whilst their speed will increase as a consequence of the expanding demand for cheap, quick and sensitive methodologies. Consequently the use of molecular methods for the identification of both species and individual animals is likely to become more widespread as their use becomes financially acceptable. Animal DNA databases are unlikely to raise the same ethical issues as human DNA databases although they would require the same rigorous standards concerning the collection, analysis and storage of the information. Retinal identification systems have been trialled with racehorses although these have suffered from practical problems and iris scans may be more effective (Cordes, 2000). It would be interesting to evaluate the usefulness of these techniques for other high-value animals. Retinal and iris identification systems will, however, only be effective whilst the animals are alive.

Stable isotope analytical techniques hold great promise for determining the geographical origin of all sorts of biological material from bones to illegal drugs. Once reliable databases become available, these techniques will undoubtedly be used more frequently. Similarly, stable isotopes could be used to establish whether an organism was collected from the wild rather than being raised in captivity or cultivated. This is particularly relevant for distinguishing meat derived from game animals raised in captivity on game farms from that from animals that were poached in the wild. Similarly, in the exotic pet trade, it could differentiate between captive born animals and those caught in the wild.

Quick quiz

(1) State four ways in which the activities of scavengers may affect the evidence in cases of murder / suspicious death.

(2) How would you distinguish between damage caused to bone by sawing and the gnawing of rats and dogs?

(3) Briefly discuss three reasons why a pet dog might eat its dead owner.

(4) Explain how pet hairs might be useful in a forensic investigation?

(5) Why is prevention of childhood cruelty to animals of relevance to criminal behaviour?

(6) What is 'bushmeat' and what are the concerns about its importation into the UK?

(7) Briefly explain how the presence of rhino horn in traditional Chinese medicines can be demonstrated.

(8) Why might one want to distinguish between farmed fish and those caught in the wild and how could this be achieved?

(9) Briefly discuss some of the concerns over the reliability of molecular-based techniques for the identification of vertebrate animals.

(10) Briefly discuss how the country of origin of animal products such as ivory could be established.

Project Work

Title

Identification of incorrect labelling of meat products.

Rationale

Food products are sometimes fraudulently mislabelled. This can have especial significance for Muslims and Jews if pork meat has been added and for Hindus if beef is the contaminant.

Method

Meat products, such as sausages and meat pastes may be purchased and tested directly or artificially contaminated to assess the sensitivity of the assay method. Pet foods would probably yield interesting results because they are often a mixture of ingredients – for example, 'chicken chunks' cat food from various sources is often labelled as containing only a 'minimum of 4% chicken'. The identity of the meat present in the sample may be assessed using standard mitochondrial DNA analysis.

Title

The utility of cat and dog hairs as forensic evidence.

Rationale

Cat and dog hairs are commonly found on clothing and might be used to link a person with a locality.

Method

Cat and dog hairs are obtained by combing from a wide variety of cats and dogs and a record made of the part of the body they are obtained from. The hairs are then analysed using a stage microscope to determine whether it is possible to identify individual animals from their hair characteristics. Sub-samples of the hairs would then be subjected to DNA extraction and analysis to determine whether this improved the identification. Another approach would be to ask volunteers to visit the homes of cat and dog owners, to sit down and have a drink and then to leave. Their clothes would then be analysed to determine the extent and location of any pet hairs they had acquired. Preferably they would never have visited the house before, some householders would be cleaner than others, and they would or would not have come into contact with the pet. One question would be whether it is impossible to visit a pet owner's home without acquiring pet hairs on the clothing and whether certain clothing is more effective at trapping pet hairs.

PART C Protists, Fungi, Plants and Microbes

9 Protists, fungi and plants

Chapter outline

Introduction
Protists
Fungi
Plants
Plant Secondary Metabolites as Sources of Drugs and Poisons
Illegal Trade in Protected Plant Species
Future Directions
Quick Quiz
Project Work

Objectives

Explain how algae and in particular diatoms, can provide an indication of drowning and also associate a person with a locality.

Discuss the potential of fungi as forensic indicators.

Describe how the characteristics of woody plants can provide forensic evidence.

Review the potential of pollen analysis for linking a person, animal or an object to a locality at a particular time of year.

Discuss how pollen samples should be sampled, processed and identified.

Describe the variety of ways in which fruit, seeds and leaves can provide forensic evidence.

Explain how plant material should be collected, identified and stored as part of a forensic investigation.

Discuss the criminal use of secondary plant metabolites and how these may be detected.

Discuss how forensic evidence of the illegal trade in protected plant species may be obtained.

Essential Forensic Biology, Second Edition   Alan Gunn
© 2009 John Wiley & Sons, Ltd

Introduction

This chapter covers three quite separate kingdoms – Protista, Fungi and Plantae – although in a forensic context they are commonly grouped together under the heading 'forensic botany'. Plants form part of our diet and the evidence of what we have eaten and when we ate it is contained within our digestive system and our faeces. In addition, we are always coming into physical contact with plants in our homes, gardens and the wider environment; the evidence of this can be seen in the damage caused to the plants and bits of them that get attached to our hair, clothing or possessions. Plants are therefore potentially useful sources of forensic evidence for associating a person, animal, vehicle or object with one another or a locality. Despite this, the forensic potential of plant-based evidence is poorly exploited and there is much work to be done in this area. Protists and fungi are also extremely common in the environment but their use in forensic science has received even less attention than that of plants.

Protists

Introduction

The kingdom Protista is an artificial grouping of single celled and simple multicellular eukaryotic organisms. Eukaryotes are organisms that have a membrane-bound nucleus and membraneous organelles such as mitochondria. The Protista is a hugely diverse assemblage and many of them are not even closely phylogenetically related to one another. Apart from the diatoms few protists are used as forensic indicators. Theoretically, strain typing of pathogenic species such as *Entamoeba histolytica* (amoebic dysentery) or the various species of malaria (*Plasmodium* spp.) could provide an indication of a person's association with a particular country or region within it but these organisms are normally considered from a medical rather than a forensic perspective. A number of free living protists also have forensic potential. For example, testate amoebae are common in many soils and mires and produce characteristic coverings called 'tests' (Fig. 9.1). The abundance and diversity of the tests could be used as an indicator of a particular geographical location and/or for comparing the similarity of two or more forensic samples.

Diatoms and other algae

The term 'algae' encompasses a wide variety of organisms ranging from single celled protists to colonial forms to huge multicellular seaweeds such as kelp. Most, but not all, algae contain chloroplasts. Although algae are ubiquitous in terrestrial and aquatic ecosystems, only the diatoms tend to be used to any a great extent in forensic analysis. Diatoms are unicellular algae belonging to the phylum Stramenopila (phylum Bacillariophyta in some textbooks) and are characterized by the possession of beautifully patterned cell walls made out of silica called 'frustules'. They can be found in fresh water and salt water, as well as the surface of moist

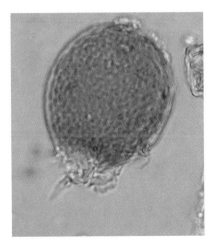

Figure 9.1   Light microscope photograph of a terrestrial testate amoeba.

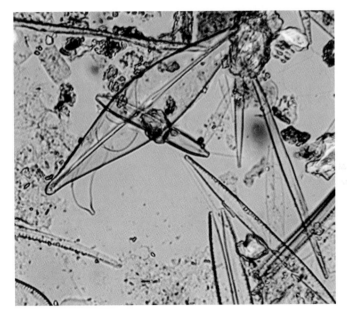

Figure 9.2   Light microscope photograph of freshwater diatoms. Note the large variety of shapes and sizes of the frustules.

terrestrial habitats (Fig. 9.2). Because they use silica extracted from their environment to build their frustules, their growth is limited by the availability of dissolved silica in the water. Each frustule is composed of two halves, which are referred to as valves, one of which is slightly bigger than the other so that the two fit together one inside the other. Many species of diatom have a characteristic frustule design – a feature that facilitates their identification. Although most frustules dissolve when the algae die, their silicaceous nature means that some are preserved, especially if the conditions are favourable. Indeed, some soils, called diatomaceous earths are composed almost entirely of fossil diatoms. There are numerous species of diatoms: more than 5600 have been described and there are probably more than 100 000 in total. Their abundance and species composition vary between locations owing to different diatoms preferring different conditions such as temperature, salinity and pH and the consequences of inter-species competition. Therefore, the presence of individual species or the species composition in a forensic sample can provide an identifying feature of a habitat or location at a particular time of year.

Diatoms and the diagnosis of drowning

Drowning results from suffocation following the immersion of the mouth and nostrils in a liquid although some pathologists would state that it also involves the aspiration of fluid into the air passages of the lungs. The mechanism of death is very complex and not simply a consequence of asphyxiation. The diagnosis of drowning at autopsy is notoriously difficult and signs of immersion merely indicate that a body was underwater (or some other fluid) for a period of time and not that the person died of drowning. Because diatoms can be found in many water sources, their presence in the lungs and other tissues has been used as an indication that the victim may have drowned (Pollanen, 1998). Diatoms can be recovered from the stomach (indicating that water was swallowed), from the lungs (indicating that it was aspirated) and also from the blood, major organs and the bone marrow (indicating the water was aspirated whilst the victim was still alive). Owing to their small size and the damage caused to the lungs during drowning, diatoms may pass through the alveoli and be swept around the body in the blood stream (Lunetta et al., 1998). However, once the blood circulation ceases, any diatoms entering the lungs with water would not be transported elsewhere. By comparing the abundance and species profile of the diatoms found in the body with that of diatoms found in the river – or wherever the victim was recovered – it is possible to provide corroborating evidence with results from the autopsy to determine whether or not they drowned and if so, whether it was at that location (Ludes et al., 1999). For example, the presence of freshwater diatoms in a body recovered at sea would suggest that the victim may have died in a river and subsequently been swept out. The possibility that a person may have died in a water body other than the one in which they were found needs to be borne in mind when deciding on extraction techniques. For example, the frustules of marine diatoms are dissolved by Soluene-350 (a solubilizing agent) whilst the frustules of freshwater diatom species are not (Sidari et al., 1999). Like all biological evidence, the results of diatom analysis need to be considered in context. For example, some workers question the sensitivity of the diatom test and the absence of diatoms does not mean drowning did not occur, nor does the presence of diatoms in the body tissues mean that it did. Although diatoms are extremely common, they are not found in all water sources and even if they were present they may not find their way into the body organs or be recovered if they did. Similarly, the abundance of diatoms means that contamination at the time the body was retrieved, during the autopsy or during laboratory analysis is always a possibility unless extreme care is taken. Furthermore, if a person repeatedly swims in a lake or the sea, it is possible that they may accumulate diatoms within their tissues over time so finding them does not mean that the person drowned there (Taylor, 1994). Diatoms may also be recovered from the tissues of persons who do not die of drowning or swim regularly. Diatoms are found in numerous man-made products ranging from building materials to the powder used in rubber gloves. There are therefore many opportunities in which diatoms may be breathed in and there is also a possibility that they might be absorbed through the gastrointestinal tract when consumed with foodstuffs – although the extent to which this occurs needs to be confirmed by further research. Consequently, diatoms can be found in the tissues of persons who died of causes other than drowning. However, it is generally accepted that finding diatoms within the bone marrow provides good corroborative evidence of drowning and identification to species level can exclude those that are contaminants (Pollanen, 1997; Pollanen et al., 1997).

Algal growth to determine length of submersion and exposure

The rate of colonization of submerged corpses by diatoms and other algae is a potential means of estimating length of submersion but there are few studies in this area (Casamatta & Verb, 2000). The amount of growth will depend upon the depth, substrate, local circumstances and time of year and therefore any predictions would have to be based upon experimentation. For example, growth would be more rapid during summer in a nutritionally rich pond than during winter in a nutrient poor mountain stream. Haefner et al. (2004) describe how algal growth can be used to determine the length of time a dead body (in their case, that of a pig) or an inanimate object was submerged in water. Rather than identify the species of algae present and their relative abundance, they determined total algal density by measuring the amount of the pigment chlorophyll a present in a sample collected in a custom-designed sampling device.

Algae also colonize exposed objects and animal remains in terrestrial environments and their growth could be assessed in a similar manner to that described above. So-called 'blue-green algae' are not algae at all – they are prokaryotic organisms (i.e. they lack an enclosed nucleus and membrane-bound organelles) and are more correctly known as cyanobacteria. Despite being extremely common in both terrestrial and aquatic ecosystems their forensic potential has received little attention, although signs of colonization of exposed bones is said to become apparent to the naked eye after 2–3 weeks under suitable conditions (Haglund et al., 1988).

Collecting, identifying and preserving diatoms and algae for forensic analysis

Many species of algae, including diatoms, exhibit seasonal growth characteristics and some are notorious for forming short-lived toxic blooms. Therefore, if an

attempt is to be made at matching the species composition recovered from a body
or object with a specific locality it may be necessary to take sequential samples
through the year and, in the case of open water, samples should be taken from both
the surface waters and the bed. In the case of a river or stream, samples should also
be taken both above and below the site at which the body was found because there
may be differences in the diatom flora and the body may have been moved by the
water currents. Owing to the small size of diatoms and their widespread occurrence,
every effort should be made to avoid the possibility of sample contamination occurring in both the field and the laboratory. All collecting equipment (e.g. plankton
nets, soil corers and collecting vials) must be scrupulously cleaned before use, and
within the laboratory the samples should be processed in clean glassware and laboratory procedures adopted to reduce the chances of contamination occurring (e.g.
working within a laminar flow cabinet). Blanks, consisting of samples of distilled
water, should be processed at the same time as a check to test whether contamination could be occurring.

Diatoms and other algae can be collected from open water using plankton nets –
these consist of a long funnel-shaped net bag mounted on a circular frame and with
a collecting vial attached to the narrower trailing end. The mesh size of the net will
determine the size of the plankton catch – planktonic algae tend to be small and
require a fine net size (0.1–0.3 mm). Algae and other microscopic organisms that
are attached to underwater substrates are sometimes referred to as 'periphyton' and
require specialized collection techniques (e.g. Haefner *et al.*, 2004). After collection,
samples may require further concentration, for example by centrifugation, although
the method needs to be chosen with care if the more delicate species of algae are to
be preserved unharmed. The samples can be observed directly using an ordinary
stage microscope although phase contrast illumination improves the amount of
detail one can see. If one is only interested in the diatoms and there is a lot of contaminating organic matter present, the samples may be air dried and subjected to
acid digestion (see below) although most ordinary algae do not have silicaceous cell
walls and would be destroyed.

Diatoms are normally extracted from tissues, soil and other solid substrates, by
heating with concentrated nitric acid for up to 48 hours (Pollanen, 1998; Hurlimann
et al., 2000). The digest is then centrifuged, the supernatant discarded, and the pellet
washed by one or two cycles of suspension in distilled water and centrifugation
after which the final pellet is observed using a stage microscope. This method relies
on the silicaceous frustules remaining after all organic matter is dissolved away – but
some workers prefer to use less dangerous reagents and have adopted techniques
based on DNA extraction (Kobayashi *et al.*, 1993) and ultrasonic digestion (Matsumoto & Fukui, 1993).

Diatoms and algae are normally identified on the basis of their morphology (e.g.
Stoermer & Smol, 2001) but this is a time consuming and often difficult exercise.
Consequently, a number of other approaches have been suggested. The complex
three-dimensional shapes of diatom frustules makes them obvious targets for computer pattern recognition software and there have been several attempts at developing automatic devices that can simultaneously recognize and count different diatom
species (e.g. Jalba *et al.*, 2005). However, even if automatic devices such as this can
be perfected, molecular evidence indicates that morphology on its own is a poor

species indicator for diatoms and the existence of 'cryptic' and 'pseudospecies' is widespread. This is an important consideration in a forensic context in which the identification of species composition is to form part of the evidence. Evans *et al.* (2007) tested the suitability of a range of genes for phylogenetic analysis of diatom species and found that the CO1 (cytochrome oxidase subunit 1) gene offered considerable potential. This is somewhat surprising because although this gene is commonly used in phylogenetic studies of animals it is considered to have limited applicability for plants (see later). Abe *et al.* (2003) attempted to identify the presence of diatoms in forensic samples using molecular techniques but although they found the approach to be sensitive it could not be used to identify sufficient species of diatoms to distinguish the locality at which a person drowned. It should be remembered that molecular techniques could not be used on diatom specimens isolated from soil or tissues using standard extraction techniques since the strong acids would destroy their DNA. Another alternative identification technique is to use Fourier-transform infrared (FT-IR) spectroscopy to distinguish between species on the basis of their chemical composition. FT-IR has proved capable of distinguishing species that are difficult to tell apart on the basis of their morphology (Vardy & Uwins, 2002) and useful for population studies (Sigee *et al.*, 2002) but it does not appear to be in widespread use for identifying diatoms.

Fungi

Mushrooms, toadstools, moulds, rusts, smuts, yeasts and other fungi are commonly thought of as plants but in fact they belong to a totally separate kingdom of their own – kingdom Fungi – and their study is a separate science – mycology. The majority of fungi are multicellular organisms and consist of long filaments called hyphae (singular hypha) that when woven together to form mats are referred to as a mycelium (plural, mycelia). The thin white strands covering the surface of a loaf of mouldy bread or a dead body (Fig. 9.3) are therefore fungal hyphae. Yeasts are single celled fungi and in addition to the well-known baker's and brewer's yeasts there are species that live on or in plants and animals. For example, *Candida albicans* occurs on our skin and in our guts and can cause infections such as vaginitis (thrush). Yeasts multiply by budding and some species produce long thin buds called pseudohyphae or 'germ tubes' but they do not produce true hyphae. A number of fungi are dimorphic and exist as either a unicellular yeast-like form or mycelial form depending upon their environment.

The cell wall of fungi is distinct from that of animals and plants in that it contains large amounts of chitin – the same substance associated with the cuticle of arthropods. Fungi form spores as part of the sexual or asexual reproductive process depending upon the species. These are usually liberated into the air although in certain species they may disperse in water or through attachment to vertebrates or invertebrates. Most fungi are saprophytic – that is they feed on dead organic matter that is broken down through the release of enzymes and the products of digestion are absorbed. A number of fungi are parasites of plants, animals or other fungi and some are predatory on nematodes and other soil organisms.

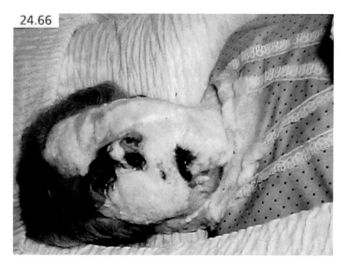

Figure 9.3 This person's body was embalmed and then buried. Several months later the body was exhumed as part of an investigation. Extensive mould formation has covered most of the face. Modern embalming techniques are only intended to preserve the body for a short period of time before burial. (Reproduced from Dolinak, D. *et al.*, (2005) *Forensic Pathology: Theory and Practice.* Copyright © 2005, Elsevier Academic Press.)

Fungi associated with dead bodies

Fungi are associated with all stages of decomposition from immediately after death to the skeletonized or mummified remains (Ishii *et al.*, 2006). From a forensic point of view they are mainly of concern from their potential to interfere with the interpretation of other forensic analyses. For example, high levels of fungal DNA can cause nonspecific inhibition of the PCR process when undertaking human STR analysis (Calacal & De Ungria, 2005). Similarly, yeasts such as *Candida albicans* which are naturally present on our skin and guts will flourish after our death and in the process produce large amounts of post mortem ethanol (Yajima *et al.*, 2006). This is of particular concern in road traffic and air accident investigations in which it is crucial to establish whether alcohol could have been involved. Unfortunately, in such cases it is common for the body to suffer horrendous injuries so that even if blood or tissue samples are taken shortly after death, contamination with gut flora is a distinct possibility. Consequently, sodium fluoride is usually added to samples to inhibit further ethanol production and the presence of yeasts such as *C. albicans* and other microbes should be considered when interpreting post mortem toxicological analyses.

Although many species of fungi are involved in decomposition of animal remains there appear to be no published studies on how fungal populations change as a body decomposes. Carter & Tibbett (2003) considered a wide range of fungi and their potential as forensic indicators, although Bunyard (2004) has commented that some of their suggestions are a bit too enthusiastic. In particular, many of the fungi they mention are small and difficult to identify. While *Hebeloma syriense* has earned itself the moniker 'the corpse finder' there is little published scientific data to confirm this. Indeed, most fungi associated with corpses are also found in other substrates containing high ammonia levels. Furthermore, the presence of a body close to the soil surface will be more readily identified by its smell, the disturbed soil, and the frantic activity of invertebrates than by searching for fungi. The usefulness of fungi in determining the PMI probably awaits the application of molecular techniques such as terminal fragment length polymorphism analysis (T-RFLP) and ribotyping (Chapter 10).

Fungi as indicators of association and locality

Fungal spores are often produced in large numbers and, like pollen (see later) can have characteristic morphologies which facilitate their identification. Because some fungi have restricted distributions, have specific ecological requirements and/or produce spores only at certain time(s) of year, their presence may be useful in associating a person or object with a locality. *Candida albicans* is not normally considered to be a sexually transmitted disease (STD) although genital infections are more commonly seen in those who are sexually active. It is possible to distinguish strains of C. *albicans* using DNA microsatellite analysis (e.g. Sampaio *et al.*, 2003) but there appear to be no attempts to link two or more people together in a forensic context through shared infections as can be done for gonorrhoea and HIV (Chapter 10).

Poisonous fungi and fungal toxins

A number of fungi produce highly toxic chemicals known as mycotoxins – this translates as fungus poisons. Mycotoxins are secondary metabolites that probably evolved to protect the fungus from being eaten and/or to prevent other organisms from exploiting the fungus's food supply. The mycotoxins are released by the fungus during its life and/or after death, and these then diffuse into the surrounding substrate. Therefore, the toxin may occur in regions in which the fungus is absent. This can make safety assessments difficult because the absence of the fungus does not mean that a substance is not contaminated with toxins. The common mould *Aspergillus flavus* (actually a species complex of 11 species) is notorious for infesting grain and other stored products and producing a range of mycotoxins called aflatoxins some of which are also carcinogenic. During Saddam Hussein's era as president of Iraq the country experimented with the use of aflatoxin as a biowarfare agent. The deliberate use of mycotoxins to poison people or domestic animals is a rare occurrence but they are a common cause of accidental poisoning. Aflatoxins have been described from marijuana as a consequence of the growth of *A. flavus* (Hazekamp, 2006). The toxins may be inhaled as they are not completely destroyed when the marijuana leaves are burnt during smoking. It is therefore possible that some of the neurological problems associated with marijuana are a consequence of contamination with mycotoxins. Fungal contamination of foodstuffs and the environment can also cause harm and even death should a person exhibit an allergic reaction to the spores. Bennett & Collins (2001) describe a case in which a man died of anaphylaxis after consuming two pancakes made from a package mix that had become heavily contaminated with several species of mould.

A number of mushrooms are poisonous and some are potentially lethal. The Roman Emperor Claudius was reportedly the victim of intentional mushroom poisoning but there are relatively few accounts of their being used to murder someone since then. In the absence of witnesses or other evidence, poisoning by any means can be difficult to detect because the forensic pathologist may have no reason to suspect that it was the cause of death and would therefore not send tissue samples for analysis. In continental Europe, where the picking of wild mushrooms is popular, accidental poisonings are common as a result of mistaking poisonous for edible species. For example, in 2000, there were 2740 reported cases of mushroom poisoning in Russia. In the UK, mushroom poisoning is more likely a consequence of people confusing wild-picked magic mushrooms, of which there are around 12 different varieties, with poisonous ones. The liberty cap mushroom (*Psilocybe semilanceata*) is the most commonly consumed 'magic mushroom' and the laws surrounding its use are decidedly odd. Growing, gathering or possessing the fungi does not contravene the law, however, any attempt at preparation (such as cutting, drying, powdering or freezing and packaging) renders it a controlled drug.

Collection and identification of fungi

The fruiting bodies of large fungi such as many mushrooms and toadstools should be collected into paper bags along with their supporting stem. If they are put into plastic bags or sealed containers they will rot very quickly, especially if the specimens are wet. Many of the larger UK fungi are relatively easy to identify from their gross morphology and spores/spore print (Phillips, 1981). If the specimens are to be preserved they should be slowly air dried but this will result in a loss of colour and their shape may become distorted. Small fungi such as moulds found growing upon a corpse can only be identified with the aid of a microscope – usually on the basis of their sexual or asexual reproductive structures. It is often helpful if these fungi are cultured in an appropriate medium (for fungi of medical importance see Winn *et al.*, 2006) – there is no 'universal medium' and some fungi cannot be cultured on artificial media. Sometimes an antibiotic is added to the culture medium to avoid overgrowth by bacteria. Spores or samples of hyphae from large fungi can be cultured in the same way. Molecular taxonomy is being used increasingly as means of identifying fungi, particularly of pathogenic species.

Plants

The kingdom Plantae includes all multicellular terrestrial organisms that carry out photosynthesis and have an embryo that is protected within the mother plant. It therefore includes organisms as diverse in size and shape as mosses, potato plants and oak trees.

Plant damage and growth as a forensic indicator

Plants are easy to damage but difficult to kill. Whenever a person, animal or object comes into contact with a plant, there is a possibility that the plant will be damaged in some way and if so, the evidence of this association will usually be preserved. For example, when a vehicle drives over grass, it leaves behind imprints from which one can determine the direction of travel and tyre characteristics. The latter, if preserved well enough, can be used to identify the vehicle responsible. Similarly, when walking across a newly mown lawn or field one usually picks up grass clippings on one's shoes and clothing and these have been used as forensic evidence (Horrocks & Walsh, 2001).

When a body or object is buried both above and below ground parts of plants will be damaged. If the growth region, known as the meristem, of a root is affected, a permanent scar will result and the number of growth rings formed after the scar will indicate how many years ago the damage was caused. The act of burial will mix up the seed bank in the soil and therefore influence the plants that germinate there. Plant succession will also be affected by the nature of the ground, exposure to sunlight, the surrounding vegetation and the time of year at which burial takes place. However, assuming that the grave surface was left bare then one can expect to observe that the plant composition is initially dominated by species that specialize in colonizing disturbed ground/bare soil, such as grasses. A typical succession pattern would be grasses and herbaceous plants, followed by shrubs and then trees. However, this does not mean that tree species would be absent until long after a grave was dug. If present in the seed bank, tree seeds will germinate and their seedlings grow as soon as the conditions are suitable. Consequently, a grave dug in the autumn may be populated by many germinating tree seedlings in the following spring. The germination and growth of these plants can provide an indication of how long it was since the burial. If a body is buried in nutritionally poor soil, its decay will probably result in the noticeably more luxuriant growth of plants upon and around the grave. When a plant, or part of one, is covered (e.g. during the course of a burial) so that it is no longer able to photosynthesize, it starts to become yellow and etiolated. However, in order to determine how long it had been in the dark one would need to know its initial condition (e.g. chlorophyll content).

Wood

The study of wood can provide information applicable to many aspects of forensic science ranging from the trade in protected species and the fraudulent sale of copies of valuable antiques to identifying the weapon(s) used in a murder. Some species of tree produce wood with a distinctive texture and smell. For example, in October 2004, 33 tons of wood from the Brazilian tree *Dalbergia nigra* (Palosanto de Rio) were impounded in Spain where it had been imported illegally for the manufacture of high quality Palosanto guitars. Police became suspicious of the increasing supply of Palosanto guitars coming onto the market despite trade in the wood being highly controlled since 1992. Following a surveillance operation, the wood, identifiable from its characteristic properties, was found hidden in batches of legally imported

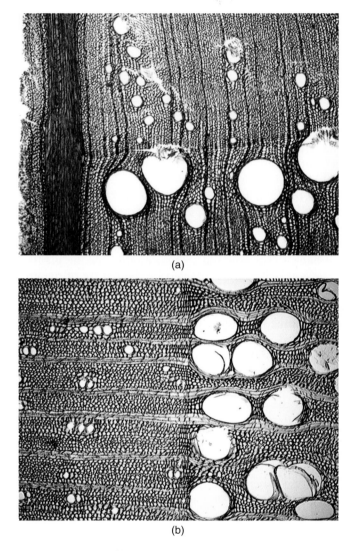

(a)

(b)

Figure 9.4 Although wood from different trees often looks very similar, its microscopic anatomy may be quite distinctive. (a) European oak (*Quercus robur*); (b) ash (*Fraxinus excelsior*). For microscopic analysis, the samples should be cut across and along the grain of the wood and then compared with reference material.

hardwood timber. Colour, texture and smell are, however, not invariably accurate means of identifying wood so microscopic analytical techniques are more normally employed (Fig. 9.4). On their own, even these techniques are seldom sufficient to enable species identification, although it may be possible to determine the genus or sub-generic grouping. A more accurate identification may be possible if more information, such as country of origin, is known. For example, Scots Pine, American red pine and several Asian red pine species cannot be distinguished from one another on the basis of microscopic anatomy but if one knew that the wood came from a tree growing in North America then the species would almost certainly be American red pine because it is the only red pine species native to the region. The feasibility of using molecular techniques for species identification from wood samples is being investigated but they are not yet in widespread use. Wood fragments can prove useful in associating an object with a crime. For example, in an Australian case, wood fragments found embedded in the skull of a murder victim were identified as coming from a species of rubber tree grown in south-east Asia. The wood from this tree is used in the production of baseball bats so the police then knew what sort of murder weapon they were looking for. Furthermore, because only certain makes of baseball bat are prepared from this type of wood, the search was narrowed down further and ultimately they were able to identify the actual bat that was used as the murder weapon and connect it to the crime.

Dendrochronology

Woody plants are not a distinct taxonomic group but they share the common characteristic of exhibiting secondary growth. That is, they increase in girth as a consequence of cell division by the lateral meristems. These plants can therefore increase the width of their trunks, branches and roots even if these are no longer increasing in length. As part of this secondary growth process, secondary xylem is produced and it is this that we more commonly call 'wood'. Xylem transmits water and minerals from the roots to the other parts of the plant but in woody plants it is only primary xylem and the youngest of the secondary xylem vessels that do this: these are found in the outermost region that is called the 'sapwood'. The majority of the secondary xylem does not transmit anything but instead provides structural support: they are found in the inner region that is called the heartwood. Secondary xylem has high lignin content and this provides the strength needed to support a large structure such as a tree.

In woody plants that have a distinct growing season, the production of xylem vessels follows one of two patterns. In ring-porous species, such as oak (*Quercus* spp.), the vessels are produced principally in the spring whilst in diffuse-porous species, such as sycamore (*Acer pseudoplatanus*) and silver birch (*Betula pendula*) the xylem vessels are produced throughout the growth period. The study of tree growth rings is known as dendrochronology and it has proved extremely useful in determining climate change over the years. Because of their more distinct banding patterns, it is the ring-porous species that are most frequently used in dendrochronology. In these species, the early xylem is characterized by large cells with thin cell walls whilst the late xylem has small cells with thicker cell walls (Fig. 9.5). The dividing line between the late xylem of one year and the early xylem of the succeeding year is seen as the line between growth rings. In temperate climates such as the UK, ring-porous trees produce only one growth ring per year and the width of the rings provides an indication of the environmental conditions at the time – thick rings reflecting good growth conditions. Dendrochronology does not involve counting tree rings but of comparing (often referred to as cross-matching or cross-dating) ring patterns between trees and between trees and wood products. Detecting

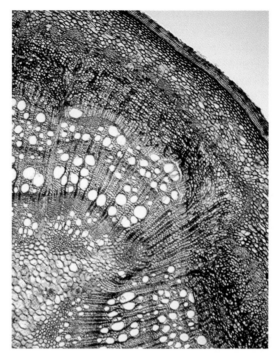

Figure 9.5 Cross-section through a 2-year-old oak (*Quercus robur*) sapling showing the formation of growth rings.

ring patterns in diffuse-ring species is often problematic although it is possible to enhance their appearance through treatment with wood rot fungi (Deflorio *et al.*, 2005).

In forensic science, the study of tree rings can provide evidence of the age of an object. Typically, the age of a piece of wood can be estimated by comparing its ring patterns with those of samples obtained from the geographic region it is claimed to have originated from. For example, if a painting is made directly onto a piece of wood it is possible to determine the date when the tree it came from was felled, provided sufficient rings are available for analysis. Consequently, it is also possible to state whether the artist to whom the work was being attributed would have been alive and at an appropriate age to have executed the work. For example, a painting attributed to a medieval artist produced on a piece of wood that came from a tree felled in 1865 could be nothing other than a fake. This is clear cut when the object in question is made out of a single piece of wood but in the case of antique furniture and musical instruments things may be more complicated owing to repairs and alterations undertaken over the years. For example, all Stradivarius violins, without exception, have had their original necks sawn off and modifications made as a consequence of changes in playing techniques. Furthermore, although dendrochronology, like most of the other forensic techniques used to determine the provenance of paintings and antiques, is good at proving falsification it cannot necessarily prove authenticity. It should also be borne in mind that fraudsters are often aware of the forensic procedures used to detect their work and will use techniques such as making picture frames out of old wood and ageing nails in salt solution. When examining furniture and similar items, it is the less visible parts (such as the back or the corner blocks) that can provide most information on the country of origin. Because these parts are not seen, they tend to be made from cheap local wood whilst the exterior ornate parts are often made from imported timber.

Dendrochronology can be used to determine when a tree was cut down or killed by comparing its tree rings with those of the same species growing in the locality. This can be important in the case of protected species when illegal logging or clearing is suspected. For example, alerce, *Fitzroya cupressoides*, is a protected species in Chile and legally classed as a National Monument but its population is threatened by illegal felling and burning. Alerce is an evergreen conifer and can live for over 3600 years but its slow growth and rate of regeneration makes it vulnerable. Despite a ban on international trade in alerce timber, the wood is still exported by people claiming that they are selling wood coming from trees that were felled or died before 1976 when Chile joined the Convention on International Trade of Endangered Species (CITES). Wolodarsky-Franke & Lara (2005) describe how dendrochronology can be used to identify instances of illegal felling when a tree stump is found. Cross-dating alerce growth rings can be difficult owing to their slow growth rate and local factors that influence their formation. The prospects of establishing a good cross-match are improved if the stump still contains sapwood and bark and if several cores can be taken for comparison with surrounding still living alerce trees. Alerce growth rings are formed between the start of the growing season in September (spring) to the end of March (autumn) (the seasons in South America are not the same as in Europe). Therefore, the 2007 growth ring would be formed between September 2007 and March/April 2008. If the 2007 growth ring was incomplete it would indicate that the tree was cut down at some point between these two dates whilst if it was fully formed it would indicate that the tree was felled during the winter period – i.e. between April and September 2008.

The study of tree rings may also be useful when a tree or sapling is found growing through a skeletonized body or within a grave plot: the number of rings indicates how long the body has rested at the site. Courtin & Fairgrieve (2004) describe a case in which the growth of a tree branch was affected by the presence of a dead body lying across it. The pressure of the body resulted in an asymmetrical pattern of growth rings and computerized quantification of these rings enabled a back calculation to when the body came to lie across the branch. Tree growth rings take time to form, so the technique could only state that the body came to rest there at some point within a 10-month period but it subsequently transpired that this encompassed the time during which the dead person disappeared. The technique could be useful in cases in which persons are left bound or hanging from a tree and more accurate evidence of the post mortem interval (e.g. physical changes, insect activity etc.) was lacking. Roots also show annual growth rings and therefore, where these have penetrated the clothing or bones, they will indicate how long the body has been present at the site, as will the depth to which roots have penetrated the body (Willey & Heilman, 1987).

Plants causing post mortem artefacts

Plants can cause a variety of post mortem artefacts and these should be distinguished from damage caused at the time of death. For example, although seemingly delicate, roots and shoots can exert enormous pressures that are sufficient to force apart the sutures of the skull. Similarly, roots growing across the surface of/through bone cause the formation of a characteristic network of intricate and ramifying grooves as a consequence of their secretions (Haglund *et al.*, 1988). The presence of plant tissues and their relationship to human remains should therefore always be taken into account.

Tool marks in wood

When a gun is fired, its bullets acquire markings that are unique to that particular weapon. In a similar manner, every hammer, saw, chisel or axe has its own unique surface properties and when used, these are faithfully preserved in substrates such as wood or bone as 'tool marks'. Any hard material causes such marks when it hits or is dragged across a softer material and their study is known as 'tool mark analysis'. Tool marks take the form of a negative impression of the object (i.e. 'the tool') that inflicted the damage from which can be deduced its shape, size, contours and rows of parallel grooves caused by the object being dragged across the surface. It is therefore possible to link a suspect with a crime by comparing the cutting surface of their tools with marks found at the scene (Chapter 5).

Case Study: The Lindbergh kidnapping case

Perhaps the most famous case in which wood analysis was central to the prosecution was that of Bruno Hauptman who was arrested for the kidnap and murder of the son of the American aviator Charles Linbergh. The crime was committed with the aid of a crude homemade wooden ladder that was left behind at the time of the kidnap. By comparing the milling marks left on the various components of the ladder it was possible to trace each of them back to their original supplier. It was also noted that one of the ladder rails was made from low grade unweathered wood, such as that used inside a house or building, and it had four distinctive square shaped nail holes. Bruno Hauptman was initially arrested when he used money from the kidnap ransom; when police searched his property they discovered that one of floorboards in the attic was eight foot shorter than the others and that the four square nail holes found in the ladder rail mentioned above were a match for the holes in one of the attic floor joists. Furthermore, the annual growth rings of the wood used in this ladder rail matched those of the short floorboard. Within Bruno Hauptman's workshop the police also found a hand plane with dull damaged blade that produced marks identical to those found on the ladder. This, together with other evidence, led to Bruno Hauptman being found guilty and he was executed in 1936.

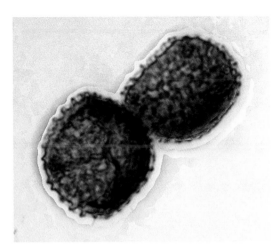

Figure 9.6 Spores of the moss, *Lycopodium clavatum*. The spores are 35–45 microns in size.

Spores

Primitive plants, such as mosses (bryophytes) and ferns (pteridophytes) do not produce seeds but they do form spores (Fig. 9.6). These spores are very small and are shed in vast numbers to be dispersed by air currents. If a spore lands in a suitable spot, it germinates and produces a gametophyte – the structure in a plant's life cycle that produces gametes.

Lycopodium spores and condoms

Spores of the club moss, *Lycopodium clavatum* are commonly used as a lubricant during the manufacture of condoms to prevent them sticking to themselves when they are rolled up; other plant products such as potato starch and corn starch are also used for the same purpose. It has therefore been suggested that the presence of these substances in vaginal swabs can be used as forensic indicators in cases of sexual assault in which the man wore a condom (Blackledge, 2005; Berkefeld, 1993); it is reported that lycopodium spores can be detected in vaginal swabs up to 4 days after sex has taken place (Keil *et al.*, 2003). This information could prove useful in a number of situations. Firstly, although not a foolproof method of avoiding leaving traces of semen, the wearing of a condom reduces the likelihood considerably. Consequently, in cases in which there is a dispute whether a sexual act took place but there is no DNA evidence, the presence of lycopodium spores or other plant evidence in a vaginal swab would indicate that the woman had had sex in the recent past with a man wearing a condom and it would therefore be worthwhile pursuing the investigation. This would also be the case if a dead woman was found and sexual assault was suspected but semen could not be detected and medical evidence was inconclusive. In addition to the presence of spores and other particulate matter, condoms also leave a 'chemical signature' formed from traces of chemicals used in the their manufacture (lubricants, spermicides etc.) that identify the brand. Men are creatures of habit and a condom-using serial sex offender is therefore likely to use the same brand in most if not all of his crimes. Therefore, identifying his brand of choice can be useful in building up a profile and in the search for associative evidence.

Pollen

Pollen is the term used to describe a collection of pollen grains. These are the male gametophytes formed by seed producing plants (i.e. gymnosperms such as pine trees and other conifers, and angiosperms such as grasses, rushes, buttercups, daisies, and lime trees). The outer layer of pollen grains is called the 'exine' and is extremely resistant to degradation. This, together with the characteristic morphology of some species of pollen means that it can be identified thousands of years after it was originally produced – a feature commonly used in archaeology and palaeobiology when reconstructing the vegetation of past landscapes. The study of pollen grains and spores is known as palynology – although researchers in this area commonly also study fungal spores and microscopic (5–500 μm) plant and animal structures collectively known as palynomorphs. Palynomorphs are living or fossil structures resistant to the strong acids etc. used to extract pollen. They include organisms such as testate amoebae, dinoflagellates, and diatoms.

Pollination

In order for fertilization to occur, the pollen grains have to be transferred from the 'male' part of the plant – the anthers – where they are produced, to the 'female' part – the stigma. This transfer process is known as pollination and when it is successful the recipient plant is said to be pollinated. In most plants the transfer is brought about between the anthers of one plant to the stigma of another by the actions of the wind or of an animal. Some plants are self pollinated and in these plants pollination may be brought about by the wind, an animal, or the plant itself. The different mechanisms are reflected in the quantity and morphology of the pollen grains and this in turn affects their usefulness as forensic indicators.

Wind pollination

Wind pollination is referred to as being anemophilous (i.e. wind-loving) and plants that affect this mechanism often produce large amounts of small pollen grains that have simple morphologies and smooth surfaces (Fig. 9.7b,c,d,f). However, some anemophilous plants, such as certain pine trees produce pollen with lateral air sacs to aid their dispersal (Fig. 9.7a). Anemophilous plants release enormous amounts of pollen owing to the low chances of any individual grain reaching its target. The pollen of wind-pollinated plants and spores of some ferns (e.g. bracken, *Pteridium*

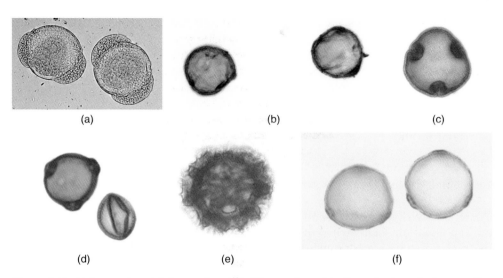

Figure 9.7 Pollen grains. (a) Scots pine tree (Pinus sylvestris). Note the lateral air sacs. Size range 55–80 μm. (b) English oak (*Quercus robur*). Typically 20.7 × 28.9 μm. (c) Lime (*Tilia vulgaris*). Typically 23.7 × 31.0 μm. (d) Birch (*Betula pendula*). Typically 19 × 22 μm. (e) Dandelion (*Taraxacum vulgare*). Typically 21.0 × 22.3 μm. (f) Cannabis (*Cannabis sativa*). Typically ~25 μm.

aquilinum), are formed in such huge numbers that any object within the general area will be contaminated with what is sometimes called the 'pollen rain'. Pollen from wind-pollinated plants is also breathed in and can be recovered from the lungs and is also swallowed and found in the gastrointestinal tract. Cannabis plants (*Cannabis sativa*) are wind pollinated and if grown indoors their pollen (Fig. 9.7f) will be found on every surface in the area. Consequently, evidence of past presence remains long after the plants are removed. Even if the grow room is thoroughly cleaned, traces of pollen can be recovered from cracks and crevices, air extractors, the tops of doors etc. Similarly, the clothes of anyone working with cannabis plants will be contaminated with the pollen. Cannabis products derived from plants grown outside will contain both cannabis pollen and that of other species of plants found in the area. This can provide an indication of the region or country of origin. By contrast, products derived from plants grown indoors in a well sealed growth room will be free of pollen from other plants (Stanley, 1992).

Animal pollination

Pollination brought about by insects or other animals is referred to as zoophilous (animal loving) or zoogamous – those specifically pollinated by insects are referred to as entomophilous (insect loving). Zoophilous plants tend to produce fewer pollen grains than anemophilous plants because it is less of a random process. Sometimes zoophilous pollen grains are covered with projections or sticky secretions that facilitate their attachment to the pollinator and this also facilitates their identification. Dandelions (*Taraxacum* spp.) are unusual plants in that, although most of them produce flowers that are attractive to insects and pollen that is morphologically suited to insect pollination (Fig. 9.7e), many of them reproduce asexually (Meirmans *et al.*, 2006).

Because they are produced in small numbers and not released into the atmosphere, zoophilous pollen tends to fall close to the parent plant or be transferred onto clothing etc. as a result of direct contact. The zoophilous pollen profile of soil or a forensic sample is therefore a good indication of the plant species composition at a particular locality.

Self-pollination

There are two types of self-pollination: autogamous and cleistogamous. Autogamous pollination occurs when the anthers and stigma develop at the same time and the pollen is transferred from the anthers to the stigma by the wind or an insect depending upon the plant species. Autogamous plants, such as the tomato, produce very small amounts of pollen. Cleistogamous plants produce flowers that never open and the pollen is transferred through the growth of the plant. Consequently, these plants produce very small amounts of pollen that lacks ornamentation and is only found in very close proximity to the parent plant. Examples include cereal grasses such as wheat (Fig. 9.8a).

Pollen dispersal

When released from plants, pollen varies in its dispersal capabilities and this can be measured as its rate of fall or 'sinking speed'. This is influenced by a combination of biological factors, such as the size and shape of the pollen and the height from which it is released, environmental factors such as the weather, and local topographical features such the presence of hedges or valleys. Maize (*Zea mays*) pollen is relatively large (90–125 μm) (Fig. 9.8c) compared with that of other wind-

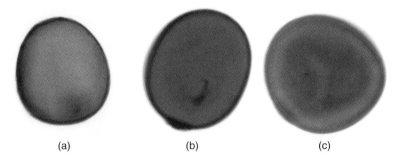

(a)   (b)   (c)

Figure 9.8 Pollen grains of Grammineae (grasses). (a) Wheat (*Triticum aestivum*). The grains lack ornamentation and are usually 42.6–61 μm in width. (b) Wild oats (*Avena fatua*). The pollen grains are extremely similar in size and appearance to those of wheat. (c) Maize (corn) (*Zea mays*). Maize produces larger pollen grains than any other grass taxon with sizes ranging from 76 to 106 μm.

pollinated grasses (usually ~40 μm) and has a fast sinking speed so the presence of large amounts in or on an item would suggest that it had spent some time situated close to a flowering maize crop. By contrast, birch (*Betula* spp.) pollen is much smaller (20–30 μm) (Fig. 9.7d) with a slow sinking speed and therefore its pollen gets distributed more widely. Many plants release their pollen over restricted time periods, so its presence can act as an indicator of the time of exposure. For example, maize pollen is released over a period of 2–14 days, although 5–8 days is more usual. Wind-pollinated plants also tend to release most of their pollen in the early morning when upward air currents are most pronounced.

Zoophilous pollen tends to fall close to the parent plant following the death of the flowers. For example, foxglove (*Digitalis pururea*) pollen is usually only found very close to the parent plant (Wiltshire, 2006b). The presence of zoophilous pollen on a person or their clothing therefore usually requires direct contact to be made with the plant or the soil in its vicinity. For example, a person may breathe in the air very close to a flower, handle a bouquet of flowers (lilies are notorious for staining clothing with their pollen) or lie on the ground or walk through a field where flowers are growing. Similarly, the presence of cleistogamous pollen on clothing etc. usually requires close contact to be made to effect transfer.

Once released and settled onto soil or an object, pollen is subject to a range of environmental factors that can redistribute it within and between habitats. For example, the pollen profile can be affected by the type of soil, digging, ploughing, differential decay of pollen, and the pollen can also be washed away by heavy rain and redeposited elsewhere.

Pollen as a forensic indicator

Because many types of pollen and spores can be readily identified, are produced at specific times of year and usually well preserved, they have forensic potential as means of associating people, animals or objects with a locality at a particular time (Mildenhall *et al.*, 2006). For example, by isolating the pollen from a person's clothing this could provide an indication of their country of origin or the country/countries through which they have passed. This is potentially useful when a person without a passport presents himself at a border crossing and attempts to claim asylum or it is believed that he may have attended a particular country or region to indulge in training in terrorist activities or to commit a crime.

Pollen analysis does, however, have its problems. It can be difficult to identify some types of pollen, especially that of grasses, to species level. Therefore, there can be problems with taxonomic resolution. Furthermore, some types of pollen degrade faster than others – for example, oak (*Quercus* spp) pollen degrades faster than lime (*Tilia* spp.). In addition, in the case of long lived plants such as trees, it may take several years for them to reach sufficient maturity to flower. Therefore their presence will not be indicated from the pollen profile until they mature. Consequently, the pollen profile in a forensic sample may not accurately represent the plant species composition of the site at which it was acquired. In general pollen profiling is most effective for identifying a location if the pollen includes examples of rare plants or unusual combinations of plant species. The analysis of pollen profiles obtained from samples taken from a body or clothing that were acquired from contact with soil etc. presents particular problems. This is because the pollen profile would include pollen released over several years and therefore its composition would be affected by a host of variables such as differential resistance to decay and redistribution etc.

Case Study: The torso in the Thames case

In September 2001, the headless, limbless torso of a 4–6-year-old boy was recovered from the banks of the River Thames, sparking one of the most detailed forensic investigations in recent years – and one in which plant evidence has proved of crucial importance in guiding the investigation. In the absence of any identifying features, detectives christened the boy 'Adam' – which helped those working on the case and kept the victim's international media profile high. A combination of stable isotope analysis and DNA studies indicated that the boy probably originated from a region of Nigeria. The next fact to be determined was whether the boy was dead before he reached the UK and if not, how long he remained alive once he got here. The child's stomach and upper intestines were empty, indicating that he had not eaten for some days before he was killed. However, it was possible to extract pollen grains from his lower intestine and these included those of plants typical of the UK and North West Europe, such as Alder, suggesting that he had been alive in the UK and therefore breathing the pollen in before he was killed. Similar pollen could not be extracted from supermarket foods indicating that the pollen was breathed in rather than consumed in the food. It takes about 72 hours for food to pass completely through the gut, so the presence of food (and pollen) in the lower intestine suggested that the boy had been alive here for about 48–72 hours. Also present in the lower intestine were fragments of the outer seed layers of the Calabar Bean (*Physostigma venenosum*). The beans of this plant are extremely toxic as they contain physostigmine (eserine) that is a potent inhibitor of cholinergic nervous transmission; a single ripe bean is capable of killing a man. It is thought that the bean(s) may have been administered to the boy to sedate him before he was killed. Calabar beans used to be employed in West African rituals in which a person suspected of a crime was forced to eat the bean: if the person died he was considered guilty, if he vomited the bean he was innocent. The case has not yet been solved and one line of investigation is the possibility that Adam was the victim of a 'muti killing' in which humans are sacrificed to propitiate spirits and to bring good luck.

Case Study: Pollen and honey fraud

Pollen may become associated with food, the most obvious example being honey (Fig. 9.9). There is an active trade in the sale of mislabelled honey in the European Union, particularly since honey from China was banned in 2002 owing to concerns about the levels of antibiotics such as chloramphenicol that were being used by beekeepers to prevent diseases among their bees. Following the ban, it is alleged that Chinese honey was smuggled into India where it was repackaged and sold on. Suspicions have also been raised about how Singapore somehow raised its honey production from virtually zero to become the world's fourth largest honey exporter at exactly the same time that Chinese honey was banned. The problem of honey containing antibiotics is by no means confined to China and they have been found in honeys from many other countries as well. The EU ban on the import of Chinese honey has been rescinded but there are still concerns about mislabelling. The presence of pollen in honey can help identify its country of origin and, in the case of single flower honey, whether this is a true reflection of its composition. For example, in July 2004 a Northumbrian beekeeper was successfully prosecuted for marketing honey as being produced in the Scottish Borders when it actually originated in Argentina. There are therefore concerns about the production of ultra-filtered honey because although this treatment would remove contaminants such as antibiotics, it would also remove the pollen that facilitates the honey's identification. Because honey is marketed as a 'natural' wholesome food there are concerns about it becoming accidentally contaminated with pollen from genetically modified (GM) crops and this will no doubt become a source of legal dispute as GM crops become widely grown.

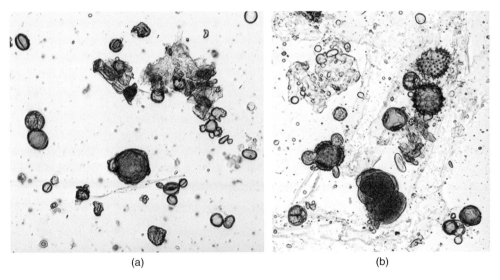

(a) (b)

Figure 9.9 Pollen analysis can indicate whether a honey's source is correctly labelled. Note the differences between these two samples. (a) Armenian wild flower honey. (b) Greek 'wild flower honey from herbs and coniferous trees'.

Case Study: Pollen as a forensic indicator in the Srebrenica massacre

During the 1990s, the Republic of Yugoslavia disintegrated as several of its regions declared independence and this resulted in a series of vicious civil wars. During the fighting, atrocities were committed by many of those involved and thousands of people were killed. One of the most notorious massacres occurred in the summer of 1995 when the town of Srebrenica fell to the Bosnian Serb army Drina corps, which was under the command of General Radislav Krstic. Subsequently, about 7000 men were led away and executed at various sites near the town and their bodies buried in seven mass graves. Some of the victims may have been buried alive. After about 3 months, the killers attempted to dispose of the evidence by digging up these 'primary' graves and dispersing the bodies among numerous 'secondary burial sites' many miles away. In this way it was hoped that the deaths would appear to be a consequence of small local skirmishes rather than a carefully planned massacre. However, a few years later, once a semblance of peace had been restored, attempts were made to bring those responsible to justice at the International Court of Human Rights at The Hague. As part of the investigations, forensic scientists compared the soils at five primary burial sites with those at 19 secondary burial sites where the bodies were found along with samples taken from the clothing and human remains and the soil surrounding them (Brown, 2006). From this they were able to prove conclusively that the majority of the bodies at the secondary sites did not die in the locality and had almost certainly been moved from the primary sites. For example, clumps of matted hay were found among some of the remains but the local vegetation indicated that this could not have arrived there as in-fill from the land surrounding the (secondary) graves. In addition, the mats of hay contained spent cartridges suggesting that the men were killed at a site where hay was being gathered or stored. Eyewitness statements indicated that many of the victims were killed at an agricultural warehouse at Kravica [referred to as Krevice in Brown (2006)] – a small village near Srebrenica. Following the massacre there were so many dead bodies at the warehouse that they were removed and buried using a bulldozer and in so doing clumps of straw and spent cartridges were buried along with the bodies. Other eyewitness accounts indicated that many men were also killed in a wheat field at about the time of harvest. Samples associated with the soils within and surrounding the victims in several of the secondary grave sites yielded large amounts of cereal pollen of the *Avena–Triticum* type and assumed to be wheat [the pollen of cultivated species of *Avena* (e.g. oats) and *Triticum* (e.g. wheat) cannot be distinguished] along with smaller quantities of maize pollen. Again, this did not match the surrounding vegetation or soil pollen profiles and as reported previously, cereals such as wheat are cleistogamous whilst maize pollen has a fast sinking speed. Therefore, the presence of large amounts of their pollen could only indicate close association and not the deposition of wind-borne grains from some distance away. This evidence, together with mineralogical analysis of the soils, indicated that many of the bodies recovered from the secondary grave sites were initially buried elsewhere and that they could be linked back to the initial primary burial sites.

Collection of pollen and other palynomorph evidence

Because pollen and other palynomorphs are so easily transferred, contamination of forensic samples needs to be rigorously guarded against. For example, ideally, samples for pollen analysis should be taken at the crime scene before other investigators disturb surrounding vegetation or, in the case of rooms, any windows are opened or excessive air currents are caused by lots of people walking around and opening doors. Similarly, if the case involves the death of someone, relatives and friends should not be permitted to leave floral tributes nearby until the specimens have been collected. Extreme care should also be taken in the laboratory to avoid cross-contamination and analysis should ideally take place in a semi-sealed room with all windows and doors closed and the sample containers should be opened and processed within a laminar flow cabinet.

Pollen may be recovered from both within and on the outer surfaces of a body, from clothing and from objects such as spades, furniture and vehicles but on each of these there are certain areas that are most likely to yield useful information. Our nasal passageways are designed to filter out suspended particulate matter from the air before it reaches the lungs and therefore nasal washings can yield not only pollen but also fibres and mineral dust particles that can also provide evidence. However, once a body starts to decay or it is buried it becomes difficult to be sure that the material retrieved is a true reflection of that breathed in or a subsequent contaminant. To avoid this, Wiltshire & Black (2006) describe a dissection method to reach the region called the cribiform plate from which the nasal passages can be washed out with minimal risk of contamination. Clearly, one cannot experiment with such techniques on humans but it would be possible to demonstrate the extraction of pollen etc. from nasal washings and relate these to environment using the heads of rabbits, sheep or pigs obtained from a friendly butcher. Pollen can also be extracted from the stomach and other parts of the gastrointestinal tract whilst on the outer body surfaces the hair and underneath the fingernails are good sources. Hairsprays and oils can make the hair sticky and help trap pollen grains. On clothing and shoes, any muddy patches are obvious places to sample. However, even if stained regions cannot be seen by the naked eye, it is still possible to extract pollen and other palynomorphs, bearing in mind the nature of the case being investigated. For example, the location of palynomorph evidence on an aggressor and their victim may be different and indicate how the assault progressed. In cases of female sexual assault that occur out of doors, the male assailant is likely to have evidence on the elbows, forearms and knees of his clothing and on the front portion of his shoes. His victim is likely to have a matching pattern of evidence on the back of her clothing and the heels of her shoes or feet. The type of clothing/footwear can strongly influence one's ability to recover pollen samples. For example, some synthetic fleece fabrics are good at picking up and retaining pollen whilst dressed cotton is rather poor (Wiltshire, 2006b). When comparing the abilities of different materials to gather pollen one should be aware that poor retrieval may be a consequence of the pollen remaining stuck within the clothing rather than not being attached in the first place. The complex patterns found on the soles of many trainers are said to be good at retaining soil and plant-based evidence. The seats, foot mats, wheel arches and air filters are good places to search for evidence in vehicles.

When attempting to link forensic samples to a particular crime scene – for example, soil samples on a spade to a grave site – it is essential that sufficient comparator (control) samples are taken from around the crime scene. This is because, as we have already seen, a wide variety of factors can affect the pollen profile and therefore the profile can vary considerably over relatively short distances. For example, even within a small area dandelion (*Taraxacum officinale*) pollen shows a very localized distribution. One way of overcoming this is to take pinch samples of small amounts of surface soil from different points within a sample area and then to combine these together for analysis. Each sample area (comparator sample) would therefore be represented by a combination of pinch samples (Wiltshire, 2006b). Precisely how many pinch samples should be combined to produce a single comparator sample and how many comparator samples should be taken would depend upon the nature of the crime scene and the questions that need to be addressed. It may not be appropriate to take comparator samples at random from around a crime scene but to focus on particular sites (Mildenhall *et al.*, 2006). For example, if the crime scene is a clearing in dense undergrowth it may be more appropriate to analyse soil samples along all the potential access routes rather than randomly 360° around the scene. If in doubt it is always better to err on the side of caution and to take more comparator samples than are needed. It is quicker and cheaper to take many samples and not analyse all of them than to have to return at a later date when there is the added risk of the pollen profile having changed. Samples should be double-bagged (e.g. a sealed container within a zip-locked plastic bag) thereby reducing the risk of the sample leaking or becoming contaminated. Those samples that are not going to be analysed immediately should be either frozen or dried to reduce the risk of bacterial or fungal growth. When collecting comparator samples it is important to also make a vegetation survey of the locality at the same time. This is because the pollen profile may not be an accurate reflection of the vegetation present at the locality. Ideally, samples of all the plants should be taken to confirm identification because some plants are not easy to recognize in the field and whilst photographs can be useful they may not have the resolution or show the key features that enable accurate identification. The number and size of evidential samples, such as those retrieved from under the fingernails of a corpse or soil found on a spade, is usually outside the control of the investigator but where multiple samples are possible the opportunity should be taken.

Although pollen samples should be amongst the first samples taken from a crime scene they may be amongst the last to be analysed. This is because the extraction of pollen grains from soil, clothing, fingernail scrapings etc. involves the use of strong alkalis and acids. Consequently, it would be impossible to obtain DNA from the samples after pollen analysis and the analysis of many other indicators (e.g. mineral content and fibres) would also be compromised. The procedure for separating out pollen grains varies with the nature of the specimen (Horrocks, 2004) – for example, whether it is a soil sample or an item of clothing. However, it usually consists of initial treatment with hot 10% w/v potassium hydroxide to dissolve any humic material and generally break up the sample matrix – this process is called deflocculation. In the case of soil samples, this is usually followed by treatment with hydrochloric acid, to remove carbonates, and then hydrofluoric acid, to remove silica. Cellulose is removed by treating the samples with glacial acetic acid in a process some authors refer to as acetylation and others as acetolysis, whilst lignin and other organic material is removed using a bleaching agent. After processing is completed the pollen is stained using 0.1% basic fuschin or 0.5% safranine and mounted onto glass microscope slides. Many workers mount their specimens in silicone oil but for forensic investigations it is necessary to make permanent preparations so that others can verify one's observations. As can be seen, the whole process involves several dangerous chemicals, especially hydrofluoric acid, and therefore requires a dedicated laboratory and a great deal of care.

As with all forensic samples, palynological evidence should be stored appropriately, carefully documented and a chain of evidence established. If it is envisaged that samples will be subjected to pollen analysis in the future it is important to guard against the possibility that contamination may occur between the time of collection and the time of analysis. For example, unless precautions are taken it is very easy for pollen grains to contaminate a sample as soon as the bag or container holding it is opened. Consequently, a record needs to be made of whenever the container is opened, by whom and it what circumstances.

Identification of pollen grains

Pollen is identified on the basis of its morphological characteristics (e.g. Hyde & Adams, 1958). This is usually done with the aid of a stage microscope and although a scanning electron microscope can provide finer detail it is not suitable for routine examination involving numerous samples. Unfortunately, even pollen grains released by a single anther can vary markedly in size and the extraction and slide preparation techniques can also influence pollen size. Consequently, pollen size cannot be used by itself as a reliable means of identification (Mildenhall *et al.*, 2006). The pollen cytoplasm usually decays within a short period of time of the grain's release and therefore DNA is unlikely to help in the identification process unless the grains were preserved in ideal circumstances (e.g. very cold or very dry conditions). Although textbook guides are available for the identification of pollen grains, to be truly reliable one needs to be able to compare the specimens with those in a reference collection of known plant species.

Interpreting pollen profiles

It is possible to express the contents of pollen and other palynomorphs within honey, a soil sample or some similar matrix as a proportion of the sample's volume or weight. This can be further refined by adding known amounts of exotic marker pollen or spores to the sample at the time of extraction and thereby enabling one to determine the recovery rate (e.g. Piana *et al.*, 2006). However, this approach is not feasible when the pollen etc. is extracted from most forensic evidential samples – such as clothing or a shoe. Indeed, the wide variety of forensic samples makes it virtually impossible to standardize protocols. Instead, the different components of the pollen profile should be expressed as percentages of the total pollen/other palynomorphs present in the sample. Usually, one would count a total of around 300 pollen grains from a sample although the number would depend upon its nature and the degree of resolution required. For example, the more grains that are counted the greater the chance of finding an extremely rare type of pollen grain but this has to be balanced against the time taken to count and identify perhaps thousands of grains. The species composition and their proportions within the samples can then be compared. There appears to be little consensus as to the most appropriate statistical method for comparing the similarity of pollen profiles in a forensic context. Although one statistical approach is to use Likelihood Ratio Analysis, this assumes that pollen is dispersed randomly and as we have seen, in many plant species their pollen exhibits a clumped distribution and environmental factors can cause pollen to be redistributed and concentrated in particular areas. Indeed, because the pollen profile is affected by so many variables it is impossible to obtain an exact match between the composition of evidential and comparator samples. One must therefore be aware of the biological and environmental factors specific to the individual case when comparing pollen profiles and consider the possible reasons why discrepancies may occur. Sometimes, it can be the presence of minor components that are most informative. For example, if similar proportions of pollen from a rare plant species are found in two samples. According to Mildenhall *et al.* (2006) the degree to which an association between a suspect/evidential sample and a crime scene can be inferred from pollen profiles is best expressed as: not supporting, inconclusive, weakly supporting, supporting, strongly supporting, very strongly supporting and conclusively proving. However, given what has been said the lattermost scenario would be a remarkably unusual occurrence. The evidence obtained from pollen profile analysis in most circumstances is therefore best considered as being circumstantial and of use in supporting other forensic tests rather than being sufficient on its own to confirm an association.

The identification of a particular pollen grain within the pollen profile of a sample indicates that the parent plant was somewhere in the vicinity. For example, a person trod on soil near to the plant or breathed in the air near to it when it was in flower. However, the absence of a particular pollen grain does not mean that the plant was not present in the vicinity. This may be because the plant's pollen was present at extremely low levels and therefore one had not analysed sufficient samples in sufficient detail or that the plant had not yet reached sufficient maturity to begin flowering or for a whole variety of environmental factors. Consequently, attempting to build a habitat survey on the basis of pollen retrieved from a small evidential sample is inherently risky. This sort of circumstance might arise, for example, if it is believed that a muddy spade retrieved from a shed was used to dig a grave but it is not known where that grave is and the suspect is not cooperating with the police. Wiltshire (2006b) relates the case of a murder victim found buried in a shallow grave among a dense Sitka spruce (*Picea sitchensis*) plantation in which the most notable plant contributing to the pollen profile within the grave and the surrounding region was not Sitka spruce but pine. The pine pollen was probably derived from the combination of that released from a single mature tree 100 metres from the grave and that remaining in the soil from the time the area was cleared to make way for the spruce plantation. The spruce trees would be all planted at about the same time and therefore although they were the dominant vegetation they would be of a similar age and not yet mature enough to produce pollen. Consequently, if the area had been sought on the basis of the pollen profile one would have searched for a pine forest not a Sitka spruce plantation. However, despite this proviso, in the absence

Table 9.1 Summary of advantages and limitations of pollen as a forensic indicator

Advantages
  (1) The combination of small size, abundance and wide variety of size and shape makes it simple to obtain a statistically valid sample from a very small amount of material. For example, 1 cm³ of soil is often sufficient and smaller sample sizes can still be useful.
  (2) Many types of pollen are resistant to decay and will last for years even under damp conditions.
  (3) Some plants are rare and/or have a restricted distribution associated with a particular habitat or country and therefore identifying their pollen can provide a reliable habitat/geographical location.
Limitations
  (1) Taxonomic resolution is often limited and it is impossible to identify the pollen of some plants to species level.
  (2) Plant species vary enormously in the amounts of pollen they produce and the extent to which this is dispersed over the surrounding area.
  (3) The pollen of some plant species decomposes faster than that of others.
  (4) After landing pollen may be redistributed within and between habitats by a wide variety of factors (e.g. flooding, digging).
  (5) The pollen profile on a forensic sample such as the sole of a shoe could represent the cumulative depositions from many different localities.

of any other evidence the pollen profile retrieved from a forensic sample does provide leads upon which an investigation can proceed.

Fruit, seeds and leaves as indicators of diet

The fruits, seeds and leaves of many plants form part of our diet. We are unable to digest the cellulose that makes up the walls of plant cells or the tough outer coating of many seeds, so these structures pass through the gut and may be identified from their characteristic morphology. Consequently an analysis of a person's stomach contents and/or faeces can provide evidence of what they were eating and therefore, potentially, where they ate it. For example, in an American case in which a young woman was murdered, some witnesses claimed that her last meal was a lunch of burger with onions and pickles that she shared with her boyfriend at a McDonald's fast food restaurant. However, when her body was found, her stomach contents included a much larger proportion of vegetable matter than this would have provided. The police therefore interviewed staff at a local Wendy's salad bar where a waitress remembered seeing the woman leave with another man after dinner. The state of degradation of the digestible components and their position within the gastrointestinal tract can provide a crude indication of how long it was since the person ate that meal but the rate of stomach emptying and passage of digesta through the gastrointestinal tract is affected by numerous factors. For example, the presence of large amounts of lipid and protein (such as that present in the traditional English breakfast of bacon and eggs) will slow gastric emptying as can medical conditions such as diabetes. Furthermore, there can be considerable variation between normal healthy subjects in the rate of transfer of food between regions of the stomach. Plant matter may be found in the body outside the gut. For example, aspiration (breathing in) of the gut contents during vomiting is a fairly common occurrence in persons suffering acute alcohol intoxication and drug overdoses and may prove fatal. In these cases, vegetable matter may be observed in the airways of the lungs either macroscopically or in histological sections.

Flowers, fruits and seeds as indicators of locality and time

Many trees produce flowers only once a year and these are often lost in large numbers over a very short period of time. The petals (e.g. cherry tree [*Prunus avium*] blossom) or catkins (e.g. birch [*Betula* spp.]) decay soon after they are shed but their presence and freshness either beneath or on top of an object (e.g. a body or vehicle) can indicate how long it has remained in that position (Fig. 9.10). Plant fruits and seeds are usually relatively large, produced in small quantities, and apart from those that are designed for aerial dispersal (e.g. dandelion seeds), they tend to fall close to their parent plant. Consequently, there is a lower risk of contamination with fruits and seeds than with pollen, and it is also possible to extract DNA from them to confirm identification and also to subject them to chemical analysis. This can be particularly important in cases of suspected fraud when attempting to determine country of origin or if contamination with GM crops is suspected. Many plants only flower once a year and they form their fruits and seeds over a restricted time period so their presence can provide good temporal evidence. Sometimes the seeds

Figure 9.10 The petals and anthers on the sole of this shoe indicate that the wearer walked underneath a laburnum tree (*Laburnum anagyroides*) and their freshness indicates that this happened only a few hours before the photograph was taken. Note how the complex tread of the trainer has retained large amounts of material. Plant evidence is seldom as obvious as this! Note also the presence of a white cat's hair and that analysis of soil particles from the shoe would provide further evidence of where this person had been walking.

may be eaten but the seedcases remain (e.g. beech mast). For example, if a body is found near a beech tree (*Fagus sylvatica*) and has mature beech mast on top of it, this would suggest that the body arrived beneath the tree before autumn: the time at which mast falls will depend on the weather and the locality. Obviously, the state of the body would need to be considered as well; if the body was skeletonized it could be several years old and therefore have accumulated more than one year's fall of mast. Many fruits are designed to be eaten by animals as part of the plant's seed dispersal strategy and they rot quickly. However, seeds tend to be very resistant to decay and many of them can survive for months in damp conditions and millennia if the conditions are dry and cold. Therefore, they can remain in the environment for a long time – and therefore, potentially, be a source of forensic evidence.

Fruits and seeds can provide good evidence linking a person, animal or object to a locality and/or a particular time of year. However, like pollen, they are most useful if they include examples of rare plants or those associated with a limited distribution or particular habitat. The large size of fruits and seeds can be an advantage as a forensic indicator but it can also restrict the number of circumstances under which they can be useful. For example, unlike pollen, it is not common to find fruits or seeds among a person's hair or under their fingernails. Furthermore, because they are not produced in large quantities, it may be difficult to find sufficient of them to produce a statistically viable sample size.

Case Study: Goosegrass fruit as a forensic indicator

A number of plants, such as goosegrass (*Galium aparine*) (Fig. 9.11), also known as 'cleavers' or 'sticky willy', wild carrot (*Daucus carota*), and lesser burdock (*Arctium minus*) disperse by producing seeds that are armed with hooks designed to catch onto the coat of passing animals. They also latch onto clothing and hair and can be extremely difficult to remove. They therefore have forensic potential in associating a person with a locality and/or their activities, especially if a reliable DNA match can be made. An interesting case has been described in Germany (http://www.strate.net/e/publications/weimar.php3) in which the fully clothed body of a young girl was found in undergrowth: she had a regular distribution of goosegrass fruit over her whole body, including the inside of her trousers and on her underwear amounting to 362 fruit in total. The number and distribution suggested that the girl did not pick up the fruit whilst playing – in which case they would have been concentrated on her legs. Goosegrass tends to grow among nettles so it is unlikely that any child would roll around on the ground where they occur. The presence of the fruit inside the trousers was considered suspicious because they are itchy and the normal reaction would be to stop whatever you are doing and remove them. This, in conjunction with other evidence, led the investigators to believe that the girl was not dressed in the trousers she was found in at the time of her death. Instead, they think that the dead girl was transported to a site close to where she was found dressed only in her underwear. There she was laid down on the ground and dressed – a trampled area where goosegrass was growing was identified as the likely spot – and this would explain the even distribution of goosegrass fruit and the presence of the fruit inside the clothes.

Figure 9.11 Goose grass (*Galium aparine*). (a) Plants growing among nettles. (b) Mature fruits showing the hairs that enable them to latch onto animal hairs and clothing. (c) Spines on the leaves and stems capture hairs and fibres from passing animals and humans.

Collection and preservation of seeds, fruits and leaves

Fresh leaves and other vegetable matter would rapidly rot if placed in the sealed plastic bags normally used for forensic samples and should therefore be put into individual paper bags that are then labelled to identify the person collecting the evidence, the date and an exhibit number. Fleshy fruits should be preserved in a fluid fixative, such as 70% w/v alcohol whilst dry fruits and seeds may be kept in small paper envelopes or sealed bottles. Leaves may be preserved by drying and pressing in a plant press before being attached to herbarium sheets for protection and storage. Alternatively, the material can be stored in a fluid fixative, as described above, although this usually leads to a loss of colour as the pigments are bleached from the sample. However, it does reduce the risk of insect and fungal damage and of contamination during storage. Faeces and gut contents should be collected into marked wide-mouthed containers and, ideally, analysed for plant material soon after collection. It can also be helpful to preserve some of the samples in 10% v/v formol–saline should further analyses be required. If the samples are extremely fluid, they should be centrifuged beforehand to concentrate the material.

Identification of plants

Botanical specimens are normally identified by traditional taxonomic techniques based on morphological features. Unfortunately, some plant species are notoriously difficult to distinguish from one another and forensic samples may consist of plant fragments that do not provide sufficient features for a reliable identification to species level. Consequently, there is increasing interest in the use of molecular techniques for plant identification. However, the techniques used successfully in molecular animal taxonomy cannot always be employed with plants. For example, in animals, the sequence of the gene coding for mitochondrial cytochrome c oxidase subunit 1 (COI) has been identified as possessing the necessary combination of conservation and divergence to facilitate discrimination between species but in higher plants this gene exhibits a comparatively slow rate of evolution that limits its suitability. Kress *et al.* (2005) have suggested two alternative regions: the nuclear internal transcribed spacer region and the plastid trnH-psbA intergenic spacer region. Using these two regions they identified sequences that could be used to barcode 99 species of plants that encompassed 80 genera and 53 families. Similarly, Ward, *et al.* (2005) have developed a series of PCR-based assays for the identification of grasses. The use of molecular techniques for the analysis of plant material in a variety of forensic contexts is reviewed in Coyle (2004).

Plant secondary metabolites as sources of drugs and poisons

Plants produce numerous chemicals that are not essential for basic growth and development that are collectively known as secondary metabolites. They help protect the plant from diseases, have structural functions, or are toxic and protect the leaves, seeds etc. from being eaten by herbivores. Plants such as hemlock (*Conium maculatum*) and foxgloves (*Digitalis purpurea*) are notorious for their poisonous properties. Even food plants can be poisonous if grown or stored under sub-optimum conditions. For example, when potato tubers (*Solanum tuberosum*) are exposed to light and turn green, they form large quantities of solanine (an alkaloidal glycoside) that is resistant to normal cooking and potentially lethal. However, green potatoes have a bitter taste and it is unlikely that a person would consume a large quantity. The use of plant toxins to poison someone intentionally is rare in the UK although accidental poisonings are more common – often the result of children consuming the berries of mistletoe (*Viscum album*), yew (*Taxus baccata*) and the seeds of laburnum (*Laburnum anagyroides*): these are seldom fatal.

Drugs are sometimes added to food either intentionally or unintentionally which may have unforeseen consequences. For example, the consumption of poppy seeds used in the baking of poppy cakes, bagels etc. can result in high levels morphine and codeine in the urine that could be mistaken for drug abuse (Selavka, 1991). This has relevance owing to the increasing use of work place drug testing, especially in the oil and transport industry. The only sure way of distinguishing between poppy seed consumption and heroin abuse is through blood tests as these are more specific. The addition of cannabis to cakes is a common practice among recreational drug users but is sometimes done with malicious intent and the cakes presented to unsus-pecting recipients as a gift. Drug intoxication can lead to accidents, especially when driving and the possibility that a person may have been the victim of 'spiked' food needs to be considered.

The 'suicide tree'

In some parts of the world plant poisoning is more common and in a study in the south-western state of Kerala in India it was estimated that 10% of all fatal poison-ings were owing to consumption of extracts prepared from the seeds of the tree *Cerbera odollam* (Gaillard *et al.*, 2004) – sometimes referred to as the 'suicide tree'. The extract contains cerberin, a highly potent toxin that is capable of inducing a fatal heart attack. At autopsy, unless the pathologist has reason to suspect poison-ing, the heart attack would be put down to natural causes. The plant extract has a bitter taste but this can be disguised with spices. Most of the victims are women and because the women often eat separately from the men in traditional households, serving separate food is made easier. Women who marry into such households are sometimes subject to enormous pressures owing to dowry disputes, the demand of exacting standards of behaviour and the running of the household, and the inability to produce a male heir. In extreme cases the woman may be murdered ('accidentally' setting themselves on fire while preparing food is remarkably common) or commit suicide. It now appears that poisoning – either homicidal or suicidal – may be far more widespread than previously thought and it is possible that they may be occur-ring in Asian communities living elsewhere in the world where pathologists would be even less likely to suspect that cerberin played a part in a person's death.

Ricin

Ricin is an extremely poisonous glycoprotein that can be isolated from the seeds of the castor bean plant *Ricinus communis* for which there is no antidote. The seeds are widely available and therefore there is considerable concern over its potential use for bioterrorism. Properly prepared castor bean oil and pomace, the castor bean seed cake used as an animal feed, are both harmless because they are processed to remove the toxins. The whole seeds are unlikely to pose a hazard because their coat is hard and impermeable thereby preventing the toxin from diffusing out and being absorbed. There is some uncertainty about the risk posed by swallowing the beans and this probably reflects variations in the toxin content and the extent (if any) to which the beans are first chewed (obviously, chewing increases the chances of the toxin being released). There is probably little risk of ricin being absorbed through the skin when handling the seeds although any dust formed through abrasion is potentially dangerous if it is breathed in. Ricin is not thought to last long in the environment and person-to-person transmission is unlikely. A number of immunologically based methods have been described for its detection (e.g. Shankar *et al.*, 2005; Shyu *et al.*, 2002).

Castor beans contain between 1 and 5% ricin and it said to be relatively easy to extract the toxin. However, without expertise and access to a suitable laboratory the risks to the person(s) extracting or handling the chemical are high. In February 2008 in Las Vegas, a man was rushed to a hospital in a critical condition and remained that way for several weeks. The cause of his illness was uncertain but a search of the motel rooms where he was staying yielded guns and an 'anarchist type textbook' along with several vials of ricin. According to local media the section of the handbook dealing with ricin was highlighted. Purified ricin could theoretically be used to contaminate food or water, or used as an aerosol (Audi *et al.*, 2005). In 1978, the Bulgarian political exile Georgi Markov was assassinated in London by being injected with ricin. As he was walking along the street he was struck on the leg by someone with an umbrella – this is a common occurrence and he thought no more about it. However, in this case the umbrella had been modified to inject a small metal sphere and within the sphere was ricin. Four days after being injected Markov died of gastroenteritis and organ failure and although ricin was never detected in his body, the coroner stated that this was almost certainly the poison that was used.

Cannabis

Among the secondary plant metabolites are various chemicals that humans have used as drugs to achieve an altered mind state. Unfortunately, this is often accompanied by addiction and damage to the user's health. Drug abuse is a major cause of criminal behaviour so, not surprisingly, the detection of drugs and drug residues often forms part of a crime scene investigation. The analysis of suspect powders, body tissues/fluids etc. for drug residues is a job for trained pharmaceutical chemists but as many of the drugs are derived from plants, these specimens would fall under the remit of a botanist or biologist. Plant taxonomy is often looked upon (incorrectly) as a rather dull subject and has been quietly dropped from many undergraduate degrees in biological sciences over the years. However, it can generate intense, acrimonious debates and even apparently simple questions, such as 'what is a species', can prove extremely difficult to answer. The taxonomy of the genus *Cannabis* is one such problem area. Some botanists consider the genus to be monotypic, that is containing only a single species, whilst others state that it is polytypic, and to contain several species or subspecies. Such distinctions were used in criminal proceedings in the past when defendants attempted to claim that they were not in possession of the illegal *Cannabis sativa* var. *indica* but another variety or subspecies. The present general consensus is that there is a single species and although it is highly variable all the varieties can cross fertilize one another. Current legislation makes attempting to avoid prosecution on the basis of a taxonomic nicety a forlorn hope. 'Skunk' is a variety of cannabis that has been developed for indoor growth and has sometimes been referred to as a quite different type of drug. However, there is little evidence that its potency is any different from normal herbal cannabis and it would be impossible to distinguish between the two using normal forensic techniques. Potency varies between individual plants, how and where the plants were grown and how long and under what conditions the cannabis was stored.

The UK laws surrounding the possession and use of cannabis plants and cannabis products are complex. For example, provided that certain strict conditions are met, such as only small numbers of plants or amounts of drug are involved and the person admits guilt, there is no need for confirmatory forensic analysis or the use of a test kit such as the Duquenois–Levine colour test: the experience of a police officer trained to recognize cannabis from its texture, physical appearance and smell is considered sufficient evidence. In more serious cases, for example the possession of large numbers of plants with intent to supply, where the defendant disputes the allegations, or the investigating officer is in any way unsure about the identification, laboratory analysis is required. Drug testing kits will prove the presence or absence of a drug but it is often useful to obtain the more detailed information that can be supplied by gas chromatography linked to mass spectrometry (GC–MS). There is currently interest in the use of DNA profiling to link samples of marijuana found on a suspect with the plants it came from. In this way it is theoretically possible to link the grower, the distributor and the user together. There are, however, potential problems because marijuana can be grown from cuttings and the resultant plants are therefore all genetically identical clones. Consequently, a match between an individual marijuana sample and an individual marijuana plant does not mean that either the plant or its grower were the source. However, there is sufficient molecular diversity to make DNA profiling a useful tool and although the techniques are still in need of refinement they could prove useful when attempting to link consortiums of growers or predict the origins of large seizures (Gilmore et al., 2007). DNA profiling is not yet applicable to cannabis resin because it contains insufficient DNA.

Under UK law, cannabis cultivation may only be undertaken under license and is either for research or medical purposes or for the commercial production of hemp – which has an extremely low psychoactive content. Illegal production of cannabis can be punished with a maximum penalty of 14 years imprisonment although the possession, sale and the use of cannabis seeds for culinary purposes is not subject to the Misuse of Drugs Act 1971.

Determining the source of drugs derived from plants

It is often necessary to determine the source of contraband drugs in order to prove importation (and therefore trafficking) and to identify the region supplying the contraband. A variety of techniques may be employed including DNA analysis, identification of insect and/or pollen contaminants and analysing the drugs for their stable isotope ratios and the presence of trace alkaloids. To be truly useful any method needs to be fast, reliable and cheap: drug seizures, unlike murders, are extremely common and therefore analytical techniques must be suitable for use on a routine basis. DNA analysis has good potential provided one has the original plant material but is no use following processing (e.g. cocaine). Measurements of organic components are often compromised because these are often affected by storage conditions and processing whilst pollen and insect analysis can be time consuming and is limited by the shortage of taxonomists capable of identifying the material.

Cocaine samples can be related to their region of origin with over 90% accuracy by analysing the ratios of ^{15}N to ^{13}C and the levels of trimethoxycocaine and truxilline (Ehleringer *et al.*, 2000). Regional differences in ^{15}N arise as a consequence of variations in the soil characteristics across the main coca growing regions of the South American Andean Ridge whilst the ^{13}C levels vary owing to local differences in the length of the wet season and overall humidity. In a similar manner, stable carbon and nitrogen isotope ratios have been shown to be useful for identifying the provenance of marijuana grown in different regions of Brazil (Shibuya *et al.*, 2007). Stable isotope analysis techniques are also proving useful in cases of food fraud such as when products are adulterated or their geographic origin is mis-stated, because it is not commercially viable to artificially adjust the isotope ratios. However, to be effective, stable isotope techniques require extensive databases of authentic reference samples and certified reference materials to monitor the accuracy of the measurements.

Illegal trade in protected plant species

The world market in the illegal trade in protected plant species is probably as widespread and lucrative as that in protected animals although it attracts far less publicity. Some plant species are now nearing extinction in the wild as a consequence of over-collection. For example, cycads are primitive plants many of which thrive in hot, dry climates that have become hugely popular for their decorative properties – and sheer rarity. Prize specimens can trade hands at over £20 000 each and an illegal trade worth millions of pounds a year is thought to operate between South Africa and North America. As a result, of world's 298 described species, over half have become endangered and some are now thought to be extinct. Specimens are often intentionally mislabelled or stated as being grown from domestic stock when in fact they were taken from the wild. Correct identification requires the services of a plant taxonomist whilst wild-collected specimens can be identified by damage caused when the plant was dug up and the typical 'wear and tear' damage that all wild plants receive, such as porcupine teeth marks in the case of South African plants. A plant's bulb and roots would also retain adherent samples of the soil from which it was obtained. This would facilitate pollen and stable isotope analysis although this would require access to suitable reference materials. However, even basic soil analysis would demonstrate whether the plant had been grown in the place it was claimed to come from.

In Britain, all native plants receive protection under the Wildlife and Countryside Act (1981) which states that it is illegal to dig up any wild plant without appropriate permission and some species, such as bluebells and snowdrops, are also protected from being picked and sold. However, this sometimes happens on a large scale and is conducted as an organized criminal activity. The act also covers algae, mosses, lichen and fungi.

Table 9.2 Summary of forensic information gained from protists, fungi and higher plants

Sample	Forensic evidence
Testate amoebae	Geographical location
	Habitat
Diatoms	Geographical location
	Habitat
	Time of year
	Cause of death (e.g. drowning)
Fungi/fungal spores	Geographical location
	Habitat
	Time of year
	Poisoning
	Illegal trade in drugs
Whole plant	Damaged plants can indicate the site of a struggle or tread marks from shoes or vehicles.
	Amount of healing can indicate time since damage occurred
	Time since burial (e.g. growth on grave site)
	Illegal trade in drugs (e.g. marijuana)
	Illegal collection and trade of protected species
Plant leaves	Geographical location
	Habitat
	Diet/last meal (e.g. fragments in gut contents or faeces)
	Illegal trade in drugs
	Illegal collection and trade of protected species
Plant roots	Time since burial (damage, repair and subsequent growth through or around skeleton or object)
Wood	Tool marks of object causing damage
	Amount of healing can indicate time since damage occurred
	Cause of death (splinters left in wound may identify murder weapon)
	Illegal trade in protected species
	Fraud (misrepresentation: object made out to be older than it really is or made from inferior wood)
Seeds	Geographical location
	Habitat
	Time of year
	Diet/last meal (in gut/faeces)
	Poisoning
	Illegal trade in drugs
	Illegal trade in protected species
	Fraud (e.g. mislabelling, contains GM material)
Pollen	Geographical location
	Habitat
	Time of year
	Poisoning (some flowers are poisonous)
	Fraud (e.g. mislabelling, contains GM material or from a different country of origin)

Future directions

The use of diatoms and other algae as well as that of fungi as forensic indicators is compromised by the problems of isolation and identification. One of the most promising ways around this would be to develop the use of microchips that allow the simultaneous identification of large numbers of organisms. This technology is already well advanced in the field of environmental microbiology and so-called 'phylochips' have been developed that can scan for numerous species of bacteria and Archaea. Phylochips that have been developed for the identification of bacteria use hundreds or even thousands of oligonucleotide probes that are immobilized to the chip surface in a high density array (e.g. DeSantis *et al.*, 2007). The probes are designed according to the 'multiple probe concept' which enables them to detect organisms at the same or different phylogenetic levels. For bacteria, the target of these probes is usually the 16S rRNA gene and specific regions of this are fragmented, amplified and fluorescently labelled before being added to the phylochip. Hybridization will occur wherever there is a 'match' (i.e. a complementary sequence) between a probe and a gene fragment and this will be identifiable by the generation of a fluorescent signal that is detected by a scanning device. The pattern of fluorescence can be used to identify the presence of particular organisms. According to Metfies *et al.* (2007) it should soon be possible to develop 'phylochips' suitable for the identification of diatoms and other algae – and, therefore, presumably, other 'difficult' organisms such as fungi.

The identification of pollen grains on the basis of their morphology is time consuming and it is difficult or impossible to distinguish between the pollen-related species of some plants. There are already preliminary studies of using analytical techniques such as Fourier transform infra red (FT-IR) spectroscopy that indicate that this could be a way around these problems (Gottardini *et al.*, 2007; Pappas *et al.*, 2003). Basically, the method works by generating a databank of spectra for the pollen of known plant species and these can then be used to identify the spectra from unknown pollen grains.

Quick quiz

(1) The presence of diatoms in which one of the following body parts would be the best indication of death by drowning: nasal sinuses, stomach, lungs, bone marrow? Explain your reasons.

(2) Following a fatal air crash why might one test the pilot's blood for the presence of both alcohol and *Candida albicans*?

(3) Briefly explain how fresh leaves and fungi should be collected for use as forensic evidence.

(4) How can an examination of tree rings be used to demonstrate whether a church panel painting is a forgery or genuine?

(5) State two ways in which the growth of roots might be used to estimate the length of time a corpse has been buried?

(6)   What is a palynomorph? Give two examples of palynomorphs.

(7)   Distinguish between anemophilous, entomophilous and cleistogamous pollination and comment on the value of pollen from such plants as forensic indicators.

(8)   Giving reasons for your choices, state three regions of a dead body you would sample to retrieve palynological evidence.

(9)   How can stable isotope analysis be used to demonstrate the provenance of drugs and foodstuffs?

(10) How is it possible to distinguish between cycads grown from domestic stock from those collected illegally in the wild?

Project work

Title

Can diatoms enter the bloodstream via the food or drinking water?

Rationale

One of the diagnoses of drowning is the discovery of diatoms in the lungs and bone marrow. However, there is limited published evidence to preclude the possibility that diatoms might also enter the body by being breathed in or through the gut.

Method

Laboratory rats or mice would be fed a diet supplemented with diatoms either in the food or the drinking water. Freshwater diatoms can be cultured relatively easily and might also be radioactively labelled to aid finding them again. After allowing the animals to feed for known amounts of time they would be sacrificed and the body organs analysed for the presence of diatoms.

Title

Determining the duration of submergence from algal colonization

Rationale

Criminals often dispose of the objects that would link them with a crime – such as a weapon, or an empty wallet – in a river or canal. It would therefore sometimes be useful to know how long an object retrieved from the water has been there.

Method

Objects of varying textures and chemical and physical properties, such as knives, screwdrivers and leather and plastic wallets would be placed at different depths in a watercourse at different times of year. They would be retrieved after set periods and the extent of the algal colonization established.

10 Bacteria and viruses

Chapter outline

Introduction
The Role of Microorganisms in the Decomposition Process
Microbial Profiles as Identification Tools
Microbial Infections and Human Behaviour
Microbial Infections That Can Be Mistaken for Signs of Criminal Activity
The Use of Microorganisms in Bioterrorism
Future Directions
Quick Quiz
Project Work

Objectives

Describe how microorganisms influence the decay process and the factors that affect the spread and growth of microorganisms in a dead body.

Discuss how microbial profiling techniques might be used to identify an individual or a location.

Describe how the transmission of HIV can, in certain circumstances, be considered a criminal act and the mechanisms by which the perpetrator may be identified.

Discuss the possibility that an underlying infectious disease may predispose an individual to commit a crime.

Critically evaluate the risks posed by the criminal use of microorganisms and the techniques used to detect their presence.

Introduction

The involvement of bacteria and viruses in legal cases is increasing as advances in technology, especially molecular biology, facilitate their identification, and courts of law become more willing to accept the reliability of DNA-based evidence. In addition, there is enhanced awareness amongst the public of the dangers posed by

Essential Forensic Biology, Second Edition   Alan Gunn
© 2009 John Wiley & Sons, Ltd

pathogenic microbes being spread deliberately or through reckless behaviour by naturally infected individuals and this has led to changes to the law. Furthermore, there has been an increase in the numbers of individuals and groups threatening to release pathogens maliciously, or simply to cause distress and gain publicity. The overwhelming majority of such threats are hoaxes but the need for rapid pathogen identification and tracing has become a priority, as has the need to prosecute those making the hoax claims.

The role of microorganisms in the decomposition process

Even when we are healthy we are host to vast numbers of microbes. Indeed, there are more of them than there are of us. It has been estimated that a typical adult human consists of 10^{14} cells but we also contain in the order of 10^{15} bacterial cells and over 10^{17} viruses. Many of the bacteria normally present upon or within our bodies are also extremely important in the decay process and anything that restricts their activity, such as low temperature, lack of oxygen or the absence of water, will prolong decomposition. In a very dry environment, especially if combined with strong air currents and extreme heat or cold, a human body will mummify and can remain in this state for hundreds, or even thousands of years. Similarly, under constantly frozen conditions a body, if undisturbed, will last for many years. Conversely, under warm, moist conditions, such as the tropics, a body will decompose extremely quickly even in the absence of invertebrates and other detritivores. There are few studies on the changes in the bacterial flora in and around a dead body (e.g. Hopkins *et al.*, 2000; Tibbett *et al.*, 2004) and, as yet, there is insufficient data to suggest how they could be used to determine the time since death. Changes in the abundance, diversity and distribution of microbes within the body are also of crucial consideration in determining the cause of death. For example, there can be a dispute about whether a person died of an ongoing disease or a fatal infection resulting from a medical procedure, diagnosing a case of fatal food poisoning, determining whether a person died of a drug overdose or an infection acquired through intravenous injection, or whether a person died of natural causes or a malicious act. Samples for microbial culture or DNA analysis should be taken from dead bodies with great care to avoid the possibility of contamination. Where numerous species of bacteria and other microbes are isolated from a sample there is a strong possibility that contamination has occurred either through the process of decay or the sampling and handling techniques employed. There is some dispute concerning the most appropriate tissues and fluids to use but the choice will be at least partly affected by the organism/ diagnosis being targeted. For example, because of the 'open access' between the lungs and the outside world, cultures isolated from post mortem lung tissue often yield false positives as a consequence of contamination. In the absence of other indicators it would be difficult to distinguish whether microbes grown from such tissues were an important factor in the cause of death, the result of nonpathogenic infections already present before death, or microbes that established themselves after death (Tsokos & Püschel, 2001).

Bacterial colonization of a dead body usually begins from the intestine, which is naturally home to large numbers coliforms (e.g. *Escherichia coli*), bacilli, and micrococci. A dead body therefore tends to decompose from the inside to the outside. By contrast, if the body is frozen immediately after death, then once it is returned to above 0 °C decomposition tends to occur from the outside to centre. This is because the outside will defrost a long time before the centre – precisely how long will depend upon the ambient temperature. As the body starts to decay, the pH becomes more acidic owing to the release of acids during autolysis (Chapter 1) and the products of bacterial fermentation, and it becomes anaerobic because all the oxygen is used up by the bacteria and the circulation system has ceased to operate. Consequently, the majority of the bacteria found in a dead body tend to be anaerobic and spore forming, such as *Clostridium perfringens*. Formerly known as *C. welchii*, this bacterium is found both within the guts and the female genital tract and surrounding skin. There are numerous strains of *C. perfringen*s and it is implicated in a range of pathogenic infections, ranging from food poisoning to cellulitis and gas gangrene. This is another good reason why care should be taken when handling dead bodies and tissue samples. Wounds whether formed before or after death are another source of entry and the bacteria spread rapidly via the blood vessels. Decay is therefore more rapid in a person who was suffering from septicaemia or other bacterial infections. By contrast, decay is delayed where there is excessive blood loss because the lack of fluid in the blood vessels makes it harder for bacteria to penetrate the body.

Just as the tissues and chemicals that constitute a body are broken down by autolytic and microbial action after death, drugs present in those tissues may also become metabolized and degraded. The speed and extent to which breakdown occurs depends upon the drug and the environmental circumstances. There is even the possibility that microbes may cause an increase in drug concentrations. For example, alcohol levels can rise following death as a consequence of microbial metabolism (Chapter 9). Similarly, gamma-hydroxybutyric acid (GHB) is naturally present at a very low level in our bodies but this level increases enormously when it is used as either a recreational or a therapeutic drug – or for criminal purposes. GHB is a class C drug that is used at low doses as an 'upper' by clubbers who often know it as 'liquid ecstasy', but at higher doses it can cause confusion and even coma and is notorious for being used to spike drinks in 'date rape'. After GHB is ingested, there is an initial increase in concentration, after which the levels decline again within a few hours. There is also a rise in the levels of GHB following death, although not as high as those seen immediately after taking the chemical as a drug. It is therefore of forensic interest to know the extent to which levels of GHB found in a body might be ascribed to natural causes as opposed to misuse. *Pseudomonas aeruginosa*, a common bacterium associated with decomposing tissues, is capable of producing GHB although not in sufficient quantities to account for all of the natural increase so there may be other bacteria and/or physiological processes contributing to the rise (Elliott *et al.*, 2004).

Microbial profiles as identification tools

The study of microbes has always been hampered by their small size and many of them are difficult or impossible to culture in the laboratory. Fortunately, advances in molecular techniques mean that it is no longer necessary to culture microbes in order to determine their presence. Furthermore, phylogenetic analysis indicates that the microbial flora is actually far more diverse than the plant or animal kingdoms. Similarly, because they are so small, their distribution was considered to be less restricted than that of plants and animals and consequently they were not thought to abide by the same taxa–area relationship. This is the ecological principle that underlies the finding that larger islands tend to have more diverse fauna and flora. However, this is now known not to be the case and microbial communities are not random (Green *et al.*, 2004; Horner-Devine *et al.*, 2004). This opens up the possibility of using the microbial genotype/ species composition of infectious agents or of forensic samples such as fluids, stains, soil and other debris found on a suspect, object or animal as a potential identifying characteristic linking them together or with the scene of a crime.

Soil microbe diversity as a forensic indicator

The study of soil is known as pedology although many workers refer to it as simply 'soil science'. As a consequence of the importance of soil characteristics for agriculture the UK has a very good mapped soil database. It is now realized that this can present a useful resource when attempting to link a person, animal or object with a geographical location. This is being expanded upon as part of the SoilFit project being coordinated by the Macauley Institute at Aberdeen (Zala, 2007). In a forensic context, soil particles are often found on shoes, clothing, vehicles, animals, plants, and objects of all kinds. If identifiable characteristics can be ascribed to the soil particles it becomes feasible to link them to a location or to link people/ animals/ objects together. However, in most cases all that can be said is that there is a match between two or more samples – and this does not necessarily mean that they are any way related.

Soil is a complex substance and with the exception of a few organic soils (e.g. peat-based soils) the bulk of it is composed of mineral material. In addition to its mineral component, soil contains varying amounts of decaying animal and vegetable matter and this supports a huge variety of microbes. Currently, most forensic analysis of soil samples is based upon their mineral and chemical composition but it is now thought that microbial evidence could also be useful. Most soil microbes cannot be cultured in the laboratory and so their presence is inferred from DNA analysis. Many workers sequence the 16S ribosomal RNA gene (16S rRNA) using a technique called terminal restriction fragment length polymorphism analysis (T-RFLP). This gene is chosen because it is found in all prokaryotes (i.e. bacteria and archaea) and it is distinct from the homologous gene, 18S rRNA, found in eukaryotes. It is therefore possible to extract and analyse the bacterial gene separately from that of plants, fungi or animals. Furthermore, although there are other genes that are unique to bacteria, the 16S rRNA gene is only about 1.5 kilobases long and this makes it particularly easy and cheap to sequence.

Case Study: Terminal restriction fragment length polymorphism analysis (T-RFLP)

First the DNA has to be isolated from the soil or forensic sample and then the 16S rRNA gene is amplified with primers as part of the PCR process – there are commercially available universal 16S rRNA primers available for bacteria. In this technique, the forward primer is fluorescently labelled but the reverse primer is not. The amplified genes are then digested with restriction enzymes that usually have a four base pair recognition sequence. This yields a variety of differently sized fragments that can be separated on the basis of their molecular weight – for example by electrophoresis. Because only the terminal fragments will be fluorescently labelled it is only these that will be detected. Each band represents a different fragment length and is indicative of a different microbial genotype. In this way a 'barcode' can be obtained for a particular soil or forensic sample.

Feasibility studies (e.g. Heath & Saunders, 2006; Horsewell *et al.*, 2002) indicate that soil microbiological profiling has real forensic potential and the subject is reviewed by Hill *et al.* (2007). However, more work needs doing to standardize and optimize the methodology and determine its strengths and limitations. In particular, some writers have suggested that because microbial profiling is based on DNA it will somehow provide a similar degree of accuracy to human DNA profiling. This is certainly not the case and, like pollen profiling or mineral analysis, soil microbial profiling is unlikely to indicate more than there is a certain degree of similarity between samples – the relevance of this is then open to question. This is because the soil microbial community is dynamic with its composition changing with time and environmental conditions – even the collection and storage of the samples can affect the profile obtained.

Determination of personal identity from microbes present in saliva

Human bites can be nasty because our saliva contains bacteria (Fig. 10.1) capable of causing a serious wound infection. Similarly, clenched fist injuries resulting from punching someone in the mouth are particularly liable to infection (Monteiro, 1995). However, these bacteria may be useful in identifying the person who inflicted the bite or linking the victim of an assault with his/her assailant. The bacterial flora, especially the streptococci naturally present in our mouths, is extremely diverse and their genomic profile appears to be characteristic for an individual. Borgula *et al.* (2003) and Rahimi *et al.* (2005), asked volunteers to bite themselves after which samples were taken over varying time periods. They demonstrated that live bacteria could be recovered from the bite site for at least 24 hours afterwards provided that it was relatively undisturbed, as well as from fabrics. Furthermore, the genomic profiles of the bacteria recovered from the bite sites provided a unique identifier of the person responsible. Obviously there is a great deal of work to be done to confirm

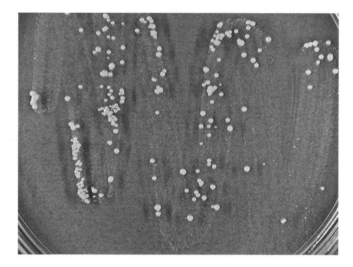

Figure 10.1 Bacteria grown from human saliva on nutrient agar. Numerous colonies are growing despite the swab being taken from the donor within two hours of him brushing his teeth and using a mouthwash.

the reliability and limitations of this technique and to agree a standardized protocol. It could prove to be a valuable tool in forensic analysis in the future, although if saliva is identified it would be most sensible to attempt to isolate human DNA before exploring alternative means of identification.

Determination of geographical origin from viral genotyping

Many of us are infected with viruses that are either nonpathogenic or remain quiescent as a result of being kept under control by our immune system. Like many other organisms, many of these viruses exhibit regional variations and it has been suggested that this could be used as a means of identifying an infected person's association with an area. For example, Ikegawa and his co-workers have indicated that in Japan it is possible to trace the origins of unidentified bodies by genotyping the JC virus (JCV) (Ikegaya *et al.*, 2007; Ikegaya & Iwase, 2004). JCV is a DNA virus belonging to the family Polyomaviridae that infects most of us during childhood and is usually asymptomatic, except in immunocompromised people. Once we are infected, the same viral strain can be found in our kidneys for the rest of our life. In addition to being found in the tissues, the virus is also excreted in the urine – this may have forensic importance where urine splashes are found at a crime scene. However, polyoma viruses are delicate and difficult to grow in culture so the urine splashes would probably have to be fresh. It is possible to extract JCV DNA from bodies that are up to 10 days old hence facilitating the identification of corpses that are already badly decayed (Ikegaya *et al.*, 2002). They also found that a related virus, BK virus, could also be isolated from kidneys and used in a similar manner.

Furthermore, because the distribution of the different genotypes of the two viruses were not the same it was possible to narrow down the likely origin of an individual. The forensic potential of JCV and BKV genotyping has not yet been assessed in other countries and would not allow a positive identification of an individual's identity but, in conjunction with other evidence, has the potential of suggesting the area where the unidentified person grew up – which would help speed up the investigation. Viral genotyping, in this instance would therefore work a bit like stable isotope ratio analysis in that it would not provide evidence of racial identity but could suggest a country or region of origin. Furthermore, like stable isotope ratio analysis, its effectiveness would depend upon the extent and reliability of databanks – in this case those of the distribution of viral genotypes within and between countries. Nevertheless, this approach could be useful when an unidentified body is found whose details do not match those of any 'missing person' and for whom there are no police records. For example, if it is suspected that the body might be that of an illegal immigrant/ someone who was trafficked illegally. Ever since the 1990s many women from Eastern Europe have been sold into prostitution in the UK and elsewhere and it remains a serious problem. Sometimes these women are killed by their clients or their pimps and it is then a difficult job to determine who they were or where they came from.

Linking a victim and a suspect through the transfer of microbial infections

Most of us have blamed a family member or friend for catching a cold but attempting to prove this scientifically would be very difficult. This is because cold viruses are usually transmitted in aerosols and therefore we can acquire them from any infectious person in our vicinity. Where the disease is sexually transmitted (STDs) the list of potential sources of infection is reduced but it can be difficult to distinguish whether person A infected person B or B infected A or whether both A and B were already infected before they had sexual contact with one another. However, if person A and person B are infected with different strains/ genotypes of the STD then it is certain that their infections are not linked. Sex offenders often have multiple victims and even if caught, they are at a high risk of re-offending once released from jail. As a group, sex offenders are more likely to be infected with STDs than the general population (e.g. Giotakos *et al.* (2003) and therefore it is potentially feasible to link victims and assailants together through STD transmission. However, apart from the transmission of HIV (see later) there are few published instances in which this has been done.

In order to be useful as a means of linking the donor and recipient of an infectious disease one needs to identify a genetic marker that exhibits three criteria. Firstly, it should exhibit sufficient variation to exclude unlinked individuals. Secondly, the variation should be capable of being identified easily and thirdly genetic variation should neither arise so slowly nor so fast that it compromises one's ability to link infected people together. In cases of child abuse there is no doubt about the direction of disease transfer and genetic typing of STDs can provide corroborating evidence of who abused the child. This can be particularly useful when the abuse takes place within families when it would not be suspicious to find the suspect adult's DNA on the child's clothes or bed linen.

Case Study: Identification of the source of gonorrhoea

Gonorrhoea is caused by the bacterium *Neisseria gonorrhoeae* and it is almost always transmitted sexually. It therefore goes without saying that a child suffering from disease has been subject to abuse. Martin *et al.* (2007) describe a case in which typing of *N. gonorrhoeae* isolated from the vagina of a young girl was used to help identify the man who abused her. There is no UK forensic DNA database for *N. gonorrhoeae* but there is a research database containing information on the bacterium's two most variable genes. This database, called the multi-antigen sequence typing database (NG-MAST), contains sequence data for over 4000 strains of *N. gonorrhoeae* but cannot be used to associate strains with particular geographical regions. In this case, the strain isolated from the young girl was found to be the same as the one identified from stained regions of the suspect's underwear (one of the symptoms of gonorrhoea is a discharge of purulent material from the genitals). Because the NG-MAST database is small and currently does not provide a reliable indication of strain distribution it could be argued that the fact that the girl and the man were infected with the same strain was circumstantial evidence and not proof of physical contact. However, the man admitted his guilt when presented with these findings and so the strength of evidence was not tested in court.

Identifying the source of an HIV infection

Human immunodefficiency virus (HIV) (Fig. 10.2) is a human retrovirus that affects the immune system and in time causes the condition 'acquired immunodefficiency syndrome' or AIDS. There are two 'species' of the virus: HIV-1 and HIV-2. HIV-1 is the more widespread of the two and the major cause of AIDS around the world whilst HIV-2 transmission is largely restricted to West Africa. Although HIV is primarily a sexually transmitted disease, it is also transmitted via blood transfusions, shared needles by intravenous drug users, accidental or malicious wounding (e.g. needlestick injuries) and an unfortunate number of infections in children arise through perinatal transmission at the time of birth. Consequently, unlike gonorrhoea, HIV infection in children may not be a consequence of abuse. An infected person can remain HIV positive but apparently healthy for months or even years before symptoms of AIDS become apparent. The virus infects and destroys cells bearing the CD4 molecule, which includes T-helper cells and a subset of blood and tissue macrophages. Following replication, the pro-viral genome becomes integrated into the host chromosome where it may remain quiescent or enter a cycle of production of progeny virus that results in the cell's destruction. The clinical manifestations of HIV infection are a result of depletion of the T-helper cell population and thereby an impairment of the body's ability to respond to other infectious agents.

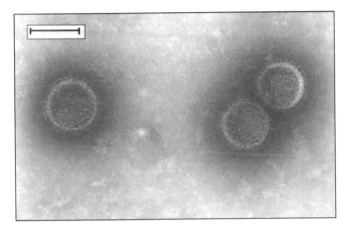

Figure 10.2 Transmission electron micrograph of negatively stained human immunodeficiency virus (HIV). The 'spikes' on the surface of the virus capsid are composed of glycoprotein molecules. The capsid contains two copies of single stranded RNA. The bar equals 100 nm. (Reproduced from Hart, T. and Shears, P. (1996) *Color Atlas of Medical Microbiology*. Copyright © 1996, Mosby Wolfe.).

HIV/AIDS is a major problem throughout the world and health workers have to treat all patient samples as though they might be infected. In the UK it is illegal to test a person for HIV without their consent because they must have counselling on being told the result – so unless a patient or person in police custody already knows that they are HIV positive and volunteers the information, or agrees to a test, there is no way of finding out. Forensic pathologists are well aware of the risks of contracting HIV and other diseases, such as hepatitis B (HBV) and hepatitis C (HCV), whilst performing an autopsy and that this risk does not disappear even if the body is many days old (Nolte & Yoon, 2003; Douceron *et al.*, 1993). Consequently, anyone handling a dead body or body parts should wear cut-resistant undergloves and appropriate protective clothing and work with extreme caution (Galloway & Snodgrass, 1998).

HIV is a single stranded RNA virus, the RNA genome being enclosed within a protein envelope (or capsid). It belongs to the family Retroviridae, genus Lentivirus. The term 'lenti' is derived from the Latin '*lentus*' which means 'slow or 'inactive' because many of the viruses belonging to this genus exhibit long incubation periods. However, like other RNA viruses, HIV is also capable of rapid multiplication under the right conditions. Once the HIV virus enters a cell, the enzyme reverse transcriptase, which is part of the virus particle, converts the viral RNA into single stranded DNA and then constructs a complementary strand of DNA for it. This DNA is then integrated into the host cell genome by the virally encoded enzyme integrase. Once integrated, the viral genome is replicated along with the host DNA by normal host cell mechanisms. The process by which the viral genome is incorporated into the host genome is the reverse of what normally happens within the cell, i.e. DNA is synthesized from an RNA template rather than RNA being synthesized from a DNA template. Hence the name of the enzyme: reverse transcriptase.

Viral reverse transcriptase exhibits a high error rate because unlike DNA polymerase it has no proof-reading ability. Consequently, the combination of speedy replication and poor proof-reading leads to a rapid rate of mutation. An infected individual therefore contains a genetically diverse HIV population – albeit one in which most viruses will share common sequences. If the host was infected on several occasions by different donors then the virus population will be even more mixed. Because HIV is rapidly evolving all the time it is impossible to obtain an exact match between the virus molecular profile from an alleged donor and recipient. Indeed, the profile in each of them will be changing constantly and the longer the time gap between infection and sampling the more variation will have evolved. Consequently any attempt at matching virus profiles between donor and recipient depends upon calculating the evolutionary relationship of their virus populations – these are known as phylogenetic tree reconstructions. There are a variety of methods for reconstructing phylogenetic trees but maximum likelihood methods (ML) tend to be used most frequently by those who work on HIV transmission chains. These use a variety of statistical techniques to determine the likelihood (probability) that a proposed phylogenetic tree and hypothetical evolutionary history would explain the observed virus population profile. By generating a variety of different phylogenetic trees using computer simulations the one with the highest likelihood (probability) can be identified. It is essential that the tree includes appropriate control samples from persons who carry the same HIV subtype as the donor and recipient and live in the same locality at the same time. This is not to incriminate other people but to ensure that the phylogenetic tree includes a reliable estimate of the evolutionary profile of the local virus population.

The profile of the virus in donor and recipient will start to diverge from the moment of transfer. The rate at which base substitutions arise varies between regions of the genome and stage of disease (for example, whether it is quiescent or rapidly progressing). Despite this, overall the substitution rate is relatively constant and can be factored into the construction of the phylogenetic tree. Furthermore, most forensic cases involve samples taken within one or a few years of the alleged transfer of infection thereby reducing the amount of divergence that could be expected. Nevertheless, phylogenetic trees cannot inform on the direction of transmission and even if two individuals are found to contain very similar viral profiles it does not mean that other people were not involved in the transmission chain. Consequently, phylogenetic trees cannot by themselves provide conclusive proof of transmission from one person to another and they must be considered alongside other evidence (Bernard *et al.*, 2007). Learn & Mullins (2004) have suggested that rather than making categorical judgements of association, statements along the lines of 'the viral sequences from person A and person B display a high level of similarity' or 'the viral sequences are compatible with the possibility that person A infected person B' should be used.

Although there are many documented cases of health workers contracting HIV infection during their work, usually through needle-stick injuries, there are few reports of patients contracting the disease from their carers. A very prominent case where this did occur was that of David Acer, a Florida dentist, who, following diagnosis of being HIV positive continued to practice without telling his patients. Even more importantly, he used poor infection control practices. For example, it was alleged that he did not always sterilize his instruments or change his gloves between patients. Several of his patients contracted HIV despite being in low risk groups and subsequent genetic analysis indicated that the HIV strains both they and David Acer carried were all closely related. Furthermore, other HIV-positive people living in the vicinity of Acer's practice had different strains of the virus – which suggests that his patients had contracted their infection from him rather than their neighbourhood (Ciesielski *et al.*, 1994). This remains an isolated case and it is still uncertain how exactly transmission from dentist to patient happened.

In the UK and several other countries, persons who know that they are HIV positive and do not inform their partner(s) before indulging in unprotected sex or bite or otherwise intentionally attempt to infect someone, are now liable to prosecution and can be sent to jail. Criminalizing the sexual transmission of HIV has proved extremely controversial, as is anything connected to HIV, its treatment and transmission. One of the arguments against is that it might prevent someone from seeking an HIV test because if you do not know that you are infected with a disease, you cannot be accused of knowingly spreading it. From a strictly legal point of view, proof of malicious intent is difficult because, where sex was consensual it relies on one person's word against another's that he/she intended to commit harm. Successful prosecutions have, however, been brought. In a landmark UK case in March 2001, a man who had known of his HIV status for almost a year was prosecuted for allowing his girlfriend to become infected. His girlfriend claimed that she was not aware of his HIV status when they began their sexual relationship and he only admitted it to her after she had discovered from a blood test that she too had become HIV positive. The man denied this version of the events but the jury did not believe him and he was jailed for 5 years. The chances of transmission of HIV (and some other sexually transmitted diseases) during rape or sexual assault are much higher than in consensual sex. This is owing to the violence resulting in genital or anal abrasions through which the virus can enter the bloodstream. Victims of assault are therefore always offered an HIV test and counselling when they report the attack. In parts of Africa, a rumour has spread that sex with virgin is a cure for the disease and this has led to a terrible increase in the number of rapes of young girls. The magical power of virgins is a longstanding and widespread belief and Howe (1950) relates a case in England (presumably, the report lacks dates and place names) in which a seven year-old girl was raped by a man suffering from gonorrhoea in the belief that this would cure of him of the disease. Despite the huge advances in forensic science, it is a sad reflection on justice in England and Wales that although the number of reported cases of rape has risen, less than 6% of these result in a conviction. This emphasizes that the successful resolution of a crime depends on more than the development of ever more sophisticated laboratory techniques.

A remarkably heartless case of deliberate HIV transmission was heard in St Charles, Missouri, USA in 1998 in which a father, who worked as a phlebotomist, stole infected blood from his place of work and injected it into his 11 month old son, who was lying sick in hospital suffering from an asthma attack. Doctors treating the child became concerned when a blood test revealed that he had somehow contracted HIV. They contacted the police who became suspicious of the father when he reportedly told his ex-girlfriend that she would never claim any child support off him because the boy would not live that long. The father was convicted on the basis of largely circumstantial evidence and received a life sentence; the child subsequently developed AIDS.

Case Study: Accidental or natural infection with HIV?

One of several interesting cases described by Learn & Mullins (2004) involved a nurse who cut herself with a broken blood collection tube whilst withdrawing blood from a child suffering from acute HIV infection. Despite thoroughly washing the wound she exhibited signs of HIV viraemia 13 days later and she claimed that the infection was a consequence of her injury. Health workers who suffer needlestick or similar injuries are always given a blood test on the day of their injury and the one taken from this nurse tested positive for anti-HIV antibodies. Consequently, her employers claimed that she was already infected with HIV – and this would limit her ability to claim compensation. The nurse stated that this was impossible because she had refrained from sex for the previous 6 years, hadn't had a blood transfusion and didn't take intravenous drugs or otherwise expose herself to high risk activities.

To resolve the case blood samples were taken from both the child and the woman and the virus profiles analysed. For controls, 28 sequences were chosen from the computer database GenBank. Three regions of the HIV genome were analysed but only the data for the *env* (envelope) V3-V5 region were presented. The *env* region codes for the precursor of proteins that are located within the virus envelope and enable the virus to attach itself and fuse to its target cells. The variability of this region makes it extremely useful for phylogenetic studies. Maximum likelihood estimates were used to develop the virus phylogenetic trees assuming the HKY (Hasagawa–Kishino–Yano) evolutionary model of nucleotide substitution. Evolutionary models are needed because base change mutations do not take place in a uniform manner along the whole genome. There are several different mathematical models available of varying degrees of complexity and different models are tested to find the most appropriate for the data. The HKY model is often used in HIV phylogenetic analysis because it allows for transitions and transversions taking place at different rates and also for the virus's skewed base composition. Base transitions are where one purine or pyramidine is substituted for another (e.g. adenine for guanine or cytosine for thymine) whilst transversions are where a purine is substituted for a pyramidine (e.g. adenine for cytosine) or a pyramidine is substituted for a purine (e.g. thymine for adenine). Transitions are far for common than transversions. HIV has high adenine, low cytosine levels and because different bases exhibit different transition/ transversion rates this needs to be taken account of when modelling evolutionary profiles. Learn & Mullins's analysis indicated the presence of two distinct virus populations – (A) and (B) – in both the nurse and the child (Fig. 10.3). Furthermore, both populations A and B were distinct from those in the control group and both the nurse and the child contained similar sequences in the two virus populations. This suggested that the viral sequences were compatible with the nurse having acquired the infection as she described. Clearly, the GenBank sequences do not provide an adequate 'control' population. However, the results were sufficiently convincing for an 'out of court' settlement to be agreed upon before more rigorous control samples were taken from the local population.

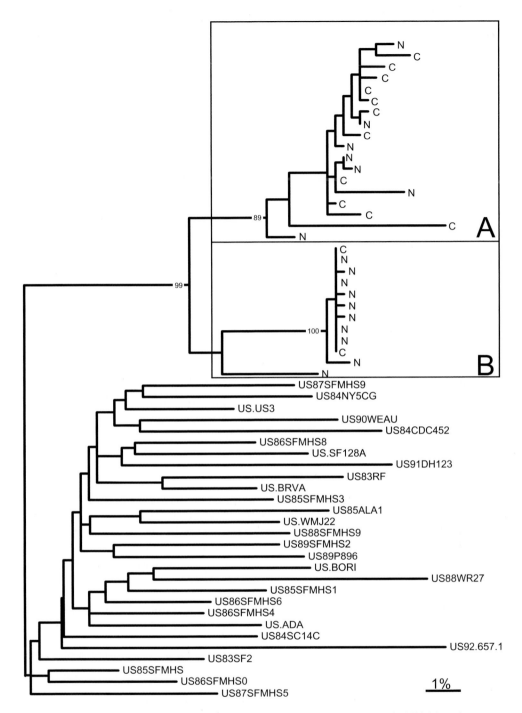

Figure 10.3 A nurse (N) claimed to have contracted HIV from an infected child (C) owing to an accident whilst taking a sample of the child's blood. This diagram represents a maximum-likelihood phylogram that reconstructs the phylogenetic relationship between the envelope (V3–V5) sequences from a nurse and the child and also 28 HIV sequences chosen at random from GenBank. The scale bar represents 1% genetic distance (0.01 substitutions per site). The numbers at the nodes represent bootstrap values (i.e. percentage of replicates supporting that group) where there was over 70% support. The phylogram indicates the presence of two distinct virus populations – (A) and (B) – in both the nurse and the child. For more details, see text. (Reproduced with permission from Learn, G.H. and Mullins, J.I. (2004) The microbial forensic use of HIV sequences, in Leitner, T. *et al.* (eds) (2005) *HIV Sequence Compendium*. Los Alamos National Laboratory, Los Alamos, NM, USA.)

Identification of microbes responsible for food poisoning

Food poisoning is extremely common and many people automatically assume that any bout of sickness or diarrhoea is a consequence of something they ate. However, some experts believe that up to 50% of such cases are actually the result of infections acquired from other sources, such as pets, contact with faeces, contact with someone who was already infected, or failure to follow simple hygiene rules such as washing hands. The most common microbes causing food poisoning in the UK are the bacteria *Campylobacter* spp. (which causes the majority of cases), *Salmonella* spp., *Escherichia coli* O157, *Clostridium perfringens* and *Listeria monocytogenes*. A number of viruses also cause food poisoning, such as Norovirus (Norwalk-like virus), rotavirus and Hepatitis A. Most cases of food poisoning, although unpleasant, are not life threatening and occur as individual or sporadic events. An outbreak is defined as occurring when two or more people fall ill having consumed a common batch of food. Occasionally food poisoning may prove fatal, especially in the very young, the elderly and the infirm. Food poisoning usually results unintentionally from the incorrect storage or cooking of food which allows the bacteria within it to survive and replicate to dangerously high levels or from poor hygiene practices which result in the transfer of bacteria from contaminated food or surfaces to previously uncontaminated food. Identifying the source of an outbreak of food poisoning requires the offending organism(s) to be cultured from the patients' faeces and questioning the patients to determine their common food source. This may prove difficult because not everyone who consumes contaminated food need fall ill and if the food has all been consumed or thrown away, it may be impossible to prove that it was the source of the infection.

The intentional contamination of food with foreign bodies (e.g. razor blades, glass) and poisons (e.g. mercury) is common practice, usually by disaffected individuals with a grudge against the manufacturer or society at large or with a view to blackmail. There are few reports of the deliberate use of microbes either through introducing them or through poor cooking or storage practices in order to cause food poisoning. However, it is a possibility where someone wishes to cause distress but not death and would be difficult to prove, though it would carry a real risk of unintentionally causing fatalities.

In addition to identifying the species of organism responsible for causing food poisoning it is also necessary to identify the strain involved. This is because different strains often vary in their pathology and susceptibility to drug treatments. It is also helpful to know the strain involved when tracing the source of the infection. Distinguishing between strains of bacteria is a time-consuming job using conventional culturing techniques and these are therefore becoming supplemented by a variety of DNA-based methods. One molecular technique that is being used with increasing frequency in the identification of bacteria responsible for food poisoning is that of ribotyping. This technique is a variant on the RFLP process and is used to examine the polymorphism in restriction sites in the rRNA gene. Usually sequences either side of the 16S rRNA gene are targeted, but sites either side of the 23S rRNA gene and within its middle are also sometimes used. Unlike the T-RFLP process it is necessary to have a pure culture of the bacterium whose strain is to be identified. The process can be automated and machines such as the Riboprinter® are commercially available (Grif *et al.*, 2003). Kishimoto *et al.* (2004) describe how ribotyping of *Staphylococcus aureus* strains isolated from the hands of kitchen workers, utensils and food items could be used to demonstrate the transmission of bacteria from specific individuals onto the food they prepared.

Case study: Intentional food poisoning by the Rajneesh cult

In 1984, members of the US-based Rajneesh cult (also referred to as Rajneshee, in the literature) attempted to influence a local Oregon election by making the voters too sick to vote. They did this by manually infected food in several restaurants with *Salmonella enterica* var *typhimurium*. Cult members obtained the bacteria via their own state-licensed medical laboratory, which in turn purchased them from commercial sources. They coated plastic gloves with the bacteria and then handled the contents of salad bars (and from the pattern of disease presumably other foodstuffs as well) in local restaurants on at least two separate occasions. It is possible that restaurant employees subsequently inadvertently contributed to the spread of the bacteria through contamination. A total of 751 people are known to have fallen sick of whom 45 were hospitalized as a result of the cult's activities but many more were probably affected since the town is situated along a tourist route and some people probably stopped by for a meal along their way. Although the possibility that the outbreak was the result of a deliberate act was considered, this was discounted for a variety of reasons (Morse & Khan, 2005; Wheelis & Sugishima, 2006). For a start, nobody claimed responsibility and no obviously disgruntled employees of the restaurants could be identified. In one of the affected restaurants, employees fell ill before their customers and this is a common pattern in food poisoning outbreaks because they are obvious sources of the infection. There appeared to be no obvious motive for the attack – at the time, nobody thought of the upcoming local election as being relevant to the case. Furthermore, the illness pattern suggested that the outbreak occurred in two waves and it was thought unlikely that anyone would disseminate pathogens in this way. Why they should think that is uncertain because poisoners often dose their victim(s) on successive occasions until the desired effect is obtained. Although the causative agent and the pattern of illness among the population were both highly unusual it was still thought possible that the illness could have arrived and spread by natural means. The causative agent was identified as *Salmonella enterica* var *typhimurium* which is seldom involved such large scale food poisonings but there had been other outbreaks involving this bacterium in the state that year. Consequently although it was unusual it was not unprecedented. Similarly, the inability to identify the original source of the infection was not unusual because the origins of many instances of food poisoning are never established.

Nevertheless, a number of features of this incident stand out that make it different from most natural food poisoning outbreaks. To begin with, one would have expected staff to fall ill before their customers in more than one of the restaurants. An initial report indicated that many employees fell ill before their customers but this was revised in later accounts (Wheelis & Sugishima, 2006). Secondly, in many of the cases, infection was linked to the salad bars and this is an unusual source of salmonellosis food poisoning (it is more commonly associated with eggs and poultry). However, it was not solely linked to this and other foods were also incriminated thereby suggesting widespread contamination. To have one infected food item is not unusual – to have several different unrelated items is distinctly odd. Furthermore, there was no obvious common factor (e.g. staff contact, food supplier, foods linked to the illness, shared water supply etc.) linking all the affected restaurants. In addition, once the salad bars were closed, the numbers of infected customers declined despite the fact that some of the employees remained at work even though they were still ill. This suggests that the employees were not the primary source of infection but that the salad bars were clearly implicated and that infection must have been occurring after the food was placed in the dining room area.

The cult's involvement only became apparent when it started to splinter and some members turned informant. Somewhat foolishly, the cult retained samples in their commune laboratory. Consequently, it was highly incriminating when it was shown that the strain isolated from people who had fallen sick and that in the commune laboratory were the same (Torok *et al.*, 1997). It was alleged that the cult had considered using the much more deadly *Salmonella typhi*, the causative agent of typhoid fever, but this idea was abandoned because it was considered that it would have been much easier to trace the source of the infection back to the cult.

Microbial infections and human behaviour

Any infectious agent that affects our nervous system can bring about changes in our ability to control our bodies. For example, the polio virus causes the destruction of motor neurons in the spinal chord resulting in paralysis that can be fatal. When the infection causes damage to those areas of the brain that control our behaviour then the result can be a dramatic change in the way in which we speak to and interact with other people. This can lead previously well-behaved persons to commit crimes, place themselves in a situation in which they could be harmed, or commit suicide. Syphilis, for example, is well known for causing a wide range of pathological symptoms amongst which is damage to the nervous system – this is called neurosyphilis. Syphilis is a sexually or congenitally transmitted disease caused by the bacterium *Treponema pallidum* (Fig. 10.4) and is thought to have been introduced into Europe by sailors returning from South America in the 1400s. Although syphilis was once a common cause of both morbidity and mortality, with the advent of modern antibiotics, it became a rare disease. However, in recent years there has been a marked rise in the number of infections in the UK and many other countries. For example, in the UK, between 1997 and 2006 the number of newly diagnosed cases in men rose by 2165% and in women by 522%. Most of the new cases have arisen through

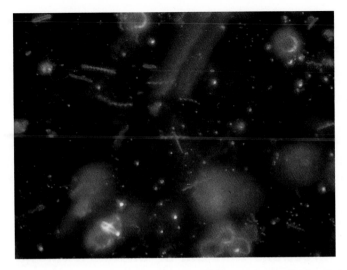

Figure 10.4 The spirochaete bacterium *Treponema pallidum* as observed using dark-ground micro-scopy (DGM). The bacterium is about 12 microns in length, has a long, thin spiral shape and exhibits rapid corkscrew-like movement. Positive identification is usually through serological tests although the bacteria may also be observed in exudates from ulcers or genital mucosa using DGM. Similar spirochaete bacteria are commonly found in the oral cavity and therefore DGM cannot be used to identify syphilis using mouth swabs. (Reproduced from Peters, W. and Gilles, H.M. (1977) *A Colour Atlas of Tropical Medicine and Parasitology*, copyright Wolfe Medical Publications Ltd.)

transmission between homosexual men. The name syphilis is derived from a character in a poem, '*Syphilis siva morbus Gallicus*' (Syphilis or the French pox) by the Italian physician Girolamo Fracastoro. In the poem, Syphilis is a swineherd (or goatherd – translators differ as to his occupation) who is afflicted with the disease by the god Apollo for lack of respect. Neurosyphilis can occur at any stage during infection although it is normally associated with the later 'tertiary' stage. In addition to paralysis, neurosyphilis can manifest itself in a variety of behavioural ways such as mood changes, irritability, grandiloquent behaviour, and dementia.

Cat scratch fever is caused by the bacterium *Bartonella henselae* and as its name suggests is transmitted to humans by young cats. It is a relatively common disease but is normally benign and self-limiting. However, sometimes it can cause a high fever, enlarged lymph nodes and changes in behaviour that may include extreme aggression. Harvey *et al.* (1991) describe a case in which a 27-year old truck driver underwent such marked changes in behaviour that he was thought to be exhibiting signs of drug abuse. As a consequence, he was sacked from his job and was eventually forcibly taken to the local hospital by the police following complaints from his wife. In a similar fashion, viral encephalitis caused by a variety of different viruses (e.g. *Herpes simplex*, Borna Disease Virus (BDV), Epstein–Barr Virus and a range of arboviruses) has been linked to odd, confused, and occasionally antisocial behaviour. Between 1915 and the late 1920s there was a pandemic of a disease called *encephalitis lethargica* that sometimes resulted in the sufferers exhibiting aggressive antisocial behaviour. It is assumed that the disease was caused by a virus but it apparently disappeared before modern virological techniques were available. It has been suggested that some of Adolf Hitler's more unpleasant personality traits, as well as the symptoms of Parkinson's Disease, could have stemmed from suffering from the after effects of *encephalitis lethargica*. Hitler may also have suffered from syphilis, allegedly caught from a Jewish prostitute in Vienna (Hayden, 2003); however, like so much concerning his behaviour and those around him, this will remain speculative. AIDS may also result in behavioural changes including loss of memory, irritability, depression and dementia.

Infections that cause antisocial behavioural changes create medico-legal problems because in many, only a small proportion of those suffering the disease exhibit mental problems, and of those a smaller percentage commit offences as a result. It is therefore a difficult clinical judgement to determine whether or not a person committed a crime as a consequence of being infected with the disease and was therefore not responsible for their actions.

Microbial infections that can be mistaken for signs of criminal activity

Some microbial infections result in the formation of skin lesions that might be mistaken for signs of physical assault (e.g. Prahlow & Linch, 2000). This is especially the case in vulnerable persons, such as the very young, the mentally disturbed and those suffering from senility, who are unable to relate what has happened to them. For example, Nields & Kessler (1998) describe a case in which a 4-year-old child who died of streptococcal toxic shock syndrome was initially thought to been a victim of child abuse. Fortunately, the bacteria were identified from samples taken at autopsy and no-one was charged with assault. Streptococcal toxic shock syndrome is caused by infection with any group A streptococcal bacteria and may result in organ failure and shock – the consequences of which can be rapidly fatal. In many cases it is not known how the bacteria enter the body but a common feature of the syndrome is the development of necrotizing fasciitis in which there is destruction of the fascia (fibrous tissues that cover and separate the muscles) and fat. Following this, the overlying skin may die, split and break and the damaged region may, at first sight be thought to have resulted from a wound that has become infected.

Case Study: The Sally Clark case

It is essential to exclude the possibility that an infectious agent might be involved in an otherwise 'unexplained' death. This message was brought painfully home in the unfortunate case of Sally Clark. Mrs Clark was tried and convicted in 1999 of murdering two of her infant sons. The first son, 11-week-old Christopher, died suddenly in 1996 and fourteen months later her second son, 8-week-old Harry, also died. The medical profession therefore became suspicious and a police investigation ended with Mrs Clark being sentenced to life in prison. The same pathologist carried out autopsies on both children. Initially he ascribed Christopher's death to a respiratory tract infection but after Harry died he revised his opinion and suggested that he had been smothered. Harry's death he ascribed to 'shaken baby syndrome'. A second pathologist also concurred that the deaths were suspicious. However, it subsequently transpired that the first pathologist had withheld the results of microbiological tests carried out as a normal part of the autopsy routine. These demonstrated that Harry was suffering from an active infection with *Staphylococcus aureus*, the bacteria being found in eight regions of his body including the cerebrospinal fluid. *S. aureus* is commonly found on the surface of the skin and in the nose of normal healthy individuals and is generally considered harmless to such people – because of this the pathologist did not submit the findings to the coroner because he considered the bacteria to be a consequence of contamination. However, *S. aureus* is also associated with skin infections such as boils and abscesses and it is capable of causing potentially fatal septicaemia and meningitis. The presence of the bacteria in Harry's body, and especially in the spinal fluid, together with his blood cell profile, indicated that Harry could have died as a result of meningitis rather than a criminal act. This resulted in Mrs Clark being freed on 29th January 2003 and her conviction being declared 'unsafe' (Dyer, 2005; Byard, 2004).

The use of microorganisms in bioterrorism

Biological warfare has been practiced in a minor and infrequent way since antiquity (Mayor, 2003) but its impact has been small compared with conventional weaponry. The development of biological warfare agents was banned under the Geneva Convention many years ago – not that that this put a stop to it! During the Cold War era, both Western governments and Eastern communist regimes devoted considerable effort to the development of agents causing human and animal diseases as a means of causing huge casualties to the enemy (Wheelis *et al.*, 2006). Following the physical and economic collapse of the Soviet Empire there has been international concern that the diseases developed by the Soviet scientists and their knowledge of how to produce and deliver them might find their way into the hands of terrorists and 'rogue governments' which bear grudges against the western world (Brumfiel, 2003). The USA is so convinced of the threats that it is currently spending billions of dollars in developing strategies to combat them. With so much money readily available it is not surprising that some scientists are talking up the threat in their search for research funds and advancement. This is not to deny that very real risks exist where state sponsored production of biological weapons is concerned but there is also a great deal of hype, which is further fuelled by media in constant search of disaster stories. In the UK, the Advisory Committee on Dangerous Pathogens, grades organisms 1–4 on the basis of the risks they pose; those in grade 4 being considered the most dangerous. The US Centres for Disease Control and Prevention (CDC) lists pathogens A–C according to their potential for use in bioterrorism, those on the A-list being considered to be the most dangerous.

A wide range of viruses and bacteria have been suggested for possible use in biological warfare and a few examples are considered here to illustrate how they are spread, identified, and traced.

The dissemination of pathogenic microorganisms

Biological warfare agents are sometimes said to be 'cheap and easy' to produce, as if all microbes are the same and that they are 'simple to deliver' to their target. Growing highly infectious, rapidly lethal bacteria or viruses usually requires Category 4 containment facilities and highly trained scientists otherwise the production staff would soon become infected and die. Even with access to the facilities of a modern biodefence laboratory, working with dangerous pathogens is a risky occupation. For example, in 2004 three researchers at Boston University's medical campus working on the tularaemia bacterium became infected (Dalton, 2005). Fortunately, the infections were not fatal and it is not normal for tularaemia to be spread by person-to-person contact. However, it was some time before it was realized that these infections had occurred and with a different pathogen it would have been possible that the disease could have unintentionally spread into the nearby community. Terrorists would also face the problems of establishing a suitable dose to incapacitate or kill the intended target(s) and a means of delivering that dose. Biological agents cannot usually be delivered in a conventional bomb because they would be destroyed in the heat of the explosion. Although several governments have developed bombs and missile warheads to deliver biological agents this represents a level of resources and skill beyond most terrorist organizations. A more low-tech approach is to release the agent in the form of an aerosol although this is not as simple as it sounds. For a start, the aerosolizer must generate droplets of around 0.5–5 µm. If they are smaller than this the droplets may not be retained in the lungs and if they are larger they tend to fall to the ground before reaching their intended target. Furthermore, the effectiveness of the aerosol will be strongly affected by environmental conditions such as humidity and air currents. For example, at low humidity the droplets will rapidly evaporate whilst the turbulent air currents associated with the high-rise buildings typical of many cities often pull air upwards and disperse it away from street level. It is theoretically possible that a disease might be spread by a suicidal 'biobomber' becoming intentionally infected and then infecting the target population by travelling around a city on crowded public transport. However, dying in a sudden explosive 'blaze of glory' is one thing but intentionally dying over a period of days from a painful incapacitating disease, in a public place and without drawing attention to oneself calls for a different and very rare type of suicidal individual. The use of pathogens by disaffected individuals or organizations is therefore most likely to be aimed at causing terror and disruption of public services rather than widespread mortality. The release of the agent would probably be publicized by the terrorists at the time of release, just as is the case with many of their conventional bombs, or there would be the threat of a release in order to blackmail the government – this would require very small amounts of agent and the delivery device need not even be very effective but would still result in widespread publicity and panic. These issues are, however, beyond the scope of this section.

It should be remembered that the majority of infectious diseases that spread through the population are a consequence of natural outbreaks or human negligence. For example, in August 2002 almost 200 people at the Barrow Arts Centre

(Leicestershire) were infected with legionnaires' disease (caused by the bacterium *Legionella pneumophila*) of whom seven died. This occurred as a consequence of the bacteria being allowed to replicate within an air conditioning unit that then sprayed them into the surrounding air. This led to charges of unlawful killing against the technical and design services manager responsible for the site and the local council.

Microbial toxins

It is not just living pathogens that could be used maliciously. A number of microorganisms produce toxins that are highly poisonous and these could potentially be employed to kill or injure an individual or community. For example, the bacterium *Clostridium botulinum* produces botulinum toxin – which is one of the most poisonous substances known – whilst a number of fungi produce chemicals known as mycotoxins some of which are potentially fatal (Chapter 9). These microbial toxins have a range of chemical structures but none of them is based on DNA. Consequently, they cannot be detected or characterized by standard molecular techniques. The microbes that produce these toxins exist naturally in the environment and many have widespread distribution. Consequently, for some of them, such as botulism, accidental contamination or poor food preparation techniques are the most likely causes of the poisoning. The number of people suffering from poisoning and the individual circumstances would arouse suspicions of malicious spreading of the toxin but the source of the infection would need to be identified as a priority regardless of the cause.

In 1990, the Aum Shrinrikyo cult in Japan made several attempts to use botulinum toxin as biological weapon by disseminating the toxin via sprays, balloons and contaminating the water supply but they were singularly unsuccessful (Wheelis & Sugishima, 2006). There are many logistical difficulties associated with the use of this organism. Firstly, one needs to find a strain that produces the toxin (not all of them do), then it has to be reared under anaerobic conditions, the toxin then has to be recovered and finally disseminated. Although state organizations have developed *C. botulinum* for use in warfare, the information to overcome these problems is not generally available.

Identifying whether a disease outbreak is natural or a consequence of a malicious act

The first task in any disease outbreak or suspected poisoning is to identify the organism or chemical(s) involved. This would involve standard medical/ veterinary/ laboratory techniques. As mentioned previously, most disease outbreaks are a consequence of natural circumstances and unless an individual or terrorist group claims to have been involved it is only through epidemiological studies that criminal activities would be suspected (Table 10.1).

Table 10.1 Clues that would provide an early indication of the malicious spreading of a microbial pathogen. Adapted from Morse & Khan (2005)

(1) The occurrence of just one case of a rare disease without any obvious explanation. For example, smallpox is now extinct in the wild so even one case anywhere in the world would be an indication of malicious release.

(2) The identification of a microorganism with an abnormal genetic profile and/ or evidence of being 'weaponized' to enhance its dispersal. For example, in the US the anthrax strain Vollum 1B was specifically developed for use in biowarfare.

(3) The occurrence of a disease that exhibits an unusual presentation. For example, inhalation anthrax (especially in several people) would be considered unusual because it normally presents as a cutaneous or gastrointestinal disease.

(4) The occurrence of a disease that is unusual or atypical for a particular population or age group. For example, an outbreak of a disease exhibiting a measles-like rash would be unusual among adults but not surprising among children.

(5) An unusual pattern of morbidity or mortality in humans/ animals that precedes or accompanies that in animals/ humans. For example, if large numbers of horses started to suffer and die of glanders-like symptoms shortly before humans started to exhibit similar symptoms. Glanders is an infectious disease caused by the bacterium *Burkholderia mallei*. It is primarily a disease of horses but humans can be infected and it has been considered for use in biowarfare.

(6) Simultaneous clusters of a rare disease or a common one with atypical symptoms in areas that did not share a common boundary within a country or in different countries. For example, if in the course of a single week in the UK cases of foot and mouth disease were reported on farms in Caernarfon, Northumberland, and Sussex.

(7) The occurrence of large levels of morbidity or mortality among humans or animals that cannot be explained.

(8) A common disease that exhibits an unusually high level of morbidity and/or mortality or the failure of the disease to respond to normal therapy. For example, anthrax infection normally responds to penicillin treatment and its lack of response to this and other common antibiotics would be considered suspicious. Similarly, flu is debilitating but it seldom causes mortality among young healthy adults. However, it should be remembered that highly pathogenic strains of microbes arise naturally (e.g. Spanish flu, SARS) and the rapid evolution of antibiotic resistance is a well-known phenomenon (e.g. methicillin resistant *Staphylococcus aureus* (MRSA) and multi-drug resistant TB).

(9) The occurrence of a disease in an unexpected geographical region or exhibiting an unusual seasonal distribution. For example, plague occurs naturally in parts of India and Africa but an outbreak in London would be highly suspicious. However, rapid mass movement, particularly plane travel, leads to infectious diseases moving rapidly around the world. For example, around 3000 cases of malaria are treated in the UK every year although the parasite has not been native here since the early 1900s.

(10) The diagnosis of several unusual or unexplained diseases in the same patient at the same time without any obvious explanation. For example, a patient suffering from both SARS and botulism would be an exceptionally unfortunate individual. However, it is not that unusual for a person to suffer from several common disease conditions at the same time and AIDS patients and those with compromised immunity often suffer a variety of concurrent infections.

(11) The occurrence of an unusual disease that afflicts a large, disparate population. For example, an outbreak of a respiratory disease in a large heterogeneous population may suggest common exposure to an inhaled pathogen or chemical agent. This might happen following the release of a microorganism or toxin on an underground system or shopping precinct.

Forensic science becomes involved in a case of suspected bioterrorism after the identification of a pathogen and the consequent need to determine where it originated (Breeze *et al.*, 2005). In the case of pathogens that are known for their biological warfare potential this might be done from a comparison of its genetic profile or stable isotope ratio (Kreutzer-Martin *et al.*, 2004) with that of cultures held legitimately by laboratories throughout the world. Ideally there should be a list of all persons who have access to the cultures in each of these laboratories. This requires a reliable database and a level of cooperation between countries that is steadily improving but far from perfect. For example, from the molecular characteristics of anthrax bacilli it would be possible to determine whether a person was suffering from a naturally acquired infection or from a variety that had been modified in the laboratory. However, for a number of possible biological agents tests are not available to enable identification to strain or sub-strain level and/or the tests are not yet fully validated.

Anthrax

Anthrax is caused by the bacterium *Bacillus anthracis* (Fig. 10.5) that owes much of its pathogenicity to the secretion of a toxin containing three proteins. One of these, protective antigen, facilitates the entry of two toxic enzymes, lethal factor and oedema factor, into the host cell. The genes coding for the toxin are found on plasmids and these have now been genetically sequenced. In addition, the full genomic sequence of several strains of *B. anthracis* is now available which will facilitate diagnosis and improve our understanding of how it causes disease. The bacteria are relatively large and rod-shaped. In clinical specimens the bacteria have a capsule and are usually seen singly or in twos. Spores are produced when growth conditions become sub-optimal and are therefore not found in living tissues. They

Figure 10.5 Light microscope photograph of anthrax bacilli stained with Loeffler's polychrome methylene blue. In culture, the bacilli often grow in chains. (Magnification ×1000) (Reproduced from Hart, T. and Shears, P. (1996) *Color Atlas of Medical Microbiology*. Copyright © 1996, Mosby Wolfe.).

are, however, formed when the organism dies or when the bacteria are deposited on the soil. By contrast, in culture anthrax bacilli form long chains, they do not have a capsule, and spores are produced. Anthrax spores are notoriously difficult to destroy and they can survive in soil for many years. In 2004 a batch of anthrax, which was shipped to researchers at the Children's Hospital and Research Centre at Oakland USA, for vaccine production was found to contain live and still lethal bacteria despite having been heat inactivated and tests being done to check that they no longer grew in culture (Anon, 2004b). The ability of the spores to survive for long periods of time coupled with the possibility of engineering the bacteria to express enhanced pathogenicity and resistance to antibiotics has led to them being considered prime candidates for biological warfare agents.

Clinically, there are three principle forms of anthrax: cutaneous, gastrointestinal and inhalation. Cutaneous anthrax results from the bacteria gaining access to the body via wounds and breaks in the skin surface and initially manifests itself as a painless papule that subsequently develops into a black necrotic ulcer. The name anthrax is derived from the Greek for coal. The lesion results in an inflammatory response and becomes surrounded by an extensive red oedema. Many cases of cutaneous anthrax eventually resolve themselves but infections can be fatal if there are complications such as invasion of the blood supply leading the bacteria being disseminated around the body. Cutaneous anthrax is the most common form of human disease and was often acquired by persons who worked with infected animals or animal products such as leather. It used to be a relatively common disease and was known as 'woolpackers' disease' in the UK from its association with the trade. Surprisingly, the fingers are seldom affected but lesions are common on the hands, arms, and neck (shaving often results in skin nicks through which the bacteria can invade). Following improvements in animal care, hygiene and the availability of an effective animal vaccine, the disease is seldom seen in humans in developed countries. In the UK there were only 14 cases between 1981 and 2000 whilst in the USA there was an average of five cases a year between 1955 and 1999. Gastrointestinal anthrax is acquired through eating food contaminated with anthrax spores and causes lesions in the gastrointestinal tract, through which the bacteria invade the rest of the body causing septicaemia and death. Naturally acquired cases of inhalation anthrax are rare but they do occur. It results from breathing in the spores and is therefore the form most likely to result from a bioterrorist attack. Anthrax spores are 1–2 μm in size and therefore of a size likely to be retained in the lungs. They are ingested by macrophages present in the alveoli, in which they germinate, and are then transported to the nearby lymph nodes and via the bloodstream around the body. Initial symptoms resemble those of a common cold but once the lungs are infected severe breathing difficulties ensue and the disease is rapidly fatal. Human-to-human transmission of anthrax is not thought to occur naturally.

The 'gold standard' for the identification of anthrax is to culture it on sheep blood agar at 37 °C under aerobic conditions. This, however, will only provide species level identification and for forensic purposes molecular and chemical tests are required to determine the strain of the infection and its likely source. Some of the molecular tests are discussed in the following case studies and Budowle et al. (2005) provide a review of the latest attempts at developing 'lab-on-a-chip' techniques for identifying anthrax strains.

Case Study: Dissemination of anthrax by the Aum Shrinikyo cult

Different strains of anthrax differ markedly in their pathogenicity, a trait used both in the development of vaccines where a lack of pathogenicity is a virtue and the development of biological warfare agents where high pathogenicity is required. Followers of the Aum Shrinikyo cult in Japan were probably not aware of this when they released anthrax spores in Kameido (which is situated near Tokyo) in 1993 (Wheelis & Sugishima, 2006). Cult members sprayed crude liquid culture medium containing a suspension of spores from the 8th floor of a building over a period of time. Nobody fell ill although several people complained of the smell. To allay the locals' complaints a cult spokesman called a press conference in which he claimed that the smell was connected to the cult's religious practices. The locals were unimpressed, their complaints continued, the police became involved and the cult ceased spraying and vacated the site. Subsequent analysis indicated that the cult's apparatus was crudely made, leaky, and produced large droplets rather than an effective aerosol. Furthermore, the spore concentration was probably too low to be effective even if a more virulent strain had been used. Their true intentions only came to light later following the cult's more successful release of Sarin gas on a Tokyo underground station which resulted in several deaths and a thorough investigation of the cult's activities. Retrospective analysis of fluid samples collected and stored by the authorities as a result of the complaints about the smell resulted in the discovery of anthrax spores from which it was possible to grow colonies of bacteria. DNA isolation and MLVA (Multiple Locus Variable number tandem repeats Analysis) genotyping of these colonies demonstrated that the anthrax belonged to the Sterne vaccine strain (Keim et al., 2001). MLVA typing is useful for strain identification of a number of bacteria (e.g. *Yersinnia pestis* [plague], *Mycobacteria tuberculosus* [TB]) although it is not suitable for all species. The technique involves identifying and analysing suitable loci (markers) on the bacterial genome (Fig. 10.6). For example, Lista et al. (2006) have identified 25 suitable markers for identifying strains of anthrax. These loci are then amplified and the PCR products are separated on the basis of their length (size) – the length being determined by the number of tandem repeats they contain. Strains can then be distinguished on the basis of the number of loci at which they differ. The Sterne34F2 vaccine strain is widely available in Japan where it is used in the preparation of animal vaccines and it is therefore probable that a cult member obtained a vial of animal vaccine from which a large number of bacteria were subsequently cultured and released. Thankfully, being nonpathogenic, it posed little risk to humans.

Case Study: The Washington anthrax letter attacks

Between September and October, 2001 letters containing anthrax spores were sent via the normal mail to media outlets and two US senators in Washington DC and New York City, USA. This happened on two separate occasions and resulted in 18 confirmed cases of anthrax from which five people died. Of these

Allele (based on number of repeats)

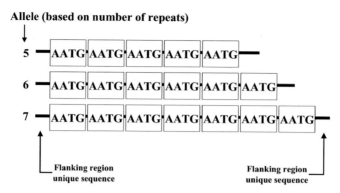

Figure 10.6 Diagrammatic representation of multiple-locus variable number tandem repeat analysis (MLVA). The three amplicons differ in size according to the number of repeat sequences they contain. The amplicons can be distinguished by, for example, electrophoresis, on the basis of their size. (Reproduced from Budowle, B. *et al.*, (2005) Genetic analysis and attribution of microbial forensics evidence. *Critical Reviews in Microbiology*, **31**, Informa Healthcare, with permission from Taylor & Francis)

cases, seven people contracted cutaneous anthrax, all of whom survived, and 11 contracted inhalation anthrax of whom only six survived. This indicates the high incidence of inhalation anthrax when contracted via a bioterrorism attack and the greater difficulty in treating this form of the disease.

The letters were addressed by hand and contained brief handwritten notes indicating the nature of their contents, threats to America and Israel and ended with 'Allah is great'. The letters were crudely written and included spelling mistakes and the intention of whoever wrote them was obviously to cast suspicions onto a Muslim extremist organization. Subsequent investigations indicated that this was highly unlikely although some people remain desperately keen to implicate the usual suspects of Iraq and Syria. The postmarks on the letters indicated that they were sent from Princeton, New Jersey and although this area included several hundred postboxes only one of them tested positive for anthrax bacilli.

The letters were tightly sealed shut but it is virtually impossible to place dried anthrax spores inside a standard envelope without contaminating their outer surface (Beecher, 2006). Consequently, several people who unwittingly handled the envelopes after they were posted as well as those who opened them became infected. This would suggest that whoever sent the letters had access to facilities that reduced personal risk when filling them and did not lick the seals! When they were opened the letters released a powdery substance. In some of the letters the powder was of a coarse brown consistency whilst in others it had a fine white appearance. Subsequent analysis indicated that the powders represented different grades of the same strain of anthrax. Somewhat embarrassingly, molecular analysis indicated that the anthrax belonged to the 'Ames' strain that was developed in America. Furthermore, whilst a number of laboratories in the US and elsewhere held this strain at the time, all could trace their stocks back to the US Army Medical Research Institute of Infectious Disease (USAMRID). The origin of the Ames strain was a sick Texan cow that died in 1980. Although it is a highly virulent strain it is reportedly mainly used in the development of vaccines. There is still confusion in the literature whether the anthrax spores were 'weaponized' – for example, being milled into a size that is small enough to remain in suspension in the air and/or coated with an additive such as silica to reduce clumping and aid aerosolization. Once aerosolized, spores can be transmitted long distances on air currents and contaminate surfaces over a wide area. Initial reports indicated that at least some of the anthrax letters contained 'weaponized' spores (Inglesby et al., 2002). However, according to Beecher (2006), a 'widely circulated misconception is that the spores were produced using additives and sophisticated engineering supposedly akin to military weapon production'. This is significant because it increases the number of individuals capable of having sent the letters. Unfortunately, Beecher did not provide any references or data to back up this assertion and until this information is made available doubts will remain (Mereish, 2007). Scanning electron microscope photographs of some of the spores indicate exceptionally pure samples with no evidence of silica particles or 'milling'. According to Ember (2006), there is a report of silica having been detected among spore samples using X-ray mass spectrometry but others have suggested that this is owing to confusing the peak for compound silica ($SiO2$) with that for the element silicon (Si). Although silicon has been reported in the spore coat of *Bacillus cereus* other workers have failed to find confirm its presence in significant amounts and its presence in anthrax spores needs to be verified. If silica was present with the spores its chemical and physical properties could be used as a 'signature' to determine where it originated and hence narrow down the list of sources. In a more recent news item (Greenemeier, 2008) it is stated that scientists demonstrated the presence of silicon/ silica within the anthrax spores rather than on their surface as long ago as 2002 but have not yet published their results. Unfortunately, the news report uses the terms silicon and silica as though they were the same. If the substance being measured was silica and it was present on the inside rather than the outside it would be further evidence that the spores were not weaponized.

Bomb pulse dating (Chapter 1) was used to demonstrate that the spores were produced sometime in the previous two years before they were despatched in the letters (Tuniz et al., 2004). Once anthrax spores are formed their metabolism effectively ceases. Therefore, they have a ^{14}C signature that indicates their year of production.

It has been reported that isotope analysis indicates that the water used in the culture of the anthrax spores originated from the north-eastern states but as yet full details of this work are not available (Ember, 2006). If this is correct then it is further proof that the attack was fully organized within the USA.

By 2008, the FBI were focusing their attentions on Bruce Ivens, a scientist at USAMRID but he committed suicide on 29 July of that year before charges could be brought. In 2007, the FBI had established that the anthrax used in the letters belonged to the same Ames strain (RMR-1029) as that used in the laboratory in which Ivens worked (Bhattacharjee & Enserink, 2008) and claimed that further evidence incriminated Ivens as the sole person involved in the crime. In order to establish the source of the anthrax used in the letters, scientists had to develop new analytical methods and full details of these and their findings have yet to be published although a brief description is provided by Enserink (2008). In short, the investigators grew anthrax cultures on agar plates from the spores contained in the letters and then identified colonies that looked different in some way from the majority. For example, they had a slightly different colour, rougher edges or were smaller. Each colony on an agar plate would represent the progeny of a single spore and therefore these 'odd' colonies could be expected to represent different phenotypes to the majority. These 'odd' colonies were then subjected to full genome analysis – a huge undertaking – to identify how their sequences differed from the majority. Molecular tests were then developed to identify these points of difference and it then became possible to match the anthrax from the letters with that from a suspect source by looking for the presence of these 'odd' phenotypes. In this way, the scientists were then able to screen anthrax samples from laboratories from within the USA and the rest of the world.

During the initial stages of the investigation, the FBI believed that they had identified Steve Hatfill, another government scientist, as the culprit but he has since been cleared of any involvement and awarded US $5.8 million in damages. It is therefore unfortunate that the strength of the case against Bruce Ivens will never be tested in a court of law.

Therefore, as 2008 draws to a close and 7 years after the letters were sent, we can be certain that the anthrax belonged to the Ames strain and in particular the RMR-1029 strain but not whether it was weaponized. We can be confident that the anthrax was produced shortly before the letters were sent and therefore it did not originate from an old stockpile that had somehow made its way onto the black market. There also appears to be some scientific proof that the anthrax was grown in the USA. In addition, unlike some terrorist bomb outrages it is highly unlikely that the attack could have been perpetrated from the kitchen of the average citizen. However, because the principal suspect has died there cannot be a court case and this means that suspicions will remain in the minds of many people. To counter this, the FBI has stated that it will reveal at least some of its evidence via peer-reviewed scientific publications in the near future. The one thing that can be certain at the end of this sorry saga is that, although there have been numerous hoaxes and false alarms concerning anthrax since the fatal letters were sent, there have been no repeats anywhere in the world. This would suggest that this style of terrorism is not easy, simple or cheap.

Plague

The devastating outbreaks of bubonic plague that afflicted the UK and other European countries during the middle ages have become irrevocably burnt onto the national psyche and the mere mention of 'plague' is sufficient to cause anxiety amongst the population. Because the last serious outbreaks in the UK took place over 200 years ago, many people erroneously believe that plague now exists only in the mists of history. However, it is still recorded in various parts of Africa and Asia and there is a natural cycle between prairie dogs in North America from which

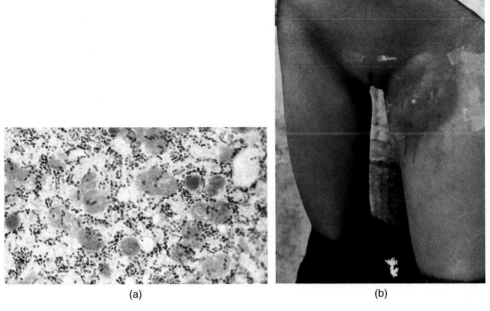

(a)   (b)

Figure 10.7 *Yersinia pestis*, the causative agent of plague. (a) Bacteria in a liver smear. The bacteria are the abundant small, round, pale lilac-stained objects. (b) Suppurating swellings (buboes) are a classic feature of bubonic plague. (Reproduced from Peters, W. and Gilles, H.M. (1977) *A Colour Atlas of Tropical Medicine and Parasitology*, copyright Wolfe Medical Publications Ltd.)

humans are occasionally infected. Although there have been suggestions that the great plagues of Black Death in the Middle Ages were due to an unidentified virus related to the Ebola and Marburg viruses (Scott & Duncan, 2004), the majority of microbiologists continue to believe that the bacterium *Yersinia pestis* was responsible. Plague exists in three principal forms, bubonic, pneumonic and septicaemic, all of which are caused by *Y. pestis* (Fig. 10.7). Bubonic plague is characterized by the formation of painful swellings (buboes) of the lymph nodes – principally the groin, axillary, and cervical nodes, and may subsequently develop into pneumonic and/or septicaemic plague. Provided that it is treated early enough, bubonic plague has a low fatality rate – approx 8%. Pneumonic plague may result directly from droplet infection or develop from bubonic plague and is rapidly fatal. Septicaemic plague may develop without any previous signs of disease or from the bubonic / pneumonic forms. It results in septic shock and has a high fatality rate – 33% or more. Plague has the distinction of being involved in the first documented use of biological warfare. When the Mongol armies encamped around the Black Sea port of Kaffa in 1346 they were afflicted by an outbreak of plague and lost so many men that they had to call off the siege. Before they left they catapulted the bodies of men who had died of plague over the city's battlement walls. Plague did break out among the defenders but whether it originated from the projected corpses or via rats outside and inside the castle walls passing the infection to one another will never be known.

Plague is usually transmitted by fleas or by droplet infection in the case of the less common pneumonic form. Fleas leave the body when it starts to cool so unless a corpse was sent airborne whilst still warm many of the fleas would have left in search of a better meal. Droplet infection from a corpse is unlikely and contaminated clothing is thought to present a low risk of infection.

Owing to changes in living conditions, the cycling of disease between rats (the normal host of the bacteria) and humans would be difficult to establish in developed countries. Therefore, releasing infected rats (which would also need to be carrying the rat flea *Xenopsylla cheopsis* to transmit the infection) would be unlikely to result in more than the death of the rats. During the Second World War, Japan maintained an establishment known as Unit 731 in China to carry out research into the use of biological warfare agents. Plague was one of several pathogens they investigated and they carried out field experiments in which they spread it in various ways – including releasing thousands of infected fleas (Harris, 2002). Although they succeeded in killing many local people they also caused the deaths of many of their own troops. The logistics of rearing of thousands of plague-infected fleas makes it an unlikely exercise for any would-be terrorist. The disease would therefore probably be released in droplet form and the bacteria would have to be in sufficient concentration for a person to breathe in 100–500 bacilli. The disease would therefore have to be released from some sort of vapour-producing device within an enclosed space. Unlike anthrax, *Y. pestis* is a nonspore forming bacteria so there is not the risk of long-term contamination. Like all terrorist weapons, however, the threat far outweighs the potential to cause deaths.

Smallpox

Smallpox is caused by a Variola virus (Fig. 10.8), and is related to cowpox virus and monkeypox virus; chickenpox virus is a herpes virus and is not related to smallpox. The term 'pox' is an old word and refers to a disease that causes the formation of pustules. Smallpox gained its common name, not in reference to the size of the pustules but to distinguish it from syphilis that was referred to as the 'great pox'. It is spread by droplet infection, by direct contact with an infectious person or from contact with clothing or bedding previously used by a person suffering from smallpox. The incubation time is approximately 3 weeks after which flu-like symptoms are expressed; a person becomes infectious 2–3 days before these symptoms appear. A few days after the onset of the disease, the patient's temperature returns to normal and they develop a vesicular rash on the face and limbs that spreads to cover the whole body. Lesions in the mouth and throat release large amounts of virus that are then aspirated in the saliva and droplets in the breath. Over the course of 7–14 days, these vesicles enlarge, rupture and start to heal over – at which point the patient ceases to be infectious. However, in about half of all cases, the fever then returns and widespread internal haemorrhage leads to death within a few days. The disfiguration and high mortality makes smallpox a very frightening disease but it is also a rare example of how a pathogen can be eradicated – the world was officially declared free of smallpox in 1979. Eradication was possible because an effective vaccine was available, everyone who is infected with the virus develops the

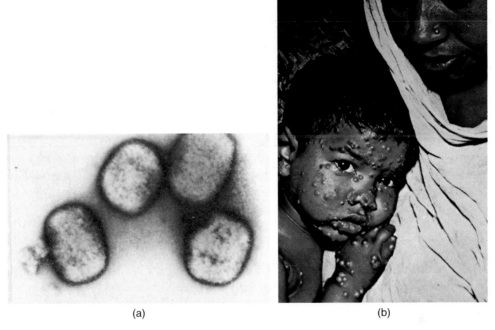

(a)   (b)

Figure 10.8 Smallpox. (a) Transmission electron micrograph of the smallpox virus. (b) Child suffering from smallpox. The skin lesions can become secondarily infected with bacteria; pneumonia and bacterial septicaemia are frequent complications. Although the lesions are fairly characteristic, confirmation of the disease depends upon laboratory identification. (Reproduced from Peters, W. and Gilles, H.M. (1977) *A Colour Atlas of Tropical Medicine and Parasitology*, copyright Wolfe Medical Publications Ltd.)

symptoms (they could therefore be identified and treated before the disease was transmitted further), and all countries in the world were willing to cooperate with the WHO eradication campaign.

Although the disease was eradicated, stocks of the virus were maintained in the USA at CDC Atlanta and in Russia at the Vektor Institute, Novoskibirsk, ostensibly so that vaccines could be developed should the disease re-emerge. There have been continuous rumours that the security of these stocks may have been breached and that other sources are held elsewhere in the world. The countries and organizations that are accused of holding illicit stocks all deny this but it is impossible to prove a negative and the consequences if the disease was released are dire. For example, in the UK, vaccination against smallpox ceased in the 1960s and therefore by 2005 anyone under 40 would not have been vaccinated whilst the immunity of older people would have declined – therefore the majority of the population would be susceptible. Computer models of how the disease would spread indicate that within 3 weeks, ten people deliberately infected from an initial 'index case' would each have inadvertently infected a further ten people (i.e. a total of 100 people). It should be remembered from the symptoms listed above that the index person would have a very short 'window of opportunity' during which they would be infectious but their diseased condition was not patently obvious from the facial rash – and they would probably be feeling very ill. Regardless of how the virus was introduced into the population, the diagnosis of smallpox anywhere in the world would be an indication of the release of illicitly held virus. The UK government has drawn up contingency plans for dealing with smallpox, as it has for all biological and chemical weapons (www.dh.gov.uk, www.hpa.org.uk), in which it is expected that the virus would spread rapidly owing to the low immunity in the population, the speed and frequency with which people travel during their daily lives, and the unfamiliarity of people, including doctors and nurses, with the symptoms of the disease. Alert levels (1–6) are to be declared by the Chief Medical Officer and if there is 'an overt release' (i.e. the terrorists gave a specific warning), the contingency plan would be led by the police. If the release is 'covert' (i.e. there was no warning but cases of disease were positively identified), the plan would be led by the Department of Health. Regional smallpox diagnosis and response groups have been established which include a smallpox diagnostic expert and smallpox management and response teams. Confirmation of the disease would require electron microscopy and PCR analysis of the virus. At Alert Level 3 (outbreak occurring in the UK), the contingency plan is to control the spread of smallpox and minimize disruption and inconvenience by identifying positive cases and isolating them in 'smallpox centres' then identifying all their possible contacts and vaccinating them. Mass vaccination of the whole population would only be considered at Alert Level 4 when there would be large multiple outbreaks across the country in which the cases could not be linked. The value of vaccinating the general population has to be balanced against the 1 in a million chances of death resulting from reactions against the vaccine itself (in a population of greater than 60 million, this is a serious consideration) and the problems of supplying sufficient doses of vaccine.

Agroterrorism

Although most of the focus on bioterrorism concerns diseases and toxins that cause human mortality, agricultural bioterrorism has the potential to cripple a country's economy. This was exemplified by the UK Foot and Mouth Disease (FMD) outbreak of 2001 in which the direct and indirect costs to the country were put at over £12 billion. There is no suggestion that this outbreak was a consequence of a malicious act but it indicates how expensive and disruptive such incidents can be.

During the twentieth century a wide variety of pathogens were investigated by governments for their potential to affect agriculture (Wheelis et al., 2006). These have included viruses, bacteria, fungi, and insects either to destroy crops or to kill or incapacitate farm animals (Tables 10.3 & 10.4). The prime focus of this research was to disrupt the enemy's economy and to reduce their ability to feed themselves. Very few of these agents were used in anger and even fewer had any significant impact. However, the use of herbicides has been more widespread – such as the use of Agent Orange (this contained n-butyl 2,4-dichlorophenoxyacetate (2,4-D) and n-butyl 2,4,5-trichlorophenoxyacetate (2,4,5-T) by the Americans in Viet Nam during the 1960s – and devastating. Agriculture is particularly vulnerable to terrorist attack because it takes place over large unsecured stretches of land that are, for the most part, impossible to police. In the case of crops, many of them are grown as monocultures over large areas and their genetic similarity makes them vulnerable to pathogen attack. Farm animals are vulnerable from being brought together and mingled when taken to market or for slaughter and they are routinely transported long distances within and between countries thereby facilitating the spread of disease. In some cases farm animals are kept indoors at high densities and although the buildings can be secured to an extent the animals' proximity to one another makes them vulnerable to infectious pathogens. Furthermore, the presence of plant or animal disease within a country may make it impossible to export produce and home consumption may drop – for example, BSE (bovine spongiform encephalopathy) in UK cattle prevented the export of meat to the European Union for many years.

Despite the inherent vulnerability of the agricultural system to malicious spreading of pests and diseases it is considered a relatively low risk as a target from terrorist groups. This is because it lacks the obvious impact of human casualties or damage to the infrastructure (e.g. an explosion). However, should such cases occur, the pathogens could be identified and traced using similar techniques to those described previously for human pathogens.

Table 10.2 A brief summary of some human pathogens that have either been investigated for their potential as biowarfare agents in the past or are currently considered likely to be spread maliciously. This is an abbreviated list and many others have been investigated for their biowarfare potential by governments

Micro-organism	Transmission
Viruses	
Smallpox	Inhalation
Hantavirus Pulmonary Syndrome (HPC)	Inhalation
West Nile virus	Mosquito bite
Ebola virus	Contact
Marburg virus	Contact, sexual transmission possible
Lassa Fever virus	Contact, sexual transmission possible
Bacteria	
Anthrax, *Bacillus anthracis*	Inhalation, ingestion, breaks in skin
Plague, *Yersinia pestis*	Flea bite, inhalation (pneumonic plague)
Salmonella food poisoning, *Salmonella typhimurium*	Ingestion
Typhoid, *Salmonella typhi*	Ingestion
Cholera, *Vibrio cholerae*	Ingestion
Tularaemia, *Francisella tularensis*	Tick bites, ingestion, contact
Brucellosis, *Brucella* spp,	Contact, ingestion
Botulism, *Clostridium botulinum*	Ingestion
Rocky Mountain spotted fever, *Rickettsia rickettsii*	Tick bite
Epidemic typhus, *Rickettsia prowazekii*	Louse bite/faeces

Table 10.3 A brief summary of some domestic animal pathogens that have either been investigated for their potential as biowarfare agents in the past or are currently considered likely to be spread maliciously. This is an abbreviated list and many others have been investigated for their biowarfare potential by governments

Micro-organism	Host	Transmission
Viruses		
Foot and mouth disease	Cattle, sheep, pigs	Contact, contamination
Rinderpest	Cattle, (sheep and pigs can be infected)	Contact, ingestion
African swine fever	Pigs	Contact, ingestion, tick vectors
Avian influenza (Fowl plague)	Birds, can be transmitted to man	Contact, contamination
Newcastle disease	Birds, can be transmitted to man	Contact, contamination
Bacteria		
Brucellosis, *Brucella* spp.	Cattle, can be transmitted to man	Contact, ingestion
Glanders, *Burkholderia mallei*	Horses + other equids, can be transmitted to man	Contact, inhalation
Melioidosis, *Burkhoderia pseudomallei*	Cattle, sheep, goats, pigs, cats + many others, can be transmitted to man	Contamination, ingestion
Psittacosis, *Chlamydia psittaci*	Birds, can be transmitted to man	Inhalation

Table 10.4 A brief summary of some plant pests and pathogens that have either been investigated for their biowarfare potential in the past or are currently considered likely to be spread maliciously

Micro-organism	Host	Transmission
Viruses		
Hoja Blanca virus	Rice	Leafhoppers (sap-sucking insects)
Bacteria		
Candidatus Liberibacter (Huanlongbing)	Citrus trees	Psyllids (sap-sucking insects)
Fungi		
Stem rust, *Puccinia gramminis*	Wheat + cereals	Air currents
Rice blast, *Piricularia oryzae*	Rice	Air currents, infested rice grain
Potato late blight, *Phytophthora infestans*	Potatoes	Air currents, infected seeds potatoes
Southern blight, *Sclerotium rolfsii*	Many broadleaved crops, tomatoes, peppers, soybean	Infected soil
Insects		
Colorado beetle *Leptinotarsa decemlineata*	Many broadleaved crops, tomatoes, peppers, soybean	Flight, contaminant within imported agricultural produce

Case Study: Foot and mouth disease in the UK

FMD virus (FMDV) is an RNA virus, containing a single strand of RNA within a capsid. Taxonomically, FMDV is classed as a picornavirus belonging to the genus Rhinovirus. It is highly contagious and infects cattle, sheep, goats, pigs and other cloven hoofed mammals. Although humans can be infected it happens very rarely. [Hand foot and mouth disease (HFMD) is a quite separate disease caused by an unrelated type of virus – an enterovirus – that commonly infects children]. FMD is spread by contamination with virus-infected saliva and faeces. It is not normally fatal in adult animals although the young may suffer high levels of mortality. Although not fatal to adults it is very debilitating and leads to a loss of production (e.g. growth, milk yield). Blisters and sores form on the feet leading to lameness and they may also be found around the nose and tongue. No cases of FMD had been diagnosed in the UK for 20 years until it was identified in a pig sent to slaughter in February 2001. However, by this time the disease was already well established in several parts of the country, its spread being aided by the large scale movement of farm livestock which is common practise nowadays. The official policy to control FMD was to slaughter all animals likely to have had contact with an infected animal regardless of whether they themselves exhibited symptoms and to restrict the movement of animals and humans over large areas of the country. After about 11 months the outbreak was over but had caused the slaughter of over 6 million animals and widespread disruption.

Like other RNA viruses, FMDV exhibits rapid multiplication and mutation rates and this can make identifying the source of an infection complicated. It is currently possible to identify the strain of an FMDV but not where it came from. The strain that caused the 2001 outbreak in the UK belonged to the type O pandemic strain (PanAsia) – one that has been spreading and causing outbreaks in many parts of the world for over 20 years. Isolates of this strain from different countries and regions within countries indicates that they are all closely related. However, a particularly close match was found to a South African isolate but whether this means that this was the source of the UK outbreak or that the UK and South African isolates shared a common source is not known (Mason et al., 2003). No evidence has been presented to explain when or how the virus was introduced into the UK although as mentioned in Chapter 8 some people suspect that bushmeat or illegal meat imports may have been to blame.

A second FMD outbreak of FMD occurred in Surrey in August to September 2007. It was brought under control more quickly than the 2001 outbreak although it too resulted in the slaughter of many animals and a lot of disruption. After the first case was diagnosed, the origins of the disease were quickly established – and proved to be highly embarrassing. The virus almost certainly came from a research and development site at Pirbright at which three organizations, the government Institute of Animal Health, Merial Animal Health Ltd, and Stabilitech Ltd all worked with FMDV to varying extents. The virus probably escaped via a leaky drainage system connecting the site to sodium hydroxide treatment tanks (that would have killed the virus). Phylogenetic analysis indicated that the FMDV responsible for the outbreak belonged to the strain

O1 BFS and was extremely closely related to that used at Pirbright. However, all three organizations at Pirbright used virus with very slight differences from one another and therefore it was not possible to categorically state which organization the virus originated from (www.hse.gov.uk/press/2007/e07032.htm). Phylogenetic analysis indicated that the viruses found at all infected farms were very closely related and it was possible to predict the sequence in which nearby farms acquired their infections. How the virus was transmitted from Pirbright to the farms is uncertain but probably involved virus infected soil and water contaminating footwear and vehicles. The first farm to be infected was only 3 miles from Pirbright with a road that directly linked it to the site. The infection was then probably transmitted between farms although at least one of them may also have been directly infected from Pirbright. This case illustrates how difficult it can be for even highly regulated officially sanctioned institutions to maintain effective biosecurity. Consequently, bioterrorists are likely to find it even harder to keep their activities secret and to prevent their microbes from escaping too early.

Future directions

The development of microbial profiling offers enormous potential as a means of linking people, animals, plants and objects to one another and/ or to a geographical region. Changes in microbial profiles could potentially be useful in estimating the PMI and/ or the length of time a body (or any other organic or inorganic object) has rested upon the soil surface. However, before this is possible the techniques will need to be refined, standardized and validated so that the evidence is sufficiently robust to be used in a court of law.

Molecular analysis of STDs offers the potential for linking assailant and victim in cases of sexual assault in which human DNA evidence is lacking or unreliable. Despite decades of sex education and health awareness campaigns, STDs remain remarkably common in the UK. Consequently, they are likely to be involved in many cases of sexual assault. As already mentioned, the appropriate standards and validations would be required and to be truly effective a national database would be needed for each STD.

Bioterrorism always captures the headlines but it will (hopefully) remain a rare event. Despite this, it is inevitable that large amounts of resources will continue to be spent on methods for detecting potential pathogens under 'field' conditions. These will probably involve refinements of the 'lab-on-a-chip' techniques. (There is always a market for paranoia and one can already purchase one's own anthrax detection kits on the internet – not that they are particularly reliable.) As much as anything, these devices are needed to quickly identify the many hoaxes and suspicious powders, packages etc. that are reported every year. The technology is also needed for identifying and tracking the spread of natural disease outbreaks (such as FMD and Bird Flu) and distinguishing such events from malicious or feckless disease transmission.

Quick quiz

(1) Why does extensive blood loss slow down the rate of decay?

(2) Why are the bacteria involved in decay predominantly anaerobic?

(3) How might soil bacteria prove useful in identifying crime suspects?

(4) Give an example of how a defendant might plead 'suffering an underlying infectious disease' as a mitigating circumstance for committing a crime.

(5) How would it be possible to identify a person recklessly spreading HIV infections?

(6) State three possible early indications that a disease might be being spread maliciously

(7) Why do anthrax spores differ in their pathogenicity?

(8) What are the three principal forms of plague?

(9) Why would even one case of smallpox anywhere in the world be evidence of illicit activity on behalf of a terrorist or government organization?

(10) With the aid of a named example, briefly discuss what is meant by the term 'agroterrorism'

Project Work

Title

Soil microbial profiles as forensic indicators

Rationale

The soil microbial flora is extremely diverse but there are few studies on its usefulness as a forensic indicator linking a person, animal or object to a location is not known.

Method

The microbial profiles would be determined by procedures such as ribotyping. Soil samples would be taken from the clothing and the soles of shoes and the microbial profiles compared with those obtained from the location where the person had been walking. They would also be compared with samples from other locations in the area and the effect of storing the samples on the microbial profile would also be assessed. For example, does the microbial profile change if the sample is not analysed for two or more days?

Title

The effect of a dead body on the underlying soil microbial flora

Rationale

It is sometimes useful to know how long a dead body has been resting at a location.

Method

The soil microbial flora might be assessed either by molecular profiling techniques as detailed above or by culturing them using traditional methods. A dead animal or tissues at varying stages of decay would be placed on the surface of the ground and the underlying soil sampled at set intervals. Changes in the microbial flora could be estimated at varying levels of detail from the total number of culturable bacteria, fungi and protozoa to studying how the molecular profile alters with time. Any changes could also be linked to other soil physical and chemical characteristics, such as pH, organic matter content, nitrogen, phosphate and protein levels as well as different soil types (e.g. comparing a heavy clay soil with a sandy loam).

References

Abbas, A. and Rutty, G.N. (2005) Ear piercing affects earprints: the role of ear piercing in human identification. *Journal of Forensic Sciences* **50**, 386–392.

Abe, S. *et al.* (2003) A novel PCR method for identifying plankton in cases of death by drowning. *Medicine Science and the Law* **43**, 23–30.

Acree, M.A. (1999) Is there a gender difference in fingerprint ridge density? *Forensic Science International* **102**, 35–44.

Adair, T.W. and Dobersen, M.J. (1999) A case of suicidal hanging staged as homicide. *Journal of Forensic Sciences* **44**, 1307–1309.

Adair, T.W. *et al.* (2006) The use of luminol to detect blood in soil one year after deposition. *International Association of Bloodstain Pattern Analysis News.* 3–7.

Adair, T.W. *et al.* (2007) Appearance of chemical burns resulting from washing of a deceased body with bleach. *Journal of Forensic Sciences* **52**, 709–711.

Adams, B.J. (2003) Establishing personal identification based on specific patterns of missing, filled, and unrestored teeth. *Journal of Forensic Sciences* **48**, 487–496.

Adams, Z.J.O. and Hall, M.J.R. (2003) Methods used for the killing and preservation of blowfly larvae, and their effect on post-mortem larval length. *Forensic Science International* **138**, 50–61.

Akiba, N. *et al.* (2007) Fluorescence spectra and images of latent fingerprints excited with a tunable laser in the ultraviolet region. *Journal of Forensic Sciences* **52**, 1103–1106.

Alasalvar, C. *et al.* (2002) Differentiation of cultured and wild sea bass (*Dicentrarchus labrax*): total lipid content, fatty acid and trace mineral composition. *Food Chemistry* **79**, 145–150.

Alberink, I. and Ruifrok, A. (2007) Performance of the FearID earprint identification system. *Forensic Science International* **166**, 145–154.

Allery. J-P. *et al.* (2001) cytological detection of spermatozoa: comparison of three staining methods. *Journal of Forensic Sciences* **46**, 349–351.

Altun, G. (2006) Planned complex suicide: report of three cases. *Forensic Science International* **157**, 83–86.

Alunni-Perret, V. *et al.* (2005) Scanning electron microscope analysis of experimental bone hacking trauma. *Jornal of Forensic Sciences* **50**, 796–801.

Ambach, W. *et al.* (1992) Corpses released from glacier ice – glaciological and forensic aspects. *Journal of Wilderness Medicine* **3**, 372–376.

Amendt, J. *et al.* (2007) Best practice in forensic entomology – standards and guidelines. *International Journal of Legal Medicine* **121**, 90–104.

Ames, C. and Turner, B. (2003) Low temperature episodes in development of blowflies: implications for postmortem interval estimation. *Medical and Veterinary Entomology* **17**, 178–186.

Essential Forensic Biology, Second Edition   Alan Gunn
© 2009 John Wiley & Sons, Ltd

Amorim, A. and Pereira, L. (2005) Pros and cons in the use of SNPs in forensic kinship investigation: a comparative analysis with STRs. *Forensic Science International* **150**, 17–21.

Anderson, G.S. (1999) Wildlife forensic entomology: determining time of death in two illegally killed black bear cubs. *Journal of Forensic Sciences* **44**, 856–859.

Anderson, G.S. (2000) Minimum and maximum development rates of some forensically important Calliphoridae (Diptera). *Journal of Forensic Sciences* **45**, 824–832.

Anon (2004a) Isotopic techniques to provide fresh evidence. *Chemistry and Industry*, Issue 19, 4th October, 2004. 14.

Anon (2004b) Live anthrax bacteria inadvertently sent to vaccine researchers. *Nature* **429**, 692.

Archer, M.S. (2003) Annual variation and departure times of carrion insects at carcasses: implications for succession studies in forensic entomology. *Australian Journal of Zoology* **51**, 569–576.

Archer, M.S. and Elgar, M.A. (1998) Cannibalism and delayed pupation in hide beetles, *Dermestes maculates* DeGeer (Coleoptera: Dermestidae) *Australian Journal of Entomology* **37**, 158–161.

Archer, M.S. and Elgar, A.A. (2003) Effects of decomposition on carcass attendance in a guild of carrion-breeding flies. *Medical and Veterinary Entomology* **17**, 263–271.

Archer, M.S. and Ranson, D.L. (2005) Potential contamination of forensic entomology samples collected in the mortuary: a case report. *Medicine Science and the Law* **45**, 89–91.

Archer, M.S. *et al.* (2006) Fly pupae and puparia as potential contaminants of forensic entomology samples from sites of body discovery. *International Journal of Legal Medicine* **120**, 364–368.

Archer, N.E. *et al.* (2005) Changes in lipid composition of latent fingerprint residue with time after deposition on a surface. *Forensic Science International* **154**, 224–239.

Arnaldos, M.I. *et al.* (2005) Estimation of postmortem interval in real cases based on experimentally obtained entomological evidence. *Forensic Science International* **149**, 57–65.

Asamura, H. *et al.* (2004) Unusual characteristic patterns of postmortem injuries. *Journal of Forensic Sciences* **49**, 1–3.

Audi, J. *et al.* (2005) Ricin poisoning. A comprehensive review. *Journal of the American Medical Association* **294**, 2342–2351.

Avila, F.W. and Goff, M.L. (1998) Arthropod succession patterns onto burnt carrion in two contrasting habitats in the Hawaiian Islands. *Journal of Forensic Sciences* **43**, 581–586.

Azam, F. and Malfatti, F. (2007) Microbial structuring of marine ecosystems. *Nature Reviews: Microbiology* **5**, 782–791.

Bal, R. (2005) How to kill with a ballpoint: credibility in Dutch forensic science. *Science Technology and Human Values* **30**, 52–75.

Bamshad, M. *et al.* (2004) Deconstructing the relationship between genetics and race. *Nature: Genetics* **5**, 598–609.

Bass, B. and Jefferson, J. (2003) *Death's Acre*. Time Warner, London.

Bauer, M. (2007) RNA in forensic science. *Forensic Science International: Genetics* **1**, 69–74.

Bauer, M. *et al.* (2003) Qunatification of mRNA degradation as a possible indicator of postmortem interval – a pilot study. *Legal Medicine* **5**, 220–227.

Beard, B.L. and Johnson, C.M. (2000) Strontium isotope composition of skeletal material can determine the birth place and geographical mobility of humans and animals. *Journal of Forensic Sciences* **45**, 1049–1061.

Beecher, D.J. (2006) Forensic application of microbiological culture analysis to identify mail intentionally contaminated with *Bacillus anthracis* spores. *Applied and Environmental Microbiology* **72**, 5304–5310.

Bender, K. *et al.* (2000) Application of mtDNA sequence analysis in forensic casework for the identification of human remains. *Foresnic Science International* **113**, 103–107.

Benecke, M. (2001) A brief history of forensic entomology. *Forensic Science International* **120**, 2–14.

Benecke, M. and Lessig, R. (2001). Child neglect and forensic entomology. *Forensic Science International* **120**, 155–159.

Bennett, A.T. and Collins, K.A. (2001) An unusual case of anaphylaxis: mold in pancake mix. *American Journal of Forensic Medicine and Pathology* **22**, 292–295.

Bennett, J.L. and Benedix, D.C. (1999) Positive identification of cremains recovered from an automobile based on presence of an internal fixation device. *Journal of Forensic Sciences* **44**, 1296–1298.

Bereuter, T.L. *et al.* (1997) Iceman's mummification – implications from infrared spectro-scopical and histological studies. *Chemistry – a European Journal* **3**, 1032–1038.

Berkefeld, K. (1993) [A possibility for verifying condom use in sex offences] *Archiv fur Kriminologie* **192**, 37–42. (in German)

Bernard, E.J. *et al.* (2007) HIV forensics: pitfalls and acceptable standards in the use of phy-logenetic analysis as evidence in criminal investigations of HIV transmission. *HIV Medicine* **8**, 382–387.

Bernd, K. *et al.* (2002) Backspatter on the firearm and hand in experimental Close-range gunshots to the head. *American Journal of Forensic Medicine and Pathology* **23**, 211–213.

Bernet, W *et al.* (2007) Bad nature, bad nurture, and testimony regarding MAOA and SLC6A4 genotyping at murder trials. *Journal of Forensic Sciences* **52**, 1–10.

Betz, P. *et al.* (1995) Frequency of blood spatters on the shooting hand and conjunctival petechiae following suicidal gunshot wounds to the head. *Forensic Science International* **76**, 47–53.

Bhattacharjee, Y. and Enserink, M. (2008) FBI releases documents on suspected anthrax poisoner. *Science*NOW Daily News. 6th August 2008. www.sciencenow.sciencemag.org/cgi/contents/full/2008/806/3

Bidmos, M.A. (2005) On the non-equivalence of documented cadaver lengths to living stature estimates based on Fully's method on bones in the Raymond A. Dart Collection. *Journal of Forensic Sciences* **50**, 501–506.

Blackledge, R. (2005) Condom trace evidence: the overlooked traces. *Forensic Nurse* www.forensicnursemag.com/articles/311feat6.html.

Bock, H. *et al.* (2000) Suicide in a lions' den. *International Journal of Legal Medicine* **114**, 101–102.

Bogenhagen, D. and Clayton, D.A. (1974) The number of mitochondrial deoxyribonucleic acid genomes in mouse L and human HeLa cells. *Journal of Biological Chemistry* **249**, 7991–7995.

Bonte, W. (1975) Tool marks in bones and cartilage. *Journal of Forensic Sciences* **20**, 315–325.

Borgula, L.M. *et al.* (2003) Isolation and genotypic comparison of oral streptococci from experimental bitemarks. *Journal of Forensic Odontostomatology* **21**, 23–30.

Bourel, B. *et al.* (1999) Effects of morphine in decomposing bodies on the development of *Lucilia sericata* (Diptera: Calliphoridae). *Journal of Forensic Sciences* **44**, 354–358.

Bourel, B. *et al.* (2000). Forensic entomology applied to a mummified corpse. *Annales de la Societe Entomologique de France* **36**, 287–290.

Bourel, B. *et al.* (2003). Flies eggs: a new method for the estimation of short-term post-mortem interval? *Forensic Science International* **135**, 27–34.

Bourel, B. *et al.* (2004) Effects of various substances on the delay of colonisation by necrophagous insects. *Proceedings of the 2ⁿᵈ Meeting of the European Association for Forensic Entomology* London, 29–30 March 2004.

Braack, L.E.O. and Retief, P.E. (1986) Dispersal, density and habitat preference of the blow-flies *Chrsyomya albiceps* (Wd.) and *Chrysomya marginalis* (Wd.) (Diptera: Calliphoridae). *Onderstepoort Journal of Veterinary Research* **53**, 13–18.

Brace, C.L. (1995) Region does not mean 'race' – reality versus convention in forensic anthropology. *Journal of Forensic Sciences* **40**, 171–175.

Brain, C.K. (1981) *The Hunters or the Hunted? An Introduction to African Cave Taphonomy.* University of Chicago Press.

Branicki, W. *et al.* (2007) Determination of phenotype associated SNPs in the MC1R gene. *Journal of Forensic Science* **52**, 349–354.

Breeze, R.G. *et al.* (2005) *Microbial Forensics.* Elsevier, Amsterdam.

Brodbeck, S.M.C. (2007) Reflections upon arteries and veins – a plea for 'spurt patterns'. *International Association of Bloodstain Pattern Analysts News.* June 4–14.

Brookes, A.J. (1999) The essence of SNPs. *Gene* **234**, 177–186.

Brown, A.G. (2006) The use of forensic botany and geology in war crimes investigations in NE Bosnia. *Forensic Science International* **163**, 204–210.

Brumfiel, G. (2003) Still out in the cold. *Nature* **423**, 678–680.

Buck, T.J. and Vidarsdottir, U.S. (2004) A proposed method for the identification of race in sub-adult skeletons: a geometric morphometric analysis of mandibular morphology. *Journal of Forensic Sciences* **49**, 1159–1164.

Budowle, B. *et al.* (1999) Mitochondrial DNA regions HVI and HVII population data. *Forensic Science International* **103**, 23–35.

Budowle, B. *et al.* (2005) Genetic analysis and attribution of microbial forensic evidence. *Critical Reviews in Microbiology* **31**, 233–254.

Bunyard, B.A. (2004) Commentary on: Carter, D.O., Tibbett, M. Taphonomic mycota: fungi with forensic potential. *Journal of Forensic Sciences* **49**, 1134.

Burragato, F. *et al.* (1998) New forensic tool for the identification of elephant or mammoth ivory. *Forensic Science International* **96**, 189–196.

Butler, J.M. (2005) *Forensic DNA Typing* 2ⁿᵈ edition. Elsevier, London, UK.

Butler, J.M. *et al.* (2003) The development of reduced size STR amplicons as a tool for the analysis of degraded DNA. *Journal of Forensic Sciences* **48**, 1054–1064.

Bux, R. *et al.* (2007) Causes and circumstances of fatal falls downstairs. *Forensic Science International* **171**, 122–126.

Byard, R.W. (2004) Unexpected infant death: lessons from the Sally Clark case. *Medical Journal of Australia* **181**, 52–54.

Byard, R.W. *et al.* (2002) Diagnostic problems associated with cadaveric trauma from animal activity. *American Journal of Forensic Medicine and Pathology* **23**, 238–244.

Byard, R.W. *et al.* (2006) The pathological features and circumstances of death of lethal crush/traumatic asphyxia in adults – a 25–year study. *Forensic Science International* **159**, 200–205.

Byrd, J.H. and Allen, J.C. (2001) Computer modelling of insect growth and its application to forensic entomology. In: Byrd, J.H. and Castner, J.L. *Forensic Entomology.* pp. 303–329. CRC Press.

Byrd, J.H. and Castner, J.L. (2001) *Forensic Entomology.* CRC Press, Boca Raton, FL.

Byers, S.N. and Myster, S. (2005) *Forensic Anthropology Laboratory Manual.* Allyn and Bacon.

Calacal, G.C. and De Ungria, M.C.A. (2005) Fungal DNA challenge in human STR typing of bone samples. *Journal of Forensic Sciences* **50**, 1394–1401.

Campobasso, C.P. *et al.* (2001) Factors affecting decomposition and Diptera colonization. *Forensic Science International* **120**, 18–27.

Campobasso, C.P. *et al.* (2004a) Post-mortem artefacts made by ants and the effects of ant activity on decomposition rates. *Proceedings of the 2ⁿᵈ Meeting of the European Association for Forensic Entomology*. London, 29–30 March 2004.

Campobasso, C.P. *et al.* (2004b) A case of *Megaselia scalaris* (Loew) (Dipt., Phoridae) breeding in a human corpse. *Aggrawal's Internet Journal of Forensic Medicine and Toxicology* **5**, 3–5.

Campobasso, C.P. *et al.* (2005) Forensic genetic analysis of insect gut contents. *American Journal of Forensic Medicine and Pathology* **26**, 161–165.

Cardinetti, B., *et al.* (2004) X-ray mapping technique: a preliminary study in discriminating gunshot residue particles from aggregates of environmental occupational origin. *Forensic Science International* **143**, 1–19.

Carson, H.J. and Esslinger, K. (2001) Carbon monoxide poisoning without Cherry-red livor. *American Journal of Forensic Medicine and Pathology* **22**, 233–235.

Carter, D.A. and Tibbett, M. (2003) Taphonomic mycota: fungi with forensic potential. *Journal of Forensic Sciences* **48**, 168–171.

Casamatta, D.A. and Verb, R.G. (2000) Algal colonization of submerged carcasses in a mid-order woodland stream. *Journal of Forensic Sciences* **45**, 1280–1285.

Catts, E.P. and Goff, M.L. (1992) Forensic entomology in criminal investigations. *Annual Review of Entomology* **37**, 253–272.

Chame, M. (2003) Terrestrial mammal feces: a morphometric summary and description. *Memorias do Instituto do Oswaldo Cruz* **98**, 71–94.

Chang, Y.M. *et al.* (2003) High failure of amelogenin test in an Indian population group. *Journal of Forensic Sciences* **48**, 1309–1313.

Chapenoire, S. and Benezech, M. (2003) Forensic Medicine in Bordeaux in the 16ᵗʰ century. *American Journal of Medical Pathology* **24**, 183–186.

Cherry, M. and Imwinkelried, E. (2006) A cautionary note about fingerprint analysis and reliance on digital technology. *Judicature* **89**, 334–338.

Ciesielski, C.A. *et al.* (1994) The 1990 Florida dental investigation – the press and the science. *Annals of Internal Medicine* **121**, 886–888.

Coleman, D.C. *et al.* (2004) *Fundamentals of Soil Ecology*. 2ⁿᵈ edn. Elsevier, Amsterdam.

Conn, C. (2008) *Forensic Toxicology, Biology and Chemistry*. Academic Press, USA.

Cooper, J.E. & Cooper, M.E. (2007) *Introduction to Veterinary and Comparative Medicine*. Blackwell Publishing. Oxford, UK.

Cordes, T. (2000) Equine identification: the state of the art. *American Association of Equine Practitioners. Annual Convention Proceedings* **46**, 300–301

Correia, A. and Pina, C. (2000) Tubercle of Carabelli: A review. *Dental Anthropology* Journal **15**, 18–21.

Courtin, G.M. and Fairgrieve, S.I. (2004) Estimation of postmortem interval (PMI) as revealed through the analysis of annual growth rings in woody tissue. *Journal of Forensic Sciences* **49**, 1–3.

Cox, T.M. *et al.* (2005) King George III and porphyria: an elemental hypothesis and investigation. *The Lancet* **366**, 332–335.

Coyle, H.M.C. (2004) *Foresic Botany*. CRC Press, Boca Raton, USA.

Coyle, H.M.C. (2007) *Nonhuman DNA Typing: Theory and Casework Applications*. CRC Press, Boca Raton, USA.

Crosby, T.K. and Watt, J.C. (1986) Entomological identification of the origin of imported cannabis. *Journal of Forensic Science* **26**, 35–44.

Dadds, M.R. *et al.* (2002) Development links between cruelty to animals and human violence. *Australian and New Zealand Journal of Criminology* **35**, 363–382.

D'Andrea, F. *et al.* (1998) Preliminary experiments on the transfer of animal hair during simulated criminal behaviour. *Journal of Forensic Sciences* **43**, 1257–1258.

Dalton, R. (2005) Infection scare inflames fight against biodefence network. *Nature* **433**, 344.

Daugman, J. (2004) How iris recognition works. *IEE Transactions on Circuits and Systems for Video Technology* **14**, 21–30

Daugman, J. and Downing, C. (2001) Epigenetic randomness, complexity and singularity of human iris patterns. *Proceedings of the Royal Society of London. Series B.* **268**, 1737–1740.

Davey, J.S. *et al.* (2007) DNA detection rates of host mtDNA in bloodmeals of human body lice (*Pediculus humanus* L., 1758). *Medical and Veterinary Entomology* **21**, 293–296.

Davies, D. *et al.* (2004) The decline of the hospital autopsy: a safety and quality issue for healthcare in Australia. *Medical Journal of Australia* **180**, 281–285.

Davies, L. (2006) Lifetime reproductive output of *Calliphora vicina* and *Lucilia sericata* in outdoor caged and field populations; flight *vs* egg production? *Medical and Veterinary Entomology* **20**, 453–458.

Davies, P.L. (2008) Residual based localization and quantification of peaks in x-ray diffractograms. *Annals of Applied Statistics*, in press.

Dawnay, N. *et al.* (2007) Validation of barcoding gene COI for use in forensic genetic species identification. *Forensic Science International* **173**, 1–6.

Day, D.M. and Wallman, J.F. (2006) Width as an alternative measurement to length for post-mortem interval estimations using *Calliphora augur* (Diptera: Calliphoridae) larvae. *Forensic Science International* **159**, 158–168.

De Greef, S. and Willems, G. (2005) Three dimensional cranio-facial reconstruction in forensic identification: latest progress and new tendencies in the 21st century. *Journal of Forensic Sciences* **50**, 12–17.

Deflorio, G. *et al.* (2005) The application of wood decay fungi to enhance annual ring detection in three diffuse-porous hardwoods. *Dendrochronologia* **22**, 123–130.

Denic, N. *et al.* (1997) Cockroach: The omnivorous scavanager – potential misinterpretation of postmortem injuries. *American Journal of Forensic Medicine and Pathology* **18**, 177–180.

Dent, B.B. *et al.* (2004) Review of human decomposition processes in soil. *Environmental Geology* **45**, 576–585.

DeSantis, T.Z. *et al.* (2007) High density universal 16S rRNA microarray analysis reveals broader diversity than typical clone library when sampling the environment. *Microbial Ecology* **53**, 371–383.

Di Martino, D. *et al.* (2004a) Single sperm cell isolation by laser microdissection. *Forensic Science International* **146S**, S151–S153.

Di Martino, D. *et al.* (2004b) Laser microdissection and DNA typing of cells from single hair follicles. *Forensic Science International* **146S**, S155–S157.

Di Nunno, N. *et al.* (2002) Self-strangulation: an uncommon but not unprecedented suicide method. *American Journal of Forensic Medicine and Pathology* **23**, 260–263.

Disney, R.H.L. (1994) *Scuttle Flies: The Phoridae.* Chapman and Hall.

Disney, R.H.L. (2008) Natural history of the scuttle fly, *Megaselia scalaris. Annual Review of Entomology* **53**, 39–60.

Disney, R.H.L. and Munk, T. (2004) Potential use of Braconidae (Hymenoptera) in forensic cases. *Medical and Veterinary Entomology* **18**, 442–444.

Dolinak, D., Matshes, E.W., and Lew, E.O. (2005) *Forensic Pathology: Theory and Practice.* Elsevier, Amsterdam.

Douceron, H. *et al.* (1993) Long lasting postmortem viability of human immunodeficiency virus – a potential risk in forensic medicine. *Forensic Science International* **60**, 61–66.

Dror, I. and Charlton, D. (2006) Why experts make errors. *Journal of Forensic Identification* **56**, 600–616.

Dror, I. *et al.* (2005) When emotions get the better of us: the effect of contextual top-down processing on matching fingerprints. *Applied Cognitive Psychology* **19**, 799–809.

Dror, I. *et al.* (2006) Contextual information renders experts vulnerable to making erroneous identifications. *Forensic Science International* **156**, 74–78.

Dupras, T.L. *et al.* (2006) *Forensic Recovery of Human Remains: Archaeological Approaches.* Taylor Francis. Boca Raton, USA.

Duric, M. *et al.* (2005) The reliability of sex determination of skeletons from forensic context in the Blakans. *Forensic Science International* **147**, 159–164.

Dyer, C. (2005) Pathologist in Sally Clark case suspended from court work. British Medical Journal **330**, 1347.

Edgar, H.J.H. (2005) Predicting race using characters of dental morphology. *Journal of Forensic Sciences* **50**, 269–273.

Edston, E. and van Hage-Hamsten, M. (2003) Death in anaphylaxis in a man with house dust mite allergy. *International Journal of Legal Medicine* **117**, 299–301.

Edwards, H.G.M. (2004) Forensic applications of Raman spectroscopy to Non-destructive analysis of biomaterials and their degradation. *Geological Society of London, Special Publications* **232**, 159–170.

Ehleringer J.R. *et al.* (2000). Tracing cocaine with stable isotopes. *Nature* **408**, 311–312.

Ehleringer J.R. *et al.* (2008) Hydrogen and oxygen isotope ratios are related to geography. *Proceedings of the National Academy of Sciences* **105**, 2788–2793.

Eichmann, C., Berger, B., Reinhold, M., Lutz, M. and Parson, W. (2004) Canine-specific STR typing of saliva traces on dog bite wounds. *International Journal of Legal Medicine* **118**, 337–342.

Elliott, S. *et al.* (2004) The possible influence of microorganisms and putrefaction in the production of GHB in post-mortem biological fluid. *Forensic Science International* **139**, 183–190.

Ember, L.R. (2006) Anthrax sleucing: science aids a nettlesome FBI criminal probe. *Chemical and Engineering News* **84**, 47–54.

Enserink, M. (2008) Full genome sequencing paved the way from spores to a suspect. *Science* **321**, 898–899.

Erdtman, G. (1969) *Handbook of Palynology.* Hafner Publishing Co.

Erzinclioglu, Z. (1996) *Blowflies.* Naturlalists' Handbooks 23, The Richmond Publishing Co. Ltd.

Erzinclioglu, Z. (2000) *Maggots, Murder and Men.* Harley Books, UK.

Evans, E.P. (1906) *The Criminal Prosecution and Capital Punishment of Animals.* Heinemann. Reprinted Faber and Faber (1988).

Evans, K.M. *et al.* (2007) An assessment of the potential diatom 'barcode' genes (cox1, rbcL, 18S and ITS rDNA) and their effectiveness in determining relationships in *Sellaphora* (Bacillariophyta). *Protist* **158**, 349–364.

Evershed, R.P. (1992) Chemical composition of a bog body adipocere. *Archaeometry* **34**, 253–265.

Fanton, L. *et al.* (2006) Criminal Burning. *Forensic Science International* **158**, 87–93.

Faria, L.D.B. *et al.* (2004) Larval predation on different instars in blowfly populations. *Brazilian Archives of Biology and Technology* **47**, 887–894.

Fedrigo, O. and Naylor, G. (2004) A gene specific DNA sequencing chip for exploring molecular evolutionary change. *Nucleic Acids Research* **32**, 1208–1213.

Fiedler, S. and Graw, M. (2003) Decomposition of buried corpses, with special reference to the formation of adipocere. *Naturwissenschaften* **90**, 291–300.

Fojtasek, L. & Kmjec, T. (2005) Time periods of GSR particles deposition after discharge-final results. *Forensic Science International* **153**, 132–135.

Foran, D.R. *et al.* (1997) Species identification from scat: an unambiguous genetic method. *Wildlife Society Bulletin* **25**, 835–839.

Forbes, S.L. *et al.* (2004) A preliminary investigation of the stages of adipocere formation. *Journal of Forensic Sciences* **49**, 1–9.

Forbes, S.L., Dent, B.B. and Stuart, B.H. (2005a) The effect of soil type on adipocere formation. *Forensic Science International* **154**, 35–43.

Forbes, S.L., Stuart, B.H. and Dent, B.B. (2005b) The effect of the burial environment on adipocere formation. *Forensic Science International* **154**, 35–43.

Forbes, S.L., Stuart, B.H. and Dent, B.B. (2005c) The effect of method of burial on adipocere formation. *Forensic Science International* **154**, 44–52.

Forbes, S.L. *et al.* (2005d) Characterization of adipocere formation in animal species. *Journal of Forensic Sciences* **50**, 633–640.

Forbes, T.R. (1985) *Surgeons at the Bailey*. Yale University Press.

Fridez, F. *et al.* (1999) Individual identification of cats and dogs using mitochondrial DNA tandem repeats? *Science and Justice* **39**, 167–171.

Gaillard, Y *et al.* (2004) *Cerbera odollam*: a 'suicide tree' and cause of death in the state of Kerala, India. *Journal of Ethnopharmacology* **95**, 123–126.

Galeotti, M. (2008) The world of the lower depths: crime and punishment in Russian history. *Global Crime* **9**, 84–107.

Galloway, A. and Snodgrass, J.J. (1998) Biological and chemical hazards of forensic skeletal analysis. *Journal of Forensic Sciences* **43**, 940–948.

Galloway, A. *et al.* (1989) Decay rates of human remains in an arid environment. *Journal of Forensic Sciences* **34**, 607–617.

Gennard, D.E. (2007) *Forensic Entomology*. Wiley, Chichester, UK.

Gheradi, M. and Constantini, G. (2004) Death, elderly neglect, and forensic entomology. *Proceedings of the 2nd Meeting of the European Association for Forensic Entomology.* London, 29–30 March 2004.

Gibbons, A. (1998) Calibrating the mitochondrial clock. *Science* **279**, 28–29.

Gill, P. (2001a) Application of low-copy number DNA profiling. *Croatian Medical Journal* **42**, 229–232.

Gill, P. (2001b) An assessment of the utility of single nucleotide polymorphisms (SNPs) for forensic purposes. *International Journal of Legal Medicine* **114**, 204–210.

Gill, P. *et al.* (1994) Identification of the remains of the Romanov family by DNA analysis. *Nature Genetics* **6**, 130–135.

Gill, P. *et al.* (2000) An investigation of the rigour of interpretation rules for STRs derived from less than 100 pg of DNA. *Forensic Science International* **112**, 17–40.

Gill, P. *et al.* (2004) An assessment of whether SNPs will replace STRs in national DNA databases – Joint considerations of the DNA working group of the European Network of Forensic Science Institutes (ENFSI) and the Scientific Working Group on DNA Analysis Methods (SWGDAM). *Science and Justice* **44**, 51–53.

Gilmore, S. *et al.* (2007) Organelle DNA haplotypes reflect crop-use characteristics and geographic origins of *Cannabis sativa*. *Forensic Science International* **172**, 179–190.

Giotakos, O. *et al.* (2003) Prevalence and risk factors of HIV, hepatitis B and hepatitis C in a forensic population of rapists and child molestors. *Epidemiology and Infection* **130**, 497–500.

Goff, M.L. (2000). *A Fly for the Prosecution: How Insect Evidence Helps Solve Crimes.* Harvard University Press.

Goff, M.L. and Lord, W.D. (2001) Entomotoxicology: insects as toxicological indicators and the impact of drugs and toxins on insect development. *In* Byrd, J.H. and Castner, J.L. (2000) *Forensic Entomology.* pp. 331–340. CRC Press, Boca Raton, USA.

Goff, M.L. *et al.* (1991) Effect of heroin in decomposing tissues on the development rate of *Boettcherisca peregrina* (Diptera: Sarcophagidae) and implications of this effect on estimations of postmortem intervals using arthropod development patterns. *Journal of Forensic Sciences* **36**, 537–542.

Goff, M.L. *et al.* (1997) Estimation of postmortem interval based on colony development time for *Anoplepsis longipes* (Hymenoptera: Formicidae). *Journal of Forensic Sciences* **42**, 1176–1179.

Gomes, L. *et al.* (2006) A review of postfeeding larval dispersal in blowflies: implications for forensic entomology. *Naturwissenschaften* **93**, 207–215.

González, W.L., *et al.* (2007) An unusual case of thermal injuries with a hot glue gun. Deliberate self-harm or maltreatment? *Forensic Science International* **167**, 53–55.

Gosline, A. (2005) Sperm clock calls time on rape. *New Scientist* **185**, 12.

Gottardini, E. *et al.* (2007) Use of Fourier transform (FT-IR) spectroscopy as a tool for pollen identification. *Aerobiologia* **23**, 211–219.

Grassberger, M. and Frank, C. (2003) Temperature-related development of the parasitoid wasp *Nasonia vitripennis* as a forensic indicator. *Medical and Veterinary Entomology* **17**, 257–262.

Grassberger, M. and Reiter, C. (2001) Effect of temperature on *Lucilia sericata* (Diptera: Calliphoridae) development with special reference to the isomegalen- and isomorphen-diagram. *Forensic Science International* **120**, 32–36.

Graves, S. *et al.* (2002) Cannibalism or violent death alone? Human remains at a small Anasazi site. In W. Haglund and M.H. Sorg (editors) *Advances in Forensic Taphonomy: Method Theory and Archaeological Perspective* pp. 310–220. CRC Press, Boca Raton, FL.

Green, J.L. *et al.* (2004) Spatial scaling of microbial eukaryote diversity. *Nature* **432**, 747–750.

Green, P.W.C. *et al.* (2003) Diet nutriment and rearing density affect the growth of the black blowfly larvae, *Phormia regina* (Diptera: Calliphoridae). *European Journal of Entomology* **100**, 39–42.

Green, S.T. and Wilson, O.F. (1996) The effect of hair colour on the incorporation of methadone in hair in rats. *Journal of Analytical Toxicology* **20**, 121–123.

Greenbaum, A.R. *et al.* (2004) Intentional burn injury: an evidence-based, clinical and forensic review. *Burns* **30**, 628–642.

Greenberg, B. (1984) Two cases of human myiasis caused by *Phaenicia sericata* (Diptera: Calliphoridae) in Chicago area hospitals. *Journal of Medical Entomology* **21**, 615.

Greenberg, B. and Kunich, J.C. (2002) *Entomology and the Law.* Cambridge University Press.

Greenemeier, L. (2008) Seven years later: electrons unlocked post-9/11 anthrax mail mystery. *Scientific American* 19th September, 2008. www.sciam.com

Greenfield, H.J. (1988) Bone consumption by pigs in a contemporary Serbian village: implications for the interpretation of a prehistoric faunal assemblage. *Journal of Field Archaeology* **15**, 473–478.

Grif, K. *et al.* (2003) Identifying and subtyping species of dangerous pathogens by automated ribotyping. *Diagnostic Microbiology and Infectious Disease* **47**, 313–320.

Grimaldi, L. *et al.* (2005) Suicide by pencil. *Journal of Forensic Sciences* **50**, 913–914.

Gungadin, S. (2007) Sex determination from fingerprint ridge density. *Internet Journal of Medical Update* **2** www.geocities.com/agnihotrimed

Haefner, J.N. *et al.* (2004) Pig decomposition in lotic aquatic systems: the potential use of algal growth in establishing a postmortem submersion interval (PMSI). *Journal of Forensic Sciences* **49**, 1–7.

Haglund, W.D. (1997) Dogs and coyotes: post-mortem involvement with human remains. pp. 367–381. In: W.D. Haglund and M.H. Sorg (editors) *Forensic Taphonomy: The Post-Mortem Fate of Human Remains*. CRC Press, FL.

Haglund, W.D. and Sperry, K. (1993) The use of hydrogen peroxide to visualize tattoos obscured by decomposition and mummification. *Journal of Forensic Sciences* **38**, 147–150.

Haglund, W.D. *et al.* (1988) Tooth mark artefacts and survival of bones in animal scavenged human skeletons. *Journal of Forensic Sciences* **33**, 985–997.

Harris, S.H. (2002) *Factories of Death*. Revised edition. Routledge. New York.

Hart, T. and Shears, P. (1996) *Color Atlas of Medical Microbiology*. Mosby Wolfe, London, UK.

Harvey, R. *et al.* (1991) Cat scratch disease: an unusual cause of combative behaviour. *American Journal of Emergency Medicine* **9**, 52–53.

Hauck, M. and Sweijd, N.A. (1999) A case study of abalone poaching in South Africa and its impact on fisheries management. *ICES Journal of Marine Science* **56**, 1024–1032.

Hayden, D. (2003) *Pox. Genius, Madness, and the Mysteries of Syphilis*. Basic Books, New York.

Hayes, E.J. and Wall, R. (1999) Age-grading adult insects: a review of techniques. *Physiological Entomology* **24**, 1–10.

Hayes, E.J. *et al.* (1998) Measurement of age and population structure in the blowfly *Lucilia sericata* (Meigen) (Diptera: Calliphoridae). *Journal of Insect Physiology* **44**, 895–901.

Hazekamp, A. (2006) An evaluation of the quality of medicinal grade cannabis in the Netherlands. *Cannabinoids* **1**, 1–9.

Heath, L.E. and Saunders, V.A. (2006) Assessing the potential of bacterial DNA profiling for forensic soil comparisons. *Journal of Forensic Sciences* **51**, 1062–1068.

Hedges, D.J. (2003) Mobile element-based assay for human gender determination. *Analytical Biochemistry* **312**, 77–79.

Hedouin, V. *et al.* (1999) Determination of drug levels in larvae of *Lucilia sericata* (Diptera: Calliphoridae) reared on rabbit carcasses containing morphine. *Journal of Forensic Sciences* **44**, 351–353.

Hellerich, U. (1992) [Tattoo pigment in regional lymph nodes – an identifying marker] *Archives Kriminologia* **190**, 163–170. (in German)

Henßge, C. and Madea, B. (2004) Estimation of time since death. *Forensic Science International* **144**, 167–175.

Henßge, C. *et al.* (2002) *The Estimation of the Time since Death in the early Postmortem Period*. 2nd edition. Arnold, London.

Herbst, J. and Haffner, H-Th. (1999) Tentative injuries to exposed skin in a homicide case. *Forensic Science International* **102**, 193–196.

Hill, J. *et al.* (2007) Soil DNA typing in forensic science. *In* Coyle, H.M.C. (editor) *Nonhuman DNA Typing: Theory and Casework Applications*. CRC Press, Boca Raton, USA.

Hillier, M.L. and Bell, L.S. (2007) Differentiating human bone from animal bone: a review of histological methods. *Journal of Forensic Sciences* **52**, 249–263.

Hira, P.R. *et al.* (2004) Myiasis in Kuwait: nosocomial infections caused by *Lucilia sericata* and *Megaselia scalaris*. *American Journal of Tropical Medicine and Hygiene* **70**, 386–389.

Hopkins, D.W. *et al.* (2000) Microbial characteristics of soils from graves: an investigation into the interface of soil microbiology and forensic science. *Applied Soil Ecology* **14**, 283–288.

Hopwood, A.J. *et al.* (1996) DNA typing from human faeces. *International Journal of Legal Medicine* **108**, 237–243.

Horner-Devine, M.C. *et al.* (2004) A taxa–area relationship for bacteria. *Nature* **432**, 750–753.

Horrocks, M. (2004) Sub-sampling and preparing forensic samples for pollen analysis. *Journal of Forensic Sciences* **49**, 1–4.

Horrocks, M. and Walsh, K.A.J. (2001) Pollen on grass clippings: putting the suspect at the scene of the crime. *Journal of Forensic Sciences* **46**, 947–949.

Horsewell, J. *et al.* (2002) Forensic comparison of soils by bacterial community DNA profiling. *Journal of Forensic Sciences* **47**, 350–353.

Howe, R.M. (1950) *Criminal Investigation* 4th edition, Sweet and Maxwell ltd, London, UK.

Hsieh, H-M. *et al.* (2003) Species identification of rhinoceros horns using the cytochrome b gene. *Forensic Science International* **136**, 1–11.

Huntington, T.E. *et al.* (2007) Maggot development during morgue storage and its effect on estimating the post mortem interval. *Journal of Forensic Sciences* **52**, 453–458.

Hurlimann, J. *et al.* (2000) Diatom detection in the diagnosis of death by drowning. *International Journal of Legal Medicine* **114**, 6–14.

Huxley, A.K. and Finnegan, M. (2004) Human remains sold to the highest bidder! A snapshot of the buying and selling of human skeletal remains on eBay ®, and Internet auction sites. *Journal of Forensic Sciences* **49**, 17–20.

Hyde, H.A. and Adams, K.F. (1958) *An Atlas of Airborne Pollen Grains*. MacMillan and Co Ltd. London, UK.

Ikegaya, H. and Iwase, H. (2004) Trial of the geographical identification using JC viral genotyping in Japan. *Forensic Science International* **139**, 169–172.

Ikegaya, H. *et al.* (2002) JC virus genotyping offers a new means of tracing the origins of unidentified cadavers. *International Journal of Legal Medicine* **116**, 242–245.

Ikegaya, H. *et al.* (2007) BK virus genotype distribution offers information of tracing the geographical origins of unidentified cadaver. *Forensic Science International* **173**, 41–46.

Ikematsu, K. *et al.* (2007) Identification of novel genes expressed in hypoxic brain condition by fluorescence differential display. *Forensic Science International* **169**, 168–172.

Ikematsu, K. *et al.* (2006) Gene response of mouse skin to pressure injury in the neck region. *Legal Medicine* **8**, 128–131.

Inglesby, T.V. *et al.* (2002) Anthrax as a biological weapon. *Journal of the American Medical Association* **287**, 2236–2252.

Inman, K. and Rudin, N. (1997) *An Introduction to Forensic DNA Analysis*. CRC Press, Boca Raton, USA.

Ireland, S. and Turner, B. (2006) The effects of larval crowding and food type on the size and development of the blowfly *Calliphora vomitoria*. *Forensic Science International* **159**, 175–181.

Irwin, J.A. *et al.* (2007) Application of low copy number STR typing to the identification of aged, degraded skeletal remains. *Journal of Forensic Sciences* **52**, 1322–1327.

Iscan, M.Y. *et al.* (1984) Age estimation from the rib by phase estimation: white males. *Journal of Forensic Sciences* **29**, 1094–1104.

Ishii, K. *et al.* (2006) Analysis of fungi detected on human cadavers. *Legal Medicine* **8**, 188–190.

Ivanov, P.L. *et al.* (1996) Mitochondrial DNA sequence heteroplasmy in the Grand Duke of Russia Georgij Romanov establishes the authenticity of the remains of Tsar Nicholas II. *Nature Genetics* **12**, 417–420.

Jackson, A.R.W. and Jackson, J.M. (2008) *Forensic Science* 2nd edition. Pearson, Prentice-Hall, Harlow, UK.

Jackson, R.M. and Raw, F. (1966) *Life in the Soil.* Edward Arnold.

Jalba, A.C. *et al.* (2005) Automatic diatom identification using contour analysis by morphological curvature scale spaces. *Machine Vision and Applications* **16**, 217–228.

James, S.H., Kish, P.E. and Sutton T.P. (2005) *Principles of Bloodstain Pattern Analysis.* CRC Press, Boca Raton, USA.

Jans, M.M.E. *et al.* (2004) Characterisation of microbial attack on archaeological bone. *Journal of Archaeological Science* **31**, 87–95.

Janssen, W. *et al.* (2005) Forensic aspects of 40 accidental autoerotic deaths in Northern Germany. *Forensic Science International* **147**, S61–S64.

Jayaprakash, P.T. (2006) Postmortem skin erosions caused by ants and their significance in crime reconstruction. *Journal of Forensic Identification* **56**, 972–999.

Jobling, M.A. and Gill, P. (2004) Encoded evidence: DNA in forensic analysis. *Nature Reviews: Genetics* **5**, 739–752.

Johnson, D.J. *et al.* (2005) STR-typing of human DNA from human fecal matter using the QIAGEN QIAmp® Stool Mini Kit. *Journal of Forensic Sciences* **50**, 802–808.

Joy, J.E. *et al.* (2006) Carrion fly (Diptera: Calliphoridae) larval colonization of sunlit and shaded pig carcasses in West Virginia. *Forensic Science International* **164**, 183–192.

Juusola, J. and Ballantyne, J. (2007) mRNA profiling for body fluid identification by multiplex quantitative RT-PCR. *Journal of Forensic Sciences* **52**, 1–10.

Kahana, T., *et al.* (1999) Marine taphonomy: adipocere formation in a series of bodies recovered from a single shipwreck. *Journal of Forensic Sciences* **44**, 897–901.

Kahana, T., *et al.* (2001) Fingerprinting the deceased: traditional and new techniques. *Journal of Forensic Sciences* **46**, 908–912.

Kahler, K. *et al.* (2003) Reanimating the dead: reconstruction of expressive faces from skull data. *ACM Transactions on Graphics* **22**, 554–561.

Karlsson, A.O. and Holmlund, G. (2007) Identification of mammal species using species-specific DNA pyrosequencing. *Forensic Science International* **173**, 16–20.

Karlsson, A.O. *et al.* (2007) DNA-testing for immigration cases: the risk of erroneous conclusions. *Forensic Science International* **172**, 144–149.

Keil, W., *et al.* (2003) Evidence of condom residues. *Forensic Science International* **136**, 261.

Keim, P. *et al.* (2001) Molecular investigation of the Aum Shinrikyo anthrax release in Kameido, Japan. *Journal of Clinical Microbiology* **39**, 4566–4577.

Kidd, K.K. *et al.* (2006) Developing a SNP panel for forensic identification of individuals. *Forensic Science International* **164**, 20–32.

Kishimoto, M. *et al.* (2004) ribotyping and a study of transmission of *Staphylococcus aureus* collected from food preparation facilities. *Journal of Food Protection* **67**, 1116–1122.

Kitano, T. *et al.* (2007) Two universal primer sets for species identification among vertebrates. *International Journal of Legal Medicine* **121**, 423–427.

Klepinger, L.L. (2006) *Fundamentals of Forensic Anthropology.* Wiley, New Jersey.

Klinbunga, S. *et al.* (2004) Species identification of the tropical abalone (*Haliotis asinine, Haliotis ovina,* and *Haliotis varia*) in Thailand using RAPD and SCAR Markers. *Journal of Biochemistry and Molecular Biology* **37**, 213–222.

Klippel, W.E. and Synstelien, J.A. (2007) Rodents as taphonomic agents: bone gnawing by brown rats and gray squirrels. *Journal of Forensic Science* **52**, 765–773.

Klotzbach, H. *et al.* (2004) Information is everything – a case report demonstrating the necessity of entomological knowledge at the crime scene. *Agrawal's Internet Journal of Forensic Medicine and Toxicology* 5, 19–21.

Knock, C. and Davison, M. (2007) Predicting the position of the source of blood stains for angled impacts. *Journal of Forensic Sciences* 52, 1044–1049.

Knudsen, P.J. (1993) Cytology in ballistics. An experimental investigation of tissue fragments on full metal jacketed bullets using routine cytological techniques. *International Journal of Legal Medicine* 106, 15–18.

Kobayashi, M. *et al.* (1993) Novel detection of plankton from lung tissue by enzymatic digestion method. *Forensic Science International* 60, 81–90.

Komar, D. and Beattie, O. (1998) Postmortem insect activity may mimic perimortem sexual assault clothing patterns. *Journal of Forensic Science* 43, 792–796.

Komar, D. and Lathrop, S. (2006) Frequencies of morphological characteristics in two contemporary forensic collections: implications for identification. *Journal of Forensic Sciences* 51, 974–978.

Kondo, T. *et al.* (2002) Ubiquitin expression in skin wounds and its application to forensic wound age determination. *International Journal of Legal Medicine* 116, 267–272.

Kosa, F. and Castellana, C. (2005) New forensic anthropological approachment for the age determination of human fetal skeletons on the base of morphometry of vertebral column. *Forensic Science International* 147, S69–S74.

Kračun, S.K. *et al.* (2007) Population substructure can significantly affect reliability of a DNA-led process of identification of mass fatality victims. *Journal of Forensic Sciences* 52, 874–878.

Kress, W.J. *et al.* (2005) Use of DNA barcodes to identify flowering plants. *Proceedings of the National Academy of Sciences of the United States of America* 102, 8369–8374.

Kreutzer-Martin, H.W. *et al.* (2004) Stable isotope ratios as a tool in microbial forensics – Part 2. Isotopic variation among different growth media as a tool for sourcing origins of bacterial cells or spores. *Journal of Forensic Sciences* 49, 961–967.

Krogman, W.M. and Iscan, M.Y. (1986) *The Human Skeleton in Forensic Medicine*. Charles C. Thomas, Springfield, Illinois, USA.

Kubiczek, P.A. and Mellen, P.F. (2004) Commentary on: Huxley, A.K. and Finnegan, M. (2004) Human remains sold to the highest bidder! A snapshot of the buying and selling of human skeletal remains on eBay ®, and Internet auction sites. *Journal of Forensic Sciences* 49, 17–20. *Journal of Forensic Sciences* 49, 1137.

Kücken, M. (2007) Models for fingerprint pattern formation. *Forensic Science International* 171, 85–96.

Kunos, C.A. *et al.* (1999) First rib metamorphosis: its possible utility for human age-at-death estimation. *American Journal of Physical Anthropology* 110, 303–323.

Kyle, C.J. and Wilson, C.C. (2007) Mitochondrial DNA identification of game and harvested freshwater fish species. *Forensic Science International* 166, 68–76.

Laloup, M. *et al.* (2003) The use of combined liquid chromatography–mass spectrometry method (LC–MS/MS) to detect nordiazepam and oxazepam in a single larva and puparium of *Calliphora vicina* (Diptera: Calliphoridae). *Forensic Science International* 136, 316–317.

Lamendin, H. *et al.* (1992) A simple technique for age estimation in adult corpses – the 2 criteria dental method. *Journal of Forensic Sciences* 37, 1373–1379.

Lantz, P-G. *et al.* (1997) Removal of PCR inhibitors from human faecal samples through the use of an aqueous two-phase system for sample preparation prior to PCR. *Journal of Microbiological Methods* 28, 159–167.

Lasczkowski, G.E. *et al.* (2002) Visualization of postmortem chondrocyte damage by vital staining and confocal laser scanning 3D microscopy. *Journal of Forensic Sciences* **47**, 663–666.

Lasseter, A.E. *et al.* (2003) Cadaver dog and handler team capabilities in the recovery of buried human remains in the southeastern United States. *Journal of Forensic Sciences* **48**, 617–621.

Lazarus, H.M. *et al.* (2001) Dangers of large exotic pets from foreign lands. *Journal of Trauma-Injury Infection and Critical Care* **51**, 1014–1015.

Learn, G.H. and Mullins, J.I. (2004) The microbial forensic use of HIV sequences. In Leitner, T. *et al.* (eds) (2005) *HIV Sequence Compendium*. Los Alamos National Laboratory, Los Alamos, NM, USA.

Leendertz, F.H. *et al.* (2004) Anthrax kills wild chimpanzees in a tropical rainforest. *Nature* **430**, 451–452.

Leikin, J.B. and Watson, W.A. (2003) Post-mortem toxicology: what the dead can and cannot tell us. *Clinical Toxicology* **41**, 47–56.

Lin, A. C-Y. *et al.* (2007) Forensic applications of infrared imaging for the detection and recording of latent evidence. *Journal of Forensic Sciences* **52**, 1148–1150.

Lin, W.F. *et al.* (2005) Identification of four *Thunnus* tuna species using mitochondrial cytochrome b gene sequences and PCR-RFLP analysis. *Journal of Food and Drug Analysis* **13**, 382–387.

Linch, C.A. *et al.* (2001) Human hair histogenesis for the mitochondrial DNA forensic scientist. *Journal of Forensic Sciences* **46**, 844–853.

Linville, J.G. *et al.* (2004) Mitochondrial DNA and STR analyses of maggot crop contents: effects of specimen preservation technique. *Journal of Forensic Sciences* **49**, 341–344.

Lista, F. *et al.* (2006) Genotyping of *Bacillus anthracis* strains based on automated 25-loci multiple locus variable-number tandem repeats analysis. *BioMed Central Microbiology* **6**, accessed at www.biomedcentral.com/1471–2180/6/33

Lo, M-C. *et al.* (2005) Analysis of heteroplasmy in hypervariable region II of mitochondrial DNA in maternally related individuals. *Annals of the New York Academy of Sciences* **1042**, 130–135.

Lockwood, R. (2000) Animal cruelty and human violence: the veterinarian's role in making connection – the American experience. *Canadian Veterinary Journal – Revue Veterinaire Canadienne* **41**, 876–878.

Lord, W.D. *et al.* (1994) The black soldier fly *Hermetia illucens* (Diptera: Stratiomyidae) as a potential measure of post-mortem interval: observations and case histories. *Journal of Forensic Sciences* **39**, 215–222.

Lord, W.D. *et al.* (1998) Isolation, amplification, and sequencing of human mitochondrial DNA obtained from a human crab louse, *Pthirus pubis* (L.), blood meals. *Journal of Forensic Sciences* **43**, 1097–1100.

Loreille, O.M. *et al.* (2007) High efficiency DNA extraction from bone by total demineralization. *Forensic Science International: Genetics* **1**, 191–195.

Lorenzini, R. (2005) DNA foresnics and the poaching of wildlife in Italy: a case study. *Forensic Science International* **153**, 218–221.

Love, J.C. and Symes, S.A. (2004) Understanding rib fracture patterns: incomplete and buckle fractures. *Journal of Forensic Sciences* **49**, 1153–1158.

Lowe, A.L. *et al.* (2001) Inferring ethnic origin by means of an STR profile. *Forensic Science International* **122**, 17–22.

Lowe, A.L. *et al.* (2002) The propensity of individuals to deposit DNA and secondary transfer of low level DNA from individuals to inert surfaces. *Forensic Science International* **129**, 25–34.

Ludes, B. *et al.* (1999) Diatom analysis in victim's tissues as an indicator of the site of drowning. *International Journal of Legal Medicine* **112**, 163–166.

Lunetta, P. *et al.* (1998) Scanning and transmission electron microscopical evidence of the capacity of diatoms to penetrate the alveo-capillary barrier in drowning. *International Journal of Legal Medicine* **111**, 229–237.

Lunetta, P. *et al.* (2002) Suicide by intracerebellar ballpoint pen. *American Journal of Forensic Medicine and Pathology* **23**, 334–337.

Madea, B. and Musshoff, F. (2007) Postmortem biochemistry. *Forensic Science International* **165**, 165–171.

Madea, B. and Rödig, A. (2006) Time of death dependent criteria in vitreous humor – accuracy of estimating the time since death. *Forensic Science International* **164**, 87–92.

Maher, J. *et al.* (2002) Evaluation of the BioSign PSA membrane test for the identification of semen stains in forensic work. *New Zealand Medical Journal* **114**, 48–49.

Mahoney, P.F. *et al.* (2005) *Ballistic Trauma* 2nd edition. Springer-Verlag, London, UK.

Mallon, W.K and Russell, M.A. (1999) Clinical and forensic significance of tattoos. *Topics in Emergency Medicine* **21**, 21–29.

Manfredi, G. *et al.* (1997) The fate of human sperm derived mtDNA in somatic cells. *American Journal of Human Genetics* **61**, 953–960.

Marko, P. *et al.* (2004) Mislabelling of a depleted reef fish. *Nature* **430**, 309–310.

Marlar, R.A. *et al.* (2000) Biochemical evidence of cannibalism at a prehistoric Puebloan site in southwestern Colorado. *Nature* **407**, 74–78.

Martin, I.M.C. *et al.* (2007) Non cultural detection and molecular genotyping of *Neiserria gonorrhoeae* from a piece of clothing. *Journal of Medical Microbiology* **56**, 487–490.

Martinez-Sanchez, A. *et al.* (2007) Geographic origin affects larval competitive ability in European populations of the blow fly *Lucilia sericata*. *Entomologia Experimentalis et Applicata* **122**, 93–98.

Mason, P.W. *et al.* (2003) Comparison of the complete genomes of Asian, African and European isolates of a recent foot and mouth disease virus type O pandemic strain (PanAsia). *Journal of General Virology* **84**, 1583–1593.

Matsuda, H. *et al.* (2004) Identification of DNA of human origin based on amplification of human specific mitochondrial cytochrome b region. *Forensic Science International* **152**, 109–114.

Matsumoto, H. and Fukui, Y. (1993) A simple method for diatom detection in drowning. *Forensic Science International* **60**, 91–95.

Mayor, A. (2003) *Greek Fire, Poison Arrows, & Scorpion Bombs. Biological and Chemical Warfare in the Ancient World*. Overlook Duckworth. Woodstock, USA.

Mejia, R. (2005) You can't rely on firearm forensics. *New Scientist* **188**, 6–7.

Meirmans, P.G. *et al.* (2006) Male sterility in triploid dandelions: asexual females vs asexual hermaphrodites. *Heredity* **96**, 45–52.

Mellen, P.F. *et al.* (1993) Experimental observations on adipocere formation. *Journal of Forensic Sciences* **38**, 91–93.

Merck, M.D. (2007) *Veterinary Forensics. Animal Cruelty Investigations*. Blackwell Publishing, Oxford, UK.

Mereish, K.A. (2007) Unsupported conclusions on the Bacillus anthracis spores. *Applied and Environmental Microbiology* **73**, 5074.

Metfies, K. *et al.* (2007) An optimization protocol for the identification of diatoms, flagellated algae and pathogenic protozoa with phylochips. *Molecular Ecology Notes* **7**, 925–936.

Mildenhall, D.C. *et al.* (2006) Forensic palynology: why do it and how it works. *Forensic Science International* **163**, 163–172.

Millington, J. (2004) development of a synthetic blood substitute for use in forensic science teaching. *LTSN Physical Sciences Development Project: Final Report.* (www.heaacademy.ac.uk)

Miras, A. *et al.* (2001) Method for determining if a corpse has been frozen: measuring the activity of short-chain 3-hydroxyacyl-CoA dehydrogenase (SCHAD). *Forensic Science International* **124**, 22–24.

Monteiro, J.A. (1995) Human and animal bite wound infections. *European Journal of Internal Medicine* **6**, 209–215.

Moran, N.C. and O'Connor, T.P. (1992) Bones that cats gnawed upon: a case study in bone modification. *Circaea* **9**, 27–34.

Morel, K. (2004) A concrete body of evidence. *The American Edge. The Official Newsletter of American Companies.* Winter 2004/05. **9**, 1–2. www.amengtest.com/news/edgWint0405.pdf

Morse, S.A. and Khan, A.S. (2005) Epidemiologic investigation for public health, biodefence, and forensic microbiology. *In.* Breeze, R.G. *et al.* (eds) *Microbial Forensics.* Elsevier, Amsterdam, Holland.

Morton, B. (2004) Marine pollution of a more ruthless kind. *Marine Pollution Bulletin* **48**, 1–2.

Mumcuoglu, K.Y *et al.* (2004). Use of lice in forensic entomology. *Journal of Medical Entomology* **41**, 803–806.

Munoz, J.J. *et al.* (2001) A new perspective in the estimation of post-mortem interval (PMI) based on vitreous [K⁺]. *Journal of Forensic Sciences* **46**, 209–214.

Munro, H.M.C. and Thrushfield, M.V. (2001) 'Battered pets': sexual abuse. *Journal of Small Animal Practice* **42**, 333–337.

Murray, C. *et al.* (2007) Identification and isolation of male cells using fluorescence in situ hybridisation and laser microdissection, for use in the investigation of sexual assault. *Forensic Science International: Genetics* **1**, 247–254.

Musshoff, F. and Madea, B. (2007) New trends in hair analysis and scientific demands on validation and technical notes. *Forensic Science International* **165**, 204–215.

Nadjem, H. *et al.* (2007) Ingestion of pointed objects in a complex suicide. *Forensic Science International* **171**, e11–e14.

Nakazono, T.O. *et al.* (2005) Successful DNA typing of urine stains using a DNA purification kit following dialfiltration. *Journal of Forensic Sciences* **50**, 860–864.

Negrusz, A. *et al.* (2001) Deposition of 7–aminoflunitrezapam and flunitrezepam in hair after a single dose of Rohypnol ®. *Journal of Forensic Sciences* **46**, 1143–1151.

Nelson, L.A. *et al.* (2007) Using COI barcodes to identify forensically and medically important blowflies. *Medical and Veterinary Entomology* **21**, 44–52.

Nields, H. and Kessler, S.C. (1998) Streptococcal toxic shock syndrome presenting as suspected child abuse. *American Journal of Forensic Medicine and Pathology* **19**, 93–97.

Nolte, K.B. and Yoon, S.S. (2003) Theoretical risk for occupational blood-borne infections in forensic pathologists. *Infection Control and Hospital Epidemiology* **24**, 772–773.

Nowak, R. (2004) Murder detectives must rethink maggot theory. *New Scientist* 3ʳᵈ April. 13.

Nuffield Council on Bioethics (2007) *The Forensic use of Bioinformation: Ethical Issues.* The Nuffield Council on Bioethics, London.

Nyhus, P.J. *et al.* (2003) Dangerous animals in captivity: *ex situ* tiger conflict and implications for private ownership of exotic animals. *Zoo Biology* **22**, 573–586.

Ohshima, T. (2000) Forensic wound examination. *Forensic Science International* **113**, 153–164.

Ohtani, S. *et al.* (2003) Differences in the D/L aspartic acid ratios in dentin among different types of teeth from the same individual and estimated age. *International Journal of Legal Medicine* **117**, 149–152.

Ohtani, S. *et al.* (2007) Age-dependent changes in the racemisation ratio of aspartic acid in human alveolar bone. *Archives of Oral Biology* **52**, 233–236.

Oxley, J.C. *et al.* (2005) Accumulation of explosives in hair. *Journal of Forensic Sciences* **50**, 826–831.

Ozdemir, M.H. *et al.* (2003) Investigating demodex in forensic autopsy cases. *Forensic Science International* **135**, 226–231.

Pamplin, C. (2004) Science on trial. *Chemistry and Industry* 6th December, 2004, 14–15.

Pappas, C.S. *et al.* (2003) New method for pollen identification by FT-IR spectroscopy. *Applied Spectroscopy* **57**, 23–27.

Parra, E.J. *et al.* (2004) Implications of correlations between skin color and genetic ancestry for biomedical research. *Nature Genetics* **36**, S54–S60.

Peeters, M. *et al.* (2002) Risk to human health from a plethora of Simian immunodeficiency viruses in primate bushmeat. *Emerging Infectious Diseases* **8**, 451–457.

Peng, Z. and Pounder, D.J. (1998) Forensic medicine in China. *American Journal of Forensic Medicine and Pathology* **19**, 368–371.

Peters, W. and Gilles, H.M. (1977) *A Colour Atlas of Tropical Medicine and Parasitology* 1st edition. Wolfe Medical Publications, London.

Petraco, N. *et al.* (2005) An ideal material for the preparation of known toolmark impressions. *Journal of Forensic Sciences* **50**, 1–4.

Petricevic, S.F. *et al.* (2006) DNA profiling of trace DNA recovered from bedding. *Forensic Science International* **159**, 21–26.

Piana, M.C. *et al.* (2006) Pollen analysis of royal jelly: contribution to analytical methods and characterisation. *Apiacta* **41**, 28–43.

Pfeiffer, I. *et al.* (2004) Forensic DNA typing of dog hair: DNA extraction and PCR amplification. *Forensic Science International* **141**, 149–151.

Phillips, C. *et al.* (2004) Nonbinary single-nucleotide polymorphism markers. *Progress in Forensic Genetics* **10**, 27–29.

Phillips, R. (1981) *Mushrooms and other fungi of Great Britain and Europe*. Pan Books. London. UK.

Phipps, M. and Petricevic, S. (2007) The tendency of individuals to transfer DNA to handled items. *Forensic Science International* **168**, 162–168.

Pickering, T.R. (2001) Carnivore voiding: a taphonomic process with the potential for the deposition of forensic evidence. *Journal of Forensic Sciences* **46**, 406–411.

Platek, S.F. *et al.* (1997) A false report of product tampering involving a rodent and soft drink can: light microscopy, image analysis and scanning electron microscopy/energy dispersive X-ray analysis. *Journal of Forensic Sciences* **42**, 1171–1175.

Pollanen, M.S. (1997) The diagnostic value of the diatom test for drowning 2. Validity: Analysis of diatoms in bone marrow and drowning medium. *Journal of Forensic Sciences* **42**, 286–290.

Pollanen, M.S. (1998) Diatoms and homicide. *Forensic Science International* **91**, 29–34.

Pollanen, M.S. *et al.* (1997) The diagnostic value of the diatom test for drowning 1. Utility: A retrospective analysis of 771 cases of drowning in Ontario, Canada. *Journal of Forensic Sciences* **42**, 281–285.

Porta, D. *et al.* (2007) A new method of reproduction of fingerprints from corpses in a bad state of preservation using latex. *Journal of Forensic Sciences* **52**, 1319–1321.

Pounder, D.J. (1991) Forensic entomo-toxicology. *Journal of the Forensic Science Society* **31**, 469–472.

Pounder, D.J. (2003) The case of Dr Shipman. *American Journal of Forensic Medicine and Pathology* **24**, 219–226.

Pragst, F. *et al.* (1999) Detection of 6-acetylmorphine in vitreous humor and cerebrospinal fluid – comparison with urinary analysis for proving heroin administration in opiate fatalities. *Journal of Analytical Toxicology* **23**, 168–172.

Prahlow, J.A. and Linch, C.A. (2000) A baby, a virus, and a rat. *American Journal of Forensic Medicine and Pathology* **21**, 127–133.

Pretty, I. and Sweet, D. (2000) Anatomical location of bitemarks and associated findings in 101 cases from the United States. *Journal of Forensic Sciences* **45**, 812–814.

Pretty, I. and Sweet, D. (2001) A look at forensic dentistry – Part 1: the role of teeth in the determination of human identity. *British Dental Journal* **190**, 359–366.

Pretty, I. and Turnbull, M.D. (2001) Lack of dental uniqueness between two bite mark suspects. *Jounal of Forensic Sciences* **46**, 1487–1491.

Preuß, J. *et al.* (2006) Dumping after homicide using setting in concrete and/or sealing with bricks – six case reports. *Forensic Science International* **159**, 55–60.

Prince, D.A. and Ubelaker, D.H. (2002) Application of Lamendin's adult dental aging technique to a diverse skeletal sample. *Journal of Forensic Sciences* **47**, 107–116.

Prinz, M. *et al.* (1993) DNA typing of urine samples following several years of storage. *International Journal of Legal Medicine* **106**, 75–79.

Quarino, L. *et al.* (2005) An ELISA method for the identification of salivary amylase. *Journal of Forensic Sciences* **50**, 873–876.

Quatrehomme, G. and Iscan, M.Y. (1997) Bevelling in exit gunshot wounds in bones. *Forensic Science International* **89**, 93–101.

Rahimi, M. *et al.* (2005) Genotypic comparison of bacteria recovered from human bite marks and teeth using arbitrarily primed PCR. *Journal of Applied Microbiology* **99**, 1265–1270.

Raupp, C.D. (1999) Treasuring, trashing or terrorizing: adult outcomes of childhood socialization about companion animals. *Society and Animals* **7**, 141–159.

Rawson, R.B. *et al.* (2000) Scanning electron microscope analysis of skin resolution as an aid in identifying trauma in forensic investigations. *Journal of Forensic Sciences* **45**, 1023–1027.

Ray, D.A. *et al.* (2005) Inference of human geographic origins using *Alu* insertion polymorphisms. *Forensic Science International* **153**, 117–124.

Reid, M.E. and Lomas-Francis, C. (1996) *The Blood Group Antigen Facts Book*. Academic Presss, San Diego, CA, USA.

Rendle, D.F. (2003) X-ray diffraction in forensic science. *The Rigaku Journal* **19**, 11–22.

Riley, W. (2006) Interpreting gang tattoos. *Corrections Today Magazine* **68**, 46–53.

Robino, C. (2006) Incestuous paternity detected by STR-typing of chorionic villi isolated from archival formalin-fixed paraffin-embedded abortion material using laser microdissection. *Journal of Forensic Sciences* **51**, 90–92.

Roeterdink, E.M. *et al.* (2004). Extraction of gunshot residues from the larvae of the forensically important blowfly *Calliphora dubia* (Macquart) (Diptera: Calliphoridae). *International Journal of Legal Medicine* **118**, 63–70.

Rognes, K. (1991) *Blowflies (Diptera, Calliphoridae) of Fennoscandia and Denmark*. Fauna Entomologica Scandinavica **24**, E.J. Brill/Scandinavian Science Press Ltd, Leiden.

Rognum, T.O. *et al.* (1991) Hypoxanthine levels in vitreous humor: evidence of hypoxia in most infants who died of sudden infant death syndrome. *Pediatrics* **87**, 306–310.

Romain, N. *et al.* (2002) Post mortem castration by a dog: a case report. *Medicine and the Law* **42**, 269–271.

Rompen, J.C. *et al.* (2000) A cause celebre: the so-called 'ballpoint murder'. *Journal of Forensic Sciences* **45**, 1144–1147.

Ropohl, D. *et al.* (1995) Postmortem injuries inflicted by domestic golden hamster: morphological aspects and evidence by DNA typing. *Forensic Science International* **72**, 81–90.

Rothschild, M.A. and Schneider, V. (1997) On the temporal onset of postmortem scavenging 'motivation' of the animal. *Forensic Science International* **89**, 57–64.

Russo, A. *et al.* (2006) Life fertility tables of *Piophila casei* L. (Diptera: Piophilidae) reared at five different temperatures. *Environmental Entomology* **35**, 194–200.

Rutty, G.N. (2005) The estimation of the time since death using temperatures recorded from the external auditory canal. Part 1: can a temperature be recorded and interpreted from this site? *Forensic Science, Medicine, and Pathology* **1**, 41–51.

Rutty, G.N. *et al.* (2000) DNA contamination of mortuary instruments and work surfaces: a significant problem in forensic practice? *International Journal of Legal Medicine* **114**, 56–60.

Sampaio, P. *et al.* (2003) Highly polymorphic microsatellite for identification of *Candida albicans* strains. *Journal of Clinical Microbiology* **41**, 552–557.

Saukko, P. and Knight, B. (2004) *Knight's Forensic Pathology* 3rd edn. Hodder Arnold, London.

Saunders, D.S. (2000) Larval diapause duration and fat metabolism in three geographical strains of the blowfly *Calliphora vicina*. *Journal of Insect Physiology* **46**, 509–517.

Saunders, D.S. and Hayward, S.A.L. (1998) Geographical and diapause-related cold tolerance in the blowfly *Calliphora vicina*. *Journal of Insect Physiology* **44**, 541–551.

Sauvageau, A. and Racette, S. (2007) Agonal sequences in a filmed suicidal hanging. Analysis of respiratory and movement responses to asphyxia by hanging. *Journal of Forensic Sciences* **52**, 957–959.

Schmidt, U. and Pollak, S. (2006) Sharp force injuries in clinical forensic medicine – findings in victims and perpetrators. *Forensic Science International* **159**, 113–118.

Schmitt, A. and Murail, P. (2004) Is the first rib a reliable indicator of age at death assessment? Test of the method developed by Kunos et al. (1999) *Homo* **54**, 207–214.

Schneider, P.M. *et al.* (1999) Forensic mtDNA hair analysis excludes a dog from having caused a traffic accident. *International Journal of Legal Medicine* **112**, 315–316.

Schroeder, H. *et al.* (2003). Use of PCR-RFLP for differentiation of calliphorid larvae (Diptera, Calliphoridae) on human corpses. *Forensic Science International* **132**, 76–81.

Schroeder, H. *et al.* (2002). Larder beetles (Coleoptera, Dermestidae) as an accelerating factor for decomposition of a human corpse. *Forensic Science International* **127**, 231–236.

Schudel, D. (2001) Screening for canine spermatozoa. *Science and Justice* **41**, 117–119.

Schumm, J.W. *et al.* (2004) Robust STR multiplexes for challenging casework samples. *Progress in Forensic Genetics ICS* **1261**, 547–549

Scott, S. and Duncan, C. (2004) *Return of the Black Death: the World's Greatest Serial Killer*. Wiley, Chichester, UK.

Sedley, S. (2005) Short cuts. *London Review of Books* **27**.

Selavka, C.M. (1991) Poppy seed ingestion as a contributing factor to opiate-positive urinalysis results: the Pacific perspective. *Journal of Forensic Sciences* **36**, 685–696.

Shankar, K. *et al.* (2005) Quantification of ricin concentrations in aqueous media. *Sensors and Actuators B: Chemical* **107**, 640–648.

Sharp, D. (1997) Infrared reveals the 'iceman's' adipocere. *The Lancet* **350**, 191.

Shepherd, R. (2003) *Simpson's Forensic Medicine* 12th edn. Hodder Arnold, London.

Sherman, R.A. and Tran, J.M.T. (1995) A simple, sterile food source for rearing the larvae of *Lucilia sericata* (Diptera: Calliphoridae). *Medical and Veterinary Entomology* **9**, 393–398.

Shibuya, E.K. *et al.* (2006) Sourcing Brazilian marijuana by applying IRMS analysis to seized samples. *Forensic Science International* **160**, 35–43.

Shyu, H.F. *et al.* (2002) Colloidal gold-based immunochromatographic assay for detection of ricin. *Toxicon* **40**, 255–258.

Sidari, L. *et al.* (1999) Diatom test with Soluene-350 to diagnose drowning in sea water. *Forensic Science International* **103**, 61–65.

Sigee, D.C. *et al.* (2002) Fourier-transform infrared spectroscopy of *Pediastrum duplex*: characterization of a micro-popultion isolated from a eutrophic lake. *European journal of Phycology* **37**, 19–26.

Simpson, E.K. *et al.* (2007) Role of orthopaedic implants and bone morphology in the identification of human remains. *Journal of Forensic Sciences* **52**, 442–448.

Sims, R.W. and Gerard, B.M. (1985) *Earthworms*. Synopses of the British Fauna. 31. E.J. Brill/Dr. W. Backhuys, London, UK.

Singh, R.R. *et al.* (2003) Application of X-ray diffraction to characterize ivory, antler and rhino horn: implications for wildlife forensics. *Forensic Science International* **136**, 376–377.

Sinha, S.K. *et al.* (2003) Development and validation of a multiplexed Y-chromosome STR genotyping system, Y-PLEX™6, for forensic casework. *Journal of Forensic Sciences* **48**, 93–103.

Skidmore, P. (1985) *The Biology of the Muscidae of the World*. Dr W. Junk Publishers.

Skinner, M. and Dupras, T. (1993) Variation in birth timing and location of the neonatal line in human enamel. *Journal of Forensic Sciences* **38**, 1383–1390.

Skopp, G. (2004) Preanalytical aspects in post-mortem toxicology. *Forensic Science International* **142**, 75–100.

Slone, D.H. and Gruner, S.V. (2007) Thermoregulation in larval aggregations of carrion feeding blow flies (Diptera: Calliphoridae). *Journal of Medical Entomology* **44**, 516–523.

Smith, K.E. and Wall, R. (1988) Estimates of population density and dispersal in the blowfly *Lucilia sericata* (Diptera: Calliphoridae). *Bulletin of Entomological Research* **88**, 65–73.

Smith, K.G.V. (1986) *A Manual of Forensic Entomology*. British Natural History Museum.

Smith, L.M. and Burgoyne, L.A. (2004) Collecting, archiving and processing DNA from wildlife samples using FTA® databasing paper. *BMC Ecology*. www.biomedcentral.com/1472–6785/4/4

Sodhi, G.S. and Kaur, J. (2001) Powder method for detecting latent fingerprints: a review. *Forensic Sciece International* **120**, 172–176.

Spalding, K. *et al.* (2005) Age written in teeth by nuclear tests. *Nature* **437**, 333–334.

Sperry, K. (1991) Tattoos and tattooing. Part 1: History and methodology. *American Journal of Forensic Medicine and Pathology* **12**, 313–319.

Spiers, E.M. (1975) The use of the dum dum bullet in colonial warfare. *Journal of Imperial and Commonwealth History* **4**, 3–14.

Stacey, R.B. (2004) Report on the erroneous fingerprint individualization in the Madrid train bombing case. *Journal of Forensic Identification* **54**, 706–718.

Stanley, E.A. (1992) Application of palynology to establish the provenance and travel history of illicit drugs. *Microscope* **40**, 149–152.

Steadman, D.W. and Worne, H. (2007) Canine scavenging of human remains in an indoor setting. *Forensic Science International* **173**, 78–82.

Stelling, M.A. and van der Peijl, G.J.Q. (2003) Analytical chemical tools in wildlife forensics. *Forensic Science International* **136**, 381–382.

Stephan, C.N. and Henneberg, M. (2001) Building faces from dry skulls: are they recognized above chance rates? *Journal of Forensic Sciences* **46**, 432–430.

Stephens, P.J. and Taff, M.L. (1987) Rectal impaction following enema with concrete mix. *American Journal of Forensic Medicine and Pathology* **8**, 179–182.

Stoermer, E.F. and Smol, J.P. (2001) *The Diatoms: Applications for the Earth and Environmental Sciences*. Cambridge University Press.

Stroud, R.K. (1998) Wildlife forensics and the veterinary practitioner. *Seminars in Avian and Exotic Pet Medicine* 7, 182–192.

Sukontason, K. *et al.* (2004) Identification of forensically important fly eggs using potassium permanganate staining technique. *Micron* 35, 391–395.

Swango, K.L. *et al.* (2006) A quantitative PCR assay for the assessment of DNA degradation in forensic samples. *Forensic Science International* 158, 14–26.

Swift, B. and Rutty, G.N. (2003) The human ear: its role in forensic practice. *Journal of Forensic Sciences* 48, 153–160.

Swift, B. *et al.* (2001) An estimation of the post-mortem interval in human skeletal remains: a radionucleotide and trace element approach. *Forensic Science International* 117, 73–87.

Sykes, L.N. *et al.* (1988) Dum-dums, hollow-points, and devastators – techniques designed to increase wounding potential of bullets. *Journal of Trauma-Injury Infection and Critical Care* 28, 618–623.

Szpila, K. *et al.* (2008). Morphology of the first instar of *Calliphora vicina*, *Phormia regina*, and *Lucilia illustris* (Diptera, Calliphoridae). *Medical and Veterinary Entomology* 22, 16–25.

Tahtouh, M. *et al.* (2007) The application of infrared chemical imaging to the detection and enhancement of latent fingerprints: method optimization and further findings. *Journal of Forensic Sciences* 52, 1089–1096.

Tailby, R. and Gant, F. (2002) The illegal market in Australian abalone. *Trends and Issues in Crime and Criminal Justice*. Number 225. Canberra: Australian Institute of Criminology.

Taylor, J.J. (1994) Diatoms and drowning – a cautionary case note. *Medicine, Science, and the Law* 34, 78–79.

Teletchea, F. *et al.* (2005) Food and forensic molecular identification: update and challenges. *Trends in Biotechnology* 23, 359–366.

Thali, M.J. *et al.* (2003) Bite mark documentation and analysis: the forensic 3D/CAD supported photogrammetry approach. *Forensic Science International* 135, 115–121.

Thompson, W.C. and Schumann, E.L. (1987) Interpretation of statistical evidence in criminal trials: the prosecutor's fallacy and the defense attorney's fallacy. *Law and Human Behavior* 11, 167–187.

Tibbett, M. *et al.* (2004) A laboratory incubation method for determining the rate of microbial degradation of skeletal muscle tissue in soil. *Journal of Forensic Sciences* 49, 1–6.

Tobe, S.S. *et al.* (2007a) Evaluation of six presumptive tests for blood, their specificity, sensitivity, and effect on high molecular weight DNA. *Forensic Science International* 52, 102–109.

Tobe, S.S. *et al.* (2007b) Successful DNA typing of a drug positive urine sample from a race horse. *Forensic Science International* 173, 85–86.

Tomberlin, J.K., *et al.* (2005) Black soldier fly (Diptera: Stratiomyidae) colonization of pig carrion in south Georgia. *Journal of Forensic Sciences* 50, 152–153.

Torok, T.J. *et al.* (1997) A large community outbreak of salmonellosis caused by intentional contamination of restaurant salad bars. *Journal of the American Medical Association* 278, 389–395.

Tracqui, A. *et al.* (1998) Suicidal hanging resulting in complete decapitation: a case report. *International Journal of Legal Medicine* 112, 55–57.

Tracqui, A. *et al.* (2004). Entomotoxicology for the forensic toxicologist: much ado about nothing? *International Journal of Legal Medicine* 118, 194–196.

Tranvik, L.J. (1997) Rapid fluorimetric assay of bacterial density in lake water and sea water. *Limnology and Oceanography* 42, 1629–1634.

Trotter, M. (1970) Estimation of stature from intact long limb bones. In. Stewart, T.D. (ed.) *Personal Identification in Mass Disasters*. Smithsonian Institution Press, Washington, DC.

Tsokos, M. and Püschel, K. (2001) Postmortem bacteriology in forensic pathology: diagnostic value and interpretation. *Legal Medicine* 3, 15–22.

Tsokos, M. *et al.* (1999) Skin and soft tissue artefacts due to postmortem damage caused by rodents. *Forensic Science International* 104, 47–57.

Tully, G. (2007) Genotype versus phenotype: human pigmentation. *Forensic Science International: Genetics* 1, 105–110.

Tun, Z. *et al.* (1999) Simultaneous detection of multiple STR loci on sex chromosomes for forensic testing of sex and identity. *Journal of Forensic Sciences* 44, 772–777.

Tuniz, C. *et al.* (2004) Sherlock Holmes counts the atoms. *Nuclear Instruments and Methods in Physics Research. B.* 213, 469–475.

Turner, C.G. and Turner, J.A. (1999) *Man Corn: Cannibalism and Violence in the American Southwest*. University of Utah Press, Salt Lake City, USA.

Ubelaker, D.H. and Buchholz, B.A. (2006) Complexities in the use of bomb-curve radiocarbon to determine time since death of human skeletal remains. *Forensic Science Communications* [Online] http://www.fbi.gov/hq/lab/fsc/backissue/jan2006/research/2006_01_research01.htm

Ullyett, K. (1963) *Crime out of Hand*. Michael Joseph, London, UK.

Van Oorschot, R.A.H. *et al.* (2005) Beware of the possibility of fingerprinting techniques transferring DNA. *Journal of Forensic Sciences* 50, 1417–1422.

Vandenberg, N. and van Oorschot, R.A.H. (2002) Extraction of human nuclear DNA from feces samples using the QIAamp DNA Stool Mini Kit. *Journal of Forensic Sciences* 47, 993–995.

Vanezis, P. *et al.* (2000) Facial reconstruction using 3–D computer graphics. *Forensic Science International* 108, 81–95.

VanLaerhoven, S.L. and Anderson, G.S. (1999) Insect succession on buried carrion in two biogeoclimatic zones of British Colombia. *Journal of Forensic Sciences* 44, 32–43.

Vardy, S. and Uwins, P. (2002) Fourier transform infrared microspectroscopy as a tool to differentiate *Nitzschia closterium* and *Nitzschia longisisima*. *Applied Spectroscopy* 56, 1545–1548.

Varetto, L. and Curto, O. (2004) Long persistence of rigor mortis at constant low temperature. *Forensic Science International* 147, 31–34.

Varnam, A. and Sutherland, J.M. (1995) *Meat and Meat Products: Technology, Chemistry and Microbiology*. Springer.

Vass, A.A. *et al.* (2002) Decomposition chemistry of human remains: a new methodology for determining postmortem interval. *Journal of Forensic Sciences* 47, 542–553.

Vege, A. *et al.* (1994) Vitreous humor hypoxanthine levels in SIDS and infectious death. *Acta Paediatrica* 83, 634–639.

Villain, A. *et al.* (2004) Windows of detection of zolpidem in urine and hair: application of drug facilitated sexual assaults. *Forensic Science International* 143, 157–161.

Vintiner, S.K. *et al.* (1992) Alleged sexual violation of a human female by a rottweiler dog. *Journal of the Forensic Science Society* 32, 357–362.

Wang, J. (2000) From DNA biosensors to gene chips. *Nucleic Acids Research* 28, 3011–3016.

Ward, D.W. *et al.* (2005) DNA barcoding Australian fish species. *Philosophical Transactions of the Royal Society. Biological Sciences* 360, 1847–1857.

Ward, J. *et al.* (2005) Molecular identification system for grasses: a novel technology for forensic botany. *Forensic Science International* 152, 121–131.

Wasser, S.K. *et al.* (2007) Using DNA to track the origin of the largest ivory seizure since the 1983 trade ban. *Proceedings of the National Academy of Sciences* 104, 4228–4233.

Watkins, W.S. *et al.* (2003) Genetic variation among world populations: inferences from 100 *Alu* insertion polymorphisms. *Genome Research* **13**, 1607–1618.

Wedel, V.L. (2007) Determination of season death using dental cementum increment analysis. *Journal of Forensic Sciences* **52**, 1334–1337.

Weir, B.S. (2007) The rarity of DNA profiles. *The Annals of Applied Statistics* **1**, 358–370.

Weitzel, M.A. (2005) A report of decomposition rates of a special burial type in Edmonton, Alberta from an experimental field study. *Journal of Forensic Sciences* **50**, 641–647.

Wells, J.D. and King, J. (2001) Incidence of precocious egg development in flies of forensic importance (Calliphoridae). *Pan-Pacific Entomologist* **77**, 235–239.

Wells, J.D. and Stevens, J.R. (2008) Application of DNA-based methods in forensic entomology. *Annual Review of Entomology* **53**, 103–120.

Wells, J.D. *et al.* (2007) Phylogenetic analysis of forensically important *Lucilia* flies based on cytochrome oxidase I: a cautionary tale for forensic species determination. *International Journal of Legal Medicine* **121**, 229–233.

Wheelis, M. and Sugishima, M. (2006) Terrorist use of biological weapons. In. M. Wheelis *et al.* (eds) *Deadly Cultures*. pp. 284–303. Harvard University Press. Massachusetts, USA.

Wheelis, M. *et al.* (2006) *Deadly Cultures*. Harvard University Press. Massachusetts, USA.

Whittaker, D.K. (1995) Forensic dentistry in the identification of victims and assailants. *Journal of Clinical Forensic Medicine* **2**, 145–151.

Whitwell, H.L. (2003) Reforming the coroner's service. *British Medical Journal* **327**, 175–176.

Wilkinson, C. (2004) *Forensic Facial Reconstruction*. Cambridge University Press, Cambridge, UK.

Will, K.W. and Rubinoff, D. (2004) Myth of the molecule: DNA barcodes for species cannot replace morphology for identification and classification. *Cladistics* **20**, 47–55.

Willey, P. and Heilman, A. (1987) Estimating time since death using plant roots and stems. *Journal of Forensic Sciences* **32**, 1264–1270.

Williams, M.C. (1996) Forensic examination of a mouse allegedly found in a previously sealed can of milk stout. *Forensic Science International* **82**, 211–215.

Willis, C. *et al.* (2001) Errors in the estimation of the distance of fall and angles of impact blood drops. *Forensic Science International* **123**, 1–4.

Willott, G.M. and Allard, J.E. (1982) Spermatozoa: their persistence after sexual intercourse. *Forensic Science International* **26**, 125–128.

Wiltshire, P.E.J. (2006a) Hair as a source of forensic evidence in murder investigations. *Forensic Science International* **163**, 241–248.

Wiltshire, P.E.J. (2006b) Consideration of some taphonomic variables of relevance to forensic palynological investigation in the United Kingdom. *Forensic Science International* **163**, 173–182.

Wiltshire, P.E.J. and Black, S. (2006) The cribiform approach to the retrieval of playnological evidence from the turbinates of murder victims. *Forensic Science International* **163**, 224–230.

Winn, W. *et al.* (2006) *Koneman's Color Atlas and Textbook of Diagnostic Microbiology*. 6th edition. Lippincott Williams and Wilkins. Baltimore, USA.

Withrow, A.G. *et al.* (2003) Extraction and analysis of human nuclear and mitochondrial DNA from electron beam irradiated envelopes. *Journal of Forensic Sciences* **48**, 1302–1308.

Wolodarsky-Franke, A. and Lara, A. (2005) The role of 'forensic' dendrochronology in the conservation of alerce (*Fitzroya cupressoides* ((Molina) Johnston)) forests in Chile. *Dendrochronologia* **22**, 235–240.

Wooldridge, J. *et al.* (2007) Flight activity of the blowflies *Calliphora vomitoria* and *Lucilia sericata* in the dark. *Forensic Science International* **172**, 94–97.

Wyss, C. and Cherix, D. (2004) Murder followed by suicide in a forest: what could be learned from a comparative field experiment. *Proceedings of the 2ⁿᵈ Meeting of the European Association for Forensic Entomology.* London, UK, 29–30 March, 2004.

Wyss, C. and Cherix, D. (2002) Beyond the limits, the case of *Calliphora* species (Diptera, Calliphoridae). *Proceedings of the 1ˢᵗ Meeting of the European Association for Forensic Entomology.* Rosny sous Bois, France, 28–30 May, 2002.

Yajima, D. *et al.* (2006) Ethanol production by *Candida albicans* in post-mortem human blood samples: effects of blood glucose level and dilution. *Forensic Science International* **164**, 116–121.

Yamamoto, M. *et al.* (2000) Forensic study of sex determination using PCR on teeth samples. *Acta Medica Okayama* **54**, 21–32.

Zala, K. (2007) Dirty science: soil forensics digs into new techniques. *Science* **318**, 386–387.

Zhu, G.H. *et al.* (2007) Puparial case hydrocarbons of *Chrysomya megacephala* as an indication of the post-mortem interval. *Forensic Science International* **169**, 1–5.

Zuccato, E. *et al.* (2005) Cocaine in surface waters: a new evidence-based tool to monitor community drug abuse. *Environmental Health: A Global Access Science* **14**, 4–14.

Zumpt, F. (1965) *Myiasis in Man and Animals in the Old World.* Butterworths, London, UK.

Index